Adolf Seilacher

Trace Fossil Analysis

Adolf Seilacher

Trace Fossil Analysis

With 75 Plates and 43 Photos

 Springer

Author

Prof. Dr. Adolf Seilacher

Yale Geology Dept.
P.O. Box 208109
New Haven, CT 06520-8109
USA

and

Engelfriedshalde 25
72076 Tübingen
Germany
E-mail: geodolf@gmx.de

Library of Congress Control Number: 2006935138

ISBN-13   978-3-540-47225-4 **Springer Berlin Heidelberg New York**

Springer is a part of Springer Science+Business Media
springer.com
© Springer-Verlag Berlin Heidelberg 2007

Cover design: Erich Kirchner, Heidelberg
Petrophaga lorioti: Copyright © Loriot
Typesetting: Stasch · Bayreuth (stasch@stasch.com)
Production: Christine Adolph, Heidelberg

Printed on acid-free paper   32/2132/CA – 5 4 3 2 1 0

Preface

This is a course book – meaning that it intends to confer not knowledge, but skill. The need for this skill becomes obvious if we look at the changing role of trace fossils during the last decades. From objects that were treated in standard paleontology text-books, at best, under "Miscellanea", together with problematica, coprolites and pseudofossils useless as index fossils, they have become subject of a special field, paleoichnology. A journal (ICHNOS), Ichnological Newsletters and regular workshops have been established, symposia are held, and the literature has increased exponentially. This success stems mainly from the intimate connection of ichnology with sedimentology and the importance that both fields have for paleoenvironmental and basin analysis, which becomes more and more important in petroleum exploration.

This useful connection, however, also had its price. In the hands of biogeologists, trace fossils easily loose their significance as unique biological documents. They are commonly treated summarily as "bioturbation", a term that was originally meant to describe the biogenic destruction of primary depositional structures and the lack of distinctive trace morphologies. In a more quantitative form, "ichnofabrics", this approach has had considerable success.

On the taxonomic side, paleoichnology is all too often considered as a field, in which there is no limit to coining new names and taxa without affiliations. If somebody would take the effort today to revise Walter Häntzschel's trace fossil volume of the "Treatise on Invertebrate Paleontology", it would probably double in size. Still a large proportion of the ichnogenera described in the meantime would probably fall under "unrecognized" or "synonyms".

The situation being as it is, this book bypasses the necessary job of taxonomic revision (the divergence between lumpers and splitters is extremely wide in this field). Rather it concentrates on the more distinctive and representative ichnogenera. It is also focussed on structures left by invertebrate animals in soft sediments, treating vertebrate tracks only marginally and leaving out hard substrate borings, bite marks on body fossils, eggs, nests, and coprolites altogether. Not that such documents would not merit attention; but by their very nature they require a different approach and should better be treated, together with the respective body fossils, by systematic paleontologists.

A word is also necessary about how this book originated and how it should be used. It grew out of courses I gave in Tübingen and many other places with the aid of representative specimens and plaster casts. In the printed version, the plates remain the core, to which the text is added in the form of extended captions. It remains to be seen, how far the style of this course can be detached from my own personal interaction and how far colleagues with different views and experiences are willing to adopt it – if only to develop alternative and possibly more adequate ideas. In no case, however, should it be done without real material, because it is the inquisitive observation that should be trained. In my own experience, drawing specimens with an old-fashioned *camera lucida* is still the most adequate method. If a form appears to be too complicated, more often than not it will become understandable after our brain has absorbed it through the pencil. Many of my drawings are new, others repeated from earlier publications; but all of them have been arranged in such a way that they make a story and wait for the addition of your own colorful scribblings. This also means that you should devote more than a glimpse to these illustrations. After all, early illiterate periods conveyed whole world views through pictures.

For the student that wants to delve further into a subject, relevant papers are arranged by plates and annotated. A glossary explains the terminology.

There follow the apologies: for not covering all ichnogenera (which can be found in the "Treatise"); for covering too much in a too condensed style (objection of my wife); and for unrealistically assuming that the time of fifteen two-hour courses could be spared for trace fossils in the curriculum of any school. But again, it is not the knowledge of forms and names that we aim at, but a method of morphological thinking in terms of processes that could easily be transferred to any other subject matter. So let us follow the motto of my Tübingen University: ATTEMPTO (Let us try!).

Acknowledgements

I collected my first trace fossils 68 years ago. Listing acknowledgements over such a long timespan amounts to writing one's own biography. My tutors in highschool times were a local physician , Dr. R. Stierlin, and a devoted forester, Dr. h.c. Otto Linck. Back at Tübingen University for the first post-war semester, the famous vertebrate paleontologist Prof. F. von Huene taught me to use the *camera lucida*, which has since remained my most important aid. Nevertheless it was the broader-minded Prof. Otto H. Schindewolf, who encouraged me to do the doctoral thesis on Mesozoic trace fossils. This brought me in contact with the Senckenberg Institute in Wilhelmshaven, where I was introduced to actuo-paleontology by Prof. Wilhelm Schäfer, who also influenced my drafting style, and with Prof. Walter Häntzschel, who then compiled the Treatise volume on "Trace Fossils and Problematica". Equally important was a first trip to Italy, where the rich university collections in Pisa and Florence made me familiar with the strange flysch ichnocoenoses and with the new ideas of Prof. C. I. Migliorini about sand being imported into deepsea basins. Later, the recognition of the *Nereites* ichnofacies lent strong support to his and P. H. Kuenen's turbidite theory, as well as to the new paradigm of plate tectonics.

Right after promotion (1951), I had the priviledge to join my teacher on his expedition to the Salt Range of Pakistan. Although our original target had been the end-Permian mass extinction, this trip introduced me to Cambrian shallow-marine ichnocoenoses, in which the various activities of trilobites play a dominant role. Later field work in Spain, in which the advise of Prof. F. Lotze has been crucial, led to the scheme of *Cruziana* ichnostratigraphy. During the following years it was successfully applied to the dating of otherwise non-fossiliferous sandstones in the deserts of North Africa, where E. Klitzsch and his team were of invaluable help. It also stood the test in a memorable excursion to the Tassili arranged by the Algerian Geological Survey in order to study the effects and the dating of the newly discovered end-Ordovician glaciation. Ichnostratigraphy was further refined in the symposia of the National Oil Company of Libya organized by M. Salem, trips to Jordan arranged by F. Bender, and to Saudi Arabia with M. Senalp.

On the American side, the interest of oil companies was more in facies relationships of trace fossils. Consulting for Jersey Research Company by the initiative of J. Campbell gave me a chance to study a broad range of ichnocoenoses through the south-western part of the country.

The idea to compound this book came with the invitation by X. T. WU to hold a course in Jao Zuo (Henan, China) with more than hundred participants. It gradually matured during similar courses throughout the world; but only when presenting it with Derek Briggs at Yale University (where I am teaching since retirement from Tübingen University), he gave me the ultimate kick to finish the manuscript.

Several people have been involved in its completion. Our son Peter Seilacher cleaned the illustrations with the computer, Gabriela Mangano, Luis Buatois (University of Saskatchewan) and Andrew Rindsberg (University of Georgia) helped in accumulating and annotating the list of references. Above all, Edith Seilacher-Drexler, my dear wife and unpaid partner for half a century, did all the word processing involved in updating and re-arranging the text.

The late Roland Goldring, Gabriela Mangano, Luis Buatois, and Andrew Rindsberg were of tremendous help in compiling and annotating the relevant references.

At various points, Wolfgang Gerber, Hans Luginsland, Werner Wetzel, Eden Volohonsky (Tübingen), Bill Sacco (New Haven) and Jens Rydell (Göteborg) supplied photographs. Hans Luginsland also produced casts in the field and in his small laboratory.

Last not least I am obliged to Springer-Verlag, whose representative Wolfgang Engel kept encouraging me throughout the years to finish this book but did not live to see it published. During the fruitful collaboration with Armin Stasch (Bayreuth) I learned about the crucial function of a scientific book designer in producing a book.

My deep gratitude goes to these people, not to forget all the inquisitive students in the courses. May their future generations share my contention that it is a wonderful thing to work as a paleodetective!

General References

Textbooks

Richter R (1925) Flachseebeobachtungen zur Paläontologie und Geologie. III–VI. Senckenberg 8(3/4): 200–224

Abel O (1935) Vorzeitliche Lebensspuren. Fischer, Jena (A classic integrative text that non-German readers will nevertheless enthuse on)

Bromley RG (1996) Trace fossils: biology, taphonomy and applications, 2nd edn. Chapman & Hall, London, 361 p (A very readable text with good appreciation of modern traces and the palaeobiology of ancient ones. Rather thin sedimentologically and does not discuss hardgrounds. Includes a glossary)

Ekdale AA, Bromley RG, Pemberton SG (1984) Ichnology – the use of trace fossils in sedimentology and stratigraphy. Society of Economic Paleontologists & Mineralogists (SEPM), Short Course 15, 317 p

Ekdale AA, Pemberton SG, Bromley RG (1984) Trace fossils. Society of Economic Paleontologists & Mineralogists (SEPM). Short Course 15, 317 p (A useful but dated text for sedimentological applications)

Frey RW (ed) (1975) The study of trace fossils. Springer-Verlag, Heidelberg, 562 p (A rather dated text)

Lockley MG (1991) Tracking dinosaurs: A new look at an ancient world. Cambridge University Press, Cambridge, 238 p

Maples CG, West RR (eds) (1992) Trace fossils: Short courses in paleontology, 5. Paleontological Society, Cincinnati

Schäfer W (1972) Ecology and palaeoecology of marine environments. Oliver and Boyd, Edinburgh, 568 p (This text, superbly translated from the German edition is in the truly scientific spirit of careful observation and interpretation. It concerns observations made predominately in the intertidal zone of the southern North Sea)

Source Books

Crimes TP, Harper JC (eds) (1970) Trace fossils. Geol J, Special Issue, 3, Seel House Press, Liverpool (A collection of 35 mostly systematic papers)

Crimes TP, Harper JC (eds) (1977) Trace fossils 2. Proceedings of an International Symposium held at Sydney, Australia. Geol J, Special Issue, 9, Seel House Press, Liverpool (A collection of 15 mostly systematic papers)

Donovan SK (ed) (1994) The palaeobiology of trace fossils. The John Hopkins University Press, Baltimore, 308 p (A collection of 11 papers on some of the aspects dealt with here, and other objects such as plant roots, coprolites and vertebrate eggs)

Gillette DD, Lockley MG (eds) (1989) Dinosaur tracks and traces. Cambridge University Press, Cambridge, 454 p (50 papers on all aspects)

Häntzschel W (1965) Vestigia Invertebratorum et Problematica. Fossilium Catalogus 1: Animalia 108, W. Junk, s'Gravenhage, 142 p (Most complete annoteted bibliography of the time)

Häntzschel W (1975) Trace fossils and problematica. In: Teichert C (ed) Treatise on invertebrate paleontology, Part W, Supplement 1. Geological Society of America and Univerity of Kansas, W1–W269 (The A–Z of trace fossils and related structures)

Kuhn O (1963) Ichnia tetrapodorum. Fossilium Catalogue, 1, Animalia 101, W. Junk, s'Gravenhage, 176 p (With 12 plates of line figures)

Lessertisseur J (1955) Traces fossiles d'activitié animale et leur significance paléobiologique. Soc Geol Fr Mem New Ser 74:1–150 (General review of invertebrate trace fossils)

Miller III W (ed) (2003) New interpretations of complex trace fossils. Palaeogeog Palaeoclim Palaeoecol (Special issue) 192, 343 p (18 papers by paleontologists and biologists concerned with the functional and fabricational context of invertebrate burrow systems. A most stimulating approach as opposed to the one of ichnofabrics)

Miller MF, Ekdale A, Picard MD (eds) (1984) Trace fossils and paleoenvironments: Marine carbonate, marginal marine terrigenous and continental terrigenous. J Paleontol 58(2):283–597 (22 papers on invertebrate trace fossils)

Noda H (1982) Check list and bibliography of trace fossils and related forms in Japan (1889–1980) and neighbourhood (1928–1980). Introduction to study of trace fossils, part 2, pp 1–80, Ibaraki, Japan

Seilacher A (1997) Fossil art. An exhibition of the Geologisches Institut Tübingen University. The Royal Tyrell Museum of Palaeontology, Drumheller, Alberta, Canada, 64 p (Large casts of trace fossils and other sedimentary structures, arranged for their appeal)

General Papers

Häntzschel W (1955) Rezente und fossile Lebensspuren, ihre Deutung und geologische Auswertung. Experientia 11(10):373–382

Richter R (1922) Flachseebeobachtungen zur Paläontologie und Geologie. III–VI. Senckenberg 4(5): 103–141 (Actuopaleontology)

Richter R (1924) Flachseebeobachtungen zur Paläontologie und Geologie. III–VI. Senckenberg 6(3/4): 119–164

Contents

Vertebrate Tracks

From the early hunter stages, man has been occupied with the recognition and interpretation of vertebrate footprints. They were also the first objects of paleoichnology. Therefore it is appropriate to start a course on trace fossils with vertebrate tracks. We shall not attempt, however, to cover all the kinds of fossil footprints that range from tracks of the first land vertebrates to those left by our bipedal australopithecine ancestors in an East African ash bed. Rather we shall concentrate on the principles of their interpretation and on the conditions that control their preservation in the fossil record.

Each footprint is a fossilized experiment in soil mechanics. For a footprint to be preserved with some degree of morphological information, the substrate must be cohesive. In dry dune sand, for instance, the print will immediately collapse and leave only a nondescript deformation – it is even less likely that such a surface impression would survive a windy day. Most vertebrate tracks were impressed in a cohesive substrate, but still there is more to them than simple casting by new sediment.

As the following plates show, there is a wide open field for future research. The experimentalist may choose to work with a model dinosaur foot and step it on a variety of substances: sand with colored layers, dry, damp or wet; clays of different wetness; mica sand; sand over clay; mud over sand; layers separated by agar films etc. Or, if you like calculating, take up your soil mechanics book! Reasonable numbers are available for the load, the area of loading and the duration of loading. The properties of the fresh sediment is the only variable we need for an environmental reconstruction.

No other trace fossils impress the public as much as dinosaur tracks. After one has seen "Jurassic Park", here is the real thing and you can almost feel the earth tremble under the steps of the giants! Modern geotourism kindles such emotion in protected track sites all over the world. In one I commonly visit with Yale students (Rocky Hill, Connecticut), a huge dome has been constructed over spectacular Triassic dinosaur tramplings. So visitors can view them from gangways in perfect artificial lighting. They can also make their own plaster cast of a real footprint, or see the dioramas of a lost world.

But visitors may also wonder why there are thousands of footprints and not a single bone or tooth of these creatures? This question touches a fundamental problem in paleontology: the fossil record is far from a collection of photographs of ancient worlds. Rather it consists of residues left-over after all possible documents have passed through a series of taphonomic filters. The "mortichnia" shown at the end of this book (Pl. 75) are the extreme exception, because the fossilization potentials of the traces and the bodily remains of the same organism diverge so much, that they are hardly ever preserved in the same kind of rocks. Generally, trace fossils are much better preserved in clastic than in argillaceous and carbonate facies, while the opposite is true for body fossils.

Beyond this basic divergence, there is selective preservation also among trace fossils. Therefore this chapter focuses on preservational biases in different ichnotopes.

Literature

Chapter I

Barbour EH (1895) Is *Daemonelix* a burrow? Am Nat 29(342): 517–527 (Spiral burrows of rodents)

Demathieu G (1977) La palichnologie des Vertébrés. Développement récent et role dans la stratigraphie du Trias. Bull. B.R.G.M. 4(3):269–278

Gillette DD, Lockley MG (eds) (1989) Dinosaur tracks and traces. Cambridge University Press, Cambridge, 454 p (This book originated from the "First International Symposium on Dinosaur Tracks and Traces" held in Albuquerque, New Mexico in 1986. Fifty papers on dinosaur paleoichnology, covering a wide variety of topics, such as history of the discipline, dinosaur behavior and paleoecology, tracksite paleoenvironmental setting, biostratigraphy, experimental neoichnology, functional morphology and techniques)

Kuhn O (1958) Die Fährten der vorzeitlichen Amphibien und Reptilien. Meisenbach, Bamberg, 64 p (Drawings of major taxa of vertebrate footprints)

Leonardi G (ed) (1987) Glossary and manual of tetrapod footprint palaeoichnology. Departamento Nacional de Produção Mineral, Brasilia, 75 p (Terminology of vertebrate paleoichnology, including a glossary of 106 terms in eight languages. Chapters on the history of the discipline and the use of statistical methods in ichnology)

Lockley MG (1991) Tracking dinosaurs: A new look at an ancient world. Cambridge University Press, Cambridge, 238 p (An engaging popular introduction to the topic of dinosaur trackways. Preservation and classification of dinosaur tracks, individual and social behavior, ecology and evolution, paleoenvironmental significance of tracksites, dinosaur bioturbation and the importance of megatracksites are among the topics covered in this book)

Lockley MG (2007) A tale of two ichnologies: The different goals and potentials of invertebrate and vertebrate (tetrapod) ichnotaxonomy and how they relate to ichnofacies analysis. Ichnos 14:39–57. (In a special issue devoted to vertebrate tracks, different principles of classification are proposed for vertebrate versus invertebrate paleoichnology)

Lockley MG, Hunt AP (1995) Dinosaur tracks and other fossil footprints of the Western United States. Columbia University Press, New York, 338 p (Valuable summary of vertebrate track-ways found in western United States aimed at a general audience. Discussion of trackways is organized according to geologic age)

Lockley MG, Meyer C (2000) Dinosaur tracks and other fossil footprints of Europe. Columbia University Press, New York, 323 p (A useful companion to Lockley and Hunt (1995); this book summarizes findings across Europe)

Martin LD, Bennett DK (1977) The burrows of the Miocene beaver *Paleocastor*, western Nebraska, USA. Palaeogeog Palaeoclim Palaeoecol 22:173–193 (Corkscrew-shaped beaver burrows comparable to *Daimonelyx*)

Pemberton SG, McCrea RT, Lockley MG (eds) (2004) William Antony Swithin Sarjeant (1935–2002): A celebration of his life and ichnological contributions. Ichnos 11, 1–386 (This special issue comprises 32 papers on different aspects of vertebrate ichnology)

Sarjeant WAS (1975) Fossil tracks and impressions of vertebrates. In: Frey RW (ed) The study of trace fossils: A synthesis of principles, problems, and procedures in ichnology. Springer-Verlag, New York, pp 283–324 (A general review of vertebrate ichnology that set the standard for the next generation of workers)

Schmidt H (1959) Die Cornberger Fährten im Rahmen der Vierfüßler-Entwicklung. Abhandlungen des Hessischen Landesamtes für Bodenforschungen 28, 137 p (Phylogenetic tree of tetrapod tracks)

Thulborn RA (1990) Dinosaur tracks. Chapman & Hall, London, 410 p (A comprehensive book on dinosaur ichnology, including preservation of dinosaur tracks, history of dinosaur paleoichnology, methods and techniques, and dinosaur behavior. Extensive coverage of the different kinds of dinosaur trackways)

Tresise G, Sarjeant WAS (1997) The tracks of triassic vertebrates: Fossil evidence from north-west England. The Stationary Office, Norwich, 204 p (Beautifully illustrated book that discusses the history of Triassic vertebrate ichnology in England)

Plate 1: *Chirotherium:* The Sherlock Holmes Approach

Baird D (1954) *Chirotherium lulli*, a pseudosuchian reptile from New Jersey. Bull Mus Comp Zool 111(4):165–192, 2 pls.

De Raaf JFM, Beets C, Kortenbout van der SLUIJS G (1965) Lower Oligocene bird-tracks from Northern Spain. Nature 207(4993): 146–148

Demathieu G, Haubold H (1972) Stratigraphische Aussagen der Tetrapodenfährten aus der terrestrischen Trias Europas. Geologie 21(7):802–836 (Ichnostratigraphy of Triassic tetrapod tracks)

Demathieu G, Haubold H (1974) Evolution und Lebensgemeinschaft terrestrischer Tetrapoden nach ihren Fährten in der Trias. Freiberger Forschh C 298:51–72

Ginsburg L, De Lapparent F, Taquet P (1968) Piste de *Chirotherium* dans le Trias du Niger. CR Acad Sci 266(D):2056–2058

Haderer FO (2001) Neues vom Handtier. Fossilien 3:172–174 (Rauisuchid *Ctenosauriscus* as potential trace maker of *Chirotherium*)

Haubold H (1984) Saurierfährten, 2nd edn. Die Neue Brehm-Bücherei. A. Ziemsen-Verlag, Wittenberg Lutherstadt, 231 p (Popular description of fossil tetrapod tracks)

Haubold H, Klein H (2000) Die dinosauroiden Fährten *Parachirotherium – Atreipus – Grallator* aus dem unteren Mittelkeuper (Obere Trias: Ladin, Karn, ?Nor) in Franken: Hallesches Jahrbuch für Geowissenschaften B22:59–85

Heyler D, Lessertisseur J (1961) Remarques dur les allures des tétrapodes paléozoiques, d'après les pistes du Permien de Lodève (Hérault). In: Coll. Internationaux du Centre National de la Recherche Sci. (ed) Problemes Actuels de Paléontologie (Évolution des Vertébrés) 104:123–134

Humboldt A von (1835) Note sur les empreintes de pieds d'un quadrupède dans les grès bigarrés de Hildburghausen. Annales des Sciences Naturelles, Serie 2(4):134–138 (Discussion of *Chirotherium*)

Jardine W (1853) The ichnology of Annandale, or illustrations of footmarks impressed on the New Red Sandstone of Cornockle Muir. WH Lizars, Edinburgh, 17 p, 13 pls.

Jux U, Pflug HD (1958) Alter und Entstehung der Triasablagerungen und ihrer Erzvorkommen am Rheinischen Schiefergebirge. Neue Wirbeltierreste und das *Chirotherium*problem. Abhandlungen des Hessischen Landesamtes für Bodenforschung 27:1–50, 3 pls. (Makers of *Chirotherium*)

Kirchner H (1927) Versteinerte Tierfährten mit besonderer Berücksichtigung der sog. *Chirotherium*-Fährten im Buntsandstein Unterfrankens. Fränkische Heimat 1–7

Krämer F (1964) Fährtenfunde von *Chirotherium barthi* KAUP der untersten Solling-Folge (Oberer Buntsandstein). Naturwissenschaften 51(1):11

Krebs B (1966) Zur Deutung der *Chirotherium*-Fährten. Natur und Museum 96(10):389–396 (Trace maker)

Leonardi G (1977) Two simple instruments for ichnological research, principally in the field of vertebrates. Dusenia 10:185–188

Leonardi G (1979) Um Glossário Comparado da Icnologia de Vertebrados em Portugues e uma História Ciencia no Brasil. A comparative glossary of vertebrate ichnology in Portuguese and the history of this science in Brasil. Glossário Comparado da Icnologia dos Vertebrados (17):4–55

Lyell C (1855) Manual of elementary geology. London, 512 p (*Chirotherium* referred to a stegocephalian)

Owen R (1842) Description of parts of the skeleton and teeth of five species of the genus *Labyrinthodon* with remarks on the probable identity of the *Cheirotherium* with this genus of extinct Batrachians. T Geol Soc Lond 6:515–543

Sarjeant WAS (1975) Fossil tracks and impressions of vertebrates. In: Frey RW (ed) The study of trace fossils: A synthesis of principles, problems, and procedures in ichnology. Springer-Verlag, New York, pp 283–324 (Summarizes the *Chirotherium* story as part of a general review of vertebrate ichnology)

Soergel W (1925) Die Fährten der Chirotheria, eine paläontologische Studie. Gustav Fischer, Jena, 92 p (Classic analysis of *Chirotherium* as trackways produced by dinosaur-like reptiles)

Plate 2: Undertracks in Wet Sands

Brown T (1999) The science and art of tracking: Nature's path to spiritual discovery. Berkley Books, New York, 219 p (Subtle effects in vertebrate tracks, including undertracks)

Courel L, Demathieu G (1976) Une Ichnofaune Reptilien remarquable dans les Grès Triasiques de Largentinière (Ardèche, France). Palaeontographica Abt A 151(4–6):194–216, 4 pls.

Courel L, Demathieu G, Gall J-C (1979) Figures sédimentaires et trace d'origine bioiogique du Trias Moyen de la Bordure Orientale du Massif Central. Signification Sédimentologique et Paléoécologique. Geobios 12(3):379–397

Hitchcock E (1858) Ichnology of New England: A report on the sandstone of the Connecticut Valley, especially its fossil footmarks. William White, Boston, 220 p, 60 pls.

Leonardi G (1981) Ichnological data on the rarity of young in North East Brazil dinosaurian populations. An. Acad. Brasil. Cienc. 53(2):345–346 (Consequence of undertrack deficiency)

Ley W (1951) Footprints in Red Sandstone. In: Dragons in Amber: Further adventures of a romantic naturalist. Sedgwick & Jackson, London, pp 53–68 (Paleontologist-turned rocket-scientist tells the *Chirotherium* story for a popular audience)

Martin AJ, Rainforth EC (2004) A theropod resting trace that is also a locomotion trace: Case study of Hitchcock's specimen AC 1/7. Geological Society of America Abstracts with Programs 36(2):96

Perkins BF, Langston W (1983) Trace fossils and paleoenvironments of selected Cretaceous localities, North-Central Texas. Annual Meeting of American Association of Petroleum Geologists, pp 1–61, 46–88 (p 57 stamped undertracks of dinosaur)

Sarjeant WAS (1975) Fossil tracks and impressions of vertebrates. In: Frey RW (ed) The study of trace fossils. Springer-Verlag, New York, pp 283–324 (Summarizes the *Chirotherium* story)

Plate 3: Tambach Ichnotope

Haubold H (1998) The Early Permian tetrapod ichnofauna of Tambach, the changing concepts in ichnotaxonomy. Hallesches Jahrbuch für Geowissenschaften (B) 20:1–16 (Ichnofauna)

Korn H (1933) Eine für die Kenntnis der Cotylosaurier des deutschen Perms bedeutsame Schwimmfährte von Tambach. Palaeobiologica 5:169–200, 1 pl. (Tambach swimming track)

Müller AH (1969) Über ein neues ichnogenus (*Tambia* ng) und andere Problematica aus dem Rotliegenden (Unterperm) von Thüringen. Deutsche Akademie der Wissenschaft in Berlin, Monatshefte 11, pp 922–931 (Diagnosis of the ichnogenus *Tambia*)

Müller AH (n.d.) Eine kombinierte Lauf- und Schwimmfährte von *Koryichnium* aus dem Oberrotliegenden von Tambach (Thüringen). Geologie 4(5):490–497

Reineck H-E, Singh IB (1986) Depositional sedimentary environments, with reference to terrigenous siliciclastic clastics (corrected 2nd printing of 2nd edn). Springer-Verlag, New York, 549 p (Text including biogenic structures)

Rindsberg AK (2005) Gas-escape structures and their paleoenvironmental significance at the Steven C. Minkin Paleozoic Footprint Site (Early Pennsylvanian, Alabama). In: Buta RD, Rindsberg AK, Kopaska-Merkel DC (eds) Pennsylvanian footprints in the Black Warrior Basin of Alabama. Alabama Paleontological Society Monograph 1, pp 177–183 (Distinguishing features of rainprints and gas-escape structures)

Seilacher A (1997) Fossil Art. An exhibition of the Geologisches Institut, Tübingen University. The Royal Tyrell Museum of Palaeontology, Drumheller, Alberta, Canada, 64 p

Steiner W, Schneider H-E (1963a) Eine neue Lauffährte mit Schwanzschleppspur aus dem Oberrotliegenden von Tambach (Thüringer Wald). Geologie 12(6):715–731 (*Herpetichnium* with sculptured tail drag)

Steiner W, Schneider H-E (1963b) Zwei Fährtenplatten aus dem Rotliegenden von Tambach in der Sammlung des Instituts für Geologie und Technische Gesteinskunde Weimar. Wiss. Z. Hochsch. f. Architektur und Bauwesen 10(1):71–87 (Tambach tetrapod trackways)

Plate 4: Coconino-Type Ichnotopes

Brown T (1999) The science and art of tracking: Nature's path to spiritual discovery. Berkley Books, New York, 219 p (Subtle effects in vertebrate tracks, including undertracks)

Haubold H, Lockley MG, Hunt AP, Lucas SG (1995) Lacertoid imprints from Permian dune sandstones. In: Lucas SG Heckert AB (eds) Footprints and facies. New Mexico Museum of Natural History and Science, Bulletin 6, pp 235–244

Huene F von (1941) Eine Fährtenplatte aus dem Stubensandstein der Tübinger Gegend. Centralbl Mineral Geol Paläontologie B 1941:138–141 (*Saurischichnus*, probably forged)

Kuban GJ (1989) Elongate dinosaur tracks. In: Gillette DD, Lockley MG (eds) Dinosaur tracks and traces. Cambridge University Press, New York, pp 57–72 (Detailed study of dinosaur tracks that were mistaken by creationists for Cretaceous giant human footprints)

Leakey MD, Robbins LM, Tuttle RH (1987) Hominid footprints. In: Leakey MD, Harris JM (eds) Laetoli, a Pliocene site in northern Tanzania. Clarendon Press, Oxford, pp 490–523 (Useful synthesis of Laetoli footprints)

Lucas SG, Lerner AJ, Hunt AP (2004) Permian tetrapod footprints from the Lucero uplift, central New Mexico, and Permian footprint biostratigraphy. In: Lucas SG, Zeigler KE (eds) Carboniferous-Permian transition. New Mexico Museum of Natural History and Science Bulletin 25, pp 291–300 (Updated and comprehensive review of Permian tetrapod footprint biostratigraphy)

McKee ED (1944) Tracks that go uphill. Plateau 16:61–72 (One of the first interpretations of trackways in Permian eolian deposits of the Coconino Sandstone)

Schmidt H (1959) Die Cornberger Fährten im Rahmen der Vierfüßler-Entwicklung. Abhandlungen des Hessischen Landesamtes für Bodenforschungen 28, 137 p (Detailed description of a Permian dune ichnotope)

Plate 5: Fish Trails

Anderson AM (1976) Fish trails from the Early Permian of South Africa. Palaeontology 19:397–409 (*Undichna*, including ichnospecies *U. bina*, *U. insolentia*, and the type *U. simplicitas*)

Buatois LA, Mángano MG, Maples CG, Lanier WP (1998) Ichnology of an Upper Carboniferous fluvioestuarine paleovalley: The Tonganoxie Sandstone, Buildex Quarry, eastern Kansas, USA. J Paleontol 72:152–180 (*Undichna* ichnospecies)

Carroll RL (1965) Lungfish burrows from the Michigan Coal Basin. Science 148:963–964

Fiege K (1951) Eine Fisch-Schwimmspur aus dem Culm bei Waldeck, mit Bemerkungen über die Lebensräume und die geographische Verbreitung der karbonischen Fische Nordwest-Europas. Neues Jahrb Geol P M (1):9–31 (Three intermittent, but not undulating parallel grooves from Lower Carboniferous referred to swimming fish)

Gibert JM de (2001) *Undichna gosiutensis*, isp. nov.: A new fish trace fossil from the Jurassic of Utah. Ichnos 8:15–22 (Diagnosis of *U. gosiutensis*)

Gibert JM de, Buatois LA, Fregenal-Martinez MA, Mangano MG, Ortegy F, Poyato-Ariza FJ, Wenz S (1999) The fish trace fossil *Undichna* from the Cretaceous of Spain. Palaeontology 42:409–427 (Analysis of two ichnospecies of *Undichna* and their potential producers in two lacustrine lagerstaetten. *U. unisulca* introduced)

Gregory MR (1981) A Miocene analogue of modern eagle ray feeding hollows. Geol Soc N Z Mtg 1

Gregory MR, Ballance PP, Gibson GW, Ayling AM (1979) On how some rays (*Elasmobranchia*) ecavate feeding depressions by Jetting Water. J Sediment Petrol 49(4):1125–1130

Higgs R (1988) Fish trails in the Upper Carboniferous of southwest England. Palaeontology 72:255–272 (Analysis of three ichnospecies of *Undichna* and potential producers from the Bude Formation. *U. britannica* and *U. consulca* diagnosed)

Linck O (1938) Schwimmfährten von Fischen im Stubensandstein. Jh Ver Vaterl Natkd Württemb 1–3. (*Undichna* in fluvial U. Triassic)

Martin AJ, Pyenson ND (2005) Behavioral significance of vertebrate trace fossils from the Union Chapel site. In: Buta RJ, Rindsberg AK, Kopaska-Merkel DC (eds) Paleozoic footprints in the Black Warrior Basin of Alabama. Alabama Paleontological Society Monograph 1, 59–73 (Earliest evidence of social behavior in amphibians and fishes)

Melchor RN, Bazán J, Fernandez MA (1993) Asociación de trazas fósiles de la facies pelitica de la Formación Agua Escondida (Carbonifero Superior?), sureste de Mendoza, Argentina. Primera Reunión Argentina de Icnología, Resúmenes y Conferencias Invitadas, 15 (*Undichna insolentia*)

Morrissey LB, Braddy SJ, Bennett JP, Marriott SB, Tarrant PR (2004) Fish trails from the Lower Old Red Sandstone of Tredomen Quarry, Powys, southeast Wales. Geol J 39:337–358 (*Undichna trisulcata*)

Simon T, Hagdorn H, Hagdorn MK, Seilacher A (2003) A swimming trace of a coelacanth fish from the lower Keuper of south-west Germany. Palaeontology 46:911–926 (*Parundichna schoelli*)

Soler-Gijón, R., Moratalla, JJ (2001) Fish and tetrapod trace fossils from the Upper Carboniferous of Puertollano, Spain. Palaeogeog Palaeoclim Palaeoecol 171:1–28 (Two ichnospecies of *Undichna* and possible tracemakers)

Trewin, NH (2000) The ichnogenus *Undichna*, with examples from the Permian of the Falkland Islands. Palaeontology 43:979–997 (Diagnosis of *U. quina*)

Turek V (1989) Fish and amphibian trace fossils from Westphalian Sediments of Bohemia. Palaeontology 32:623–634 (Diagnosis of *U. radnicensis*)

Wisshak M, Volohonsky E, Blomeier D (2004) Acanthodian fish trace fossils from the Early Devonian of Spitsbergen. Acta Palaeontol Pol 49:629–634 (Diagnosis of *U. septemsulcata*)

■

Earliest ichnology textbook: Stone Age petroglyphs, Akakus Mts., Libya

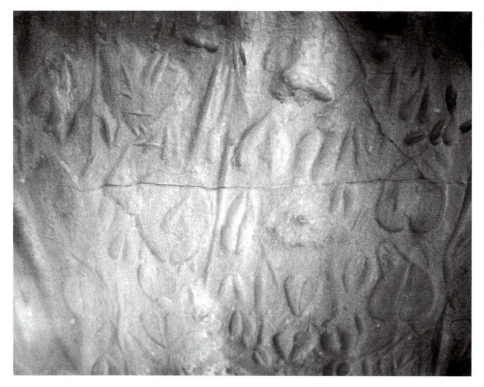

■
Chirotherium.
Buntsandstein, Hildburghausen
(Tübingen Museum)

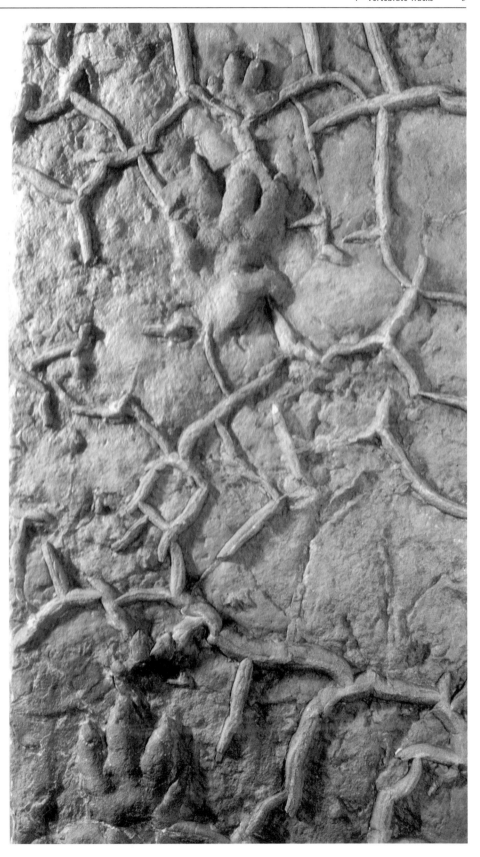

1

Plate 1
Chirotherium: The Sherlock Holmes Approach

Although tetrapod footprints had been found before in the Triassic of Massachusetts and in the Permian and Triassic of Great Britain, the discovery of *Chirotherium* in the Buntsandstein (Lower Triassic) of Germany in 1833 marks the beginning of scientific paleoichnology. The name *Chirotherium* refers to the shape of the individual footprint: as in a human hand, one recognizes five segmented fingers including an outspread thumb. With such a detailed footprint morphology it is obvious in which direction the trackway must be read. Within each trackway there is also a regular alternation of smaller and larger footprints, of which the larger ones can be referred to the rear legs, because they sometimes overstep the smaller prints. So the obvious question was: Who dunit?

Outstanding scientists of the time, such as Alexander von Humboldt, participated in the initial hot discussion over this discovery, but soon it became clear that the makers could be neither ape- nor bear-like mammals, because the thumb is on the wrong (i.e. the outer) side of the impression. The interpretation of Richard Owen and Charles Lyell appeared more plausible, because they referred the track to stegocephalian amphibians – the only group of vertebrates whose bones were known from the Buntsandstein. But Lyell's reconstruction struggles with the thumb placement by introducing a cross-legged gait. It also fails to explain why the front legs, being smaller and weighted by an enormous skull, did not become more deeply impressed than the rear ones.

The modern interpretation of *Chirotherium* as tracks of sauropod reptiles, already suggested by the turn of the century, became established with the classic analysis of W. Soergel. He started from the assumption that for a comfortable stepping angle of about 40° the legs of the *Chirotherium* animal had to be rather high. Its approximate trunk length could also be estimated from the distance between the midpoints of synchronous front- and rear-leg paces, because in narrow-gauge walking, shoulder and pelvis should be above these points. But as with equal weight distribution the smaller front feet should have made a deeper impression, Soergel hypothesized a small head and a long and heavy tail for counterbalance. Such features could be expected in the ancestry of bipedal dinosaurs.

The footprints themselves provide additional and more direct evidence. They suggest that only the fingers were in contact with the substrate. Also we can determine the number of bones (phalanges) in each finger from constrictions in the impression. In perfect preservation one may even recognize that the skin was covered with horny scales – a distinct reptilian feature. The "*phantom*" picture that Soergel derived from this analysis was later matched by skeletons found in other areas, where facies conditions favored the preservation of bones but not of tracks (Haderer reconstruction).

In the meantime, Soergel's basic procedure has been highly elaborated. The majority of known vertebrate tracks can now be attributed to certain groups of amphibians, reptiles, birds, and mammals at least at the ordinal level, so that one can design a "family tree" of tetrapod footprints, congruent to the one derived from skeletal remains.

A visit to the zoo makes us aware of the many kinds of gaits that have evolved in terrestrial tetrapods. Tracks in the snow also tell us that the various modes of locomotion are well reflected in the track patterns. Such actualistic knowledge can be readily applied to fossil trackways. They show for instance, that several groups of dinosaurs acquired bipedality later in the Triassic. But it is also possible to distinguish between running and walking tracks made by the same organism, or between the wide gauge in lizard-like and the narrow gauge in mammal-like locomotion. It is beyond the scope of this course to go into details. Obviously, there is a lot of detective work to be done by dinosaur trackers.

■ *Chirotherium*, overstepping mud-crack fills

Plate 1 · *Chirotherium*: The Sherlock Holmes Approach   7

Chirotherium
L. Triassic, Germany

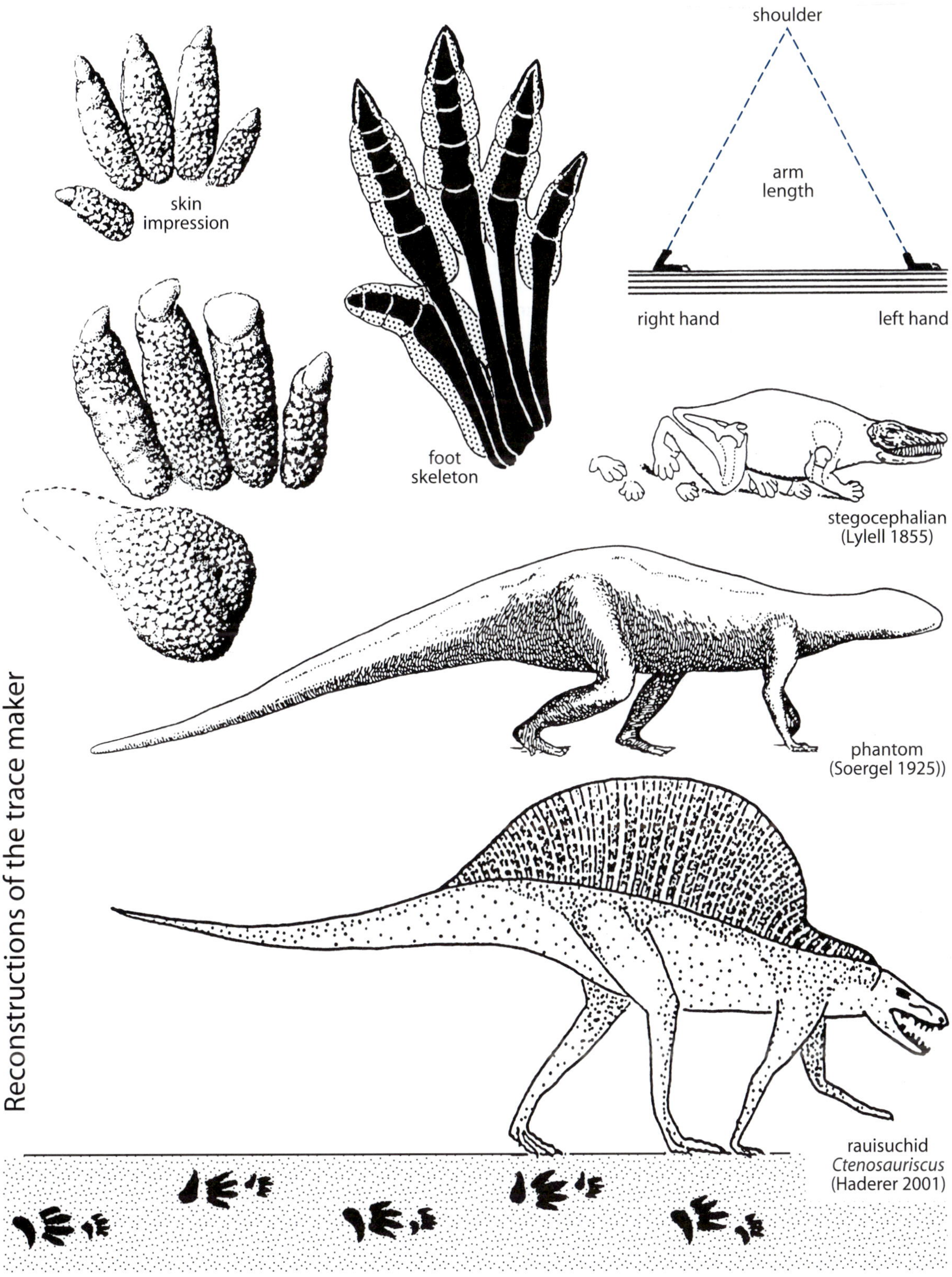

skin
impression

foot
skeleton

shoulder

arm
length

right hand   left hand

stegocephalian
(Lylell 1855)

phantom
(Soergel 1925))

rauisuchid
Ctenosauriscus
(Haderer 2001)

Reconstructions of the trace maker

2

Plate 2
Undertracks in Wet Sands

Buntsandstein-Type Ichnotopes. The preservation of footprints requires *(a)* a substrate of the right plasticity, *(b)* sediment acting as a casting agent and *(c)* the absence of erosion prior to casting. But as sand sedimentation reflects a high-energy event preceded by erosion, the footprint can be preserved only under very special conditions. Such a situation would be a flood pool in the desert, in which suspended clay had settled to form a mud veneer. As the pool dried out, footprints could form and become hardened before a new layer of sand is blown over them.

This preservational model appears to fit the common association of *Chirotherium* foot prints with polygonal mud cracks. The strange morphology of the crack fills (with a median seam and bulging sides, like the burrows of some worms or echinoids) is clearly due to secondary deformation. As the mud layer was only a few centimeters thick, the cracks extended all through. Thus their sandy infill could fuse along the "seam" with the underlying sand. Only when the mud became compacted, did the non-compactable sand-wedge squeeze out into a sausage.

More irritating is the relationship of the footprints to the crack fillings. According to the simple model, they would have been impressed when the mud was still wet, i.e. before the cracks could develop. But neither did the footprints influence the crack pattern, nor were they visibly deformed by the opening of the cracks. On the contrary we find that the footprints intersect the crack casts! Thus they must have been made after the cracks had become sand-filled. In other words, what we see are *undertracks* pressed through a thin sand cover into the cracked mud layer, which by that time had become plastic again by rain or flooding. In the present context, the undertrack concept is also important for understanding other types of Buntsandstein footprints that probably stem from the same tracemaker.

The "**flatfooted**" variant still resembles *Chirotherium* in general outline, but the margins form a vertical cliff, while the sole of the impression does not even express the toes. It was clearly made in softer mud, but the preservation of marginal slip traces suggests that it was also pressed through a protective sand cover. A third kind of impressions occurs not in trackways, but as isolated **dig traces** with groups of up to four claw traces. It reflects the digging – probably of the same creature – to a lower level, where undertracks had a still better chance to be preserved. In a regional sense, these three preservational types of *Chirotherium* characterize different sedimentary regimes within the braided river systems of the German Buntsandstein.

Very strange features have been observed in *Chirotherium*-like footprints from the Middle Triassic of France, where the terrestrial facies extends into the Lower Muschelkalk (Muschelsandstein). Some of the footprints are preserved as positive (**inverted**) epireliefs, i.e. they stick out from the sandstone like casts, although we deal with a top surface! As dogs running along a wet beach may produce "inverted" tracks, this phenomenon has been referred to the suction exerted by a flat sole that is quickly lifted off a wet sand surface. But this effect would have been even more pronounced (and not restricted to high speeds), if the medium around the foot were not a film of water, but sticky mud. Although experimental tests have yet to be made, it may be reasonably assumed that inverted footprints in the fossil record were made on sand overlain by several centimeters of wet mud. Associated footprints surrounded by concentric "**toe webs**", as well as featureless "**bolide**" impressions, support this interpretation.

Connecticut-Type Ichnotopes. A quite different group of track *lagerstaetten* is exemplified by the Upper Triassic sandstones of the Connecticut Valley (New England). Most famous is the old collection of Eduard Hitchcock in the Pratt Museum of Amherst College, Massachusetts. This museum contains his famous "**fossil volume**", in which impressions can be followed through successive bedding planes over a thickness that may exceed the diameter of a single footprint. But unlike in the vertical repetition of arthropod tracks (next chapter), the dinosaur foot did not pierce through all this thickness (penetrative undertracks), but stamped identical copies from the top by bed-to-bed deformation (compressive undertracks). This means that, in contrast to the "pressure bulb" of soil mechanics, the shock load of the dinosaur foot produced a "**pressure prism**" in the sediment underneath; i.e. the sediment was not isotropic. Was there a deformational anisotropy caused by horizontal mica flakes or biomats, or is a pressure cylinder typical for shock deformation of wet sand?

The soil mechanics approach can also be extended to smaller features. In the figured Connecticut Valley undertrack, toe impressions show a fine **microfault pattern** along the toes. Closer inspection, including thin sections, reveals that there forms an antithetic fault system in response to tensional deformation – a phenomenon that we shall again come across in asterosomid worm burrows (Pl. 41).

The Connecticut Valley undertracks also reflect the differential loading of the toes during one step. In the figured slab, the top surface still shows a clear three-toed contour, although it was already below the sediment/water interface. In the next copy, however, no more than the front part of the middle toe is expressed by longitudinal fault ridges (**undertrack deficiency**). Only the claw actually penetrated to this level, because during the final push the load became concentrated on the large middle toe and at the last moment onto its pointed claw.

Plate 2 · Undertracks in Wet Sands   9

Chirotherium: Preservational Variants

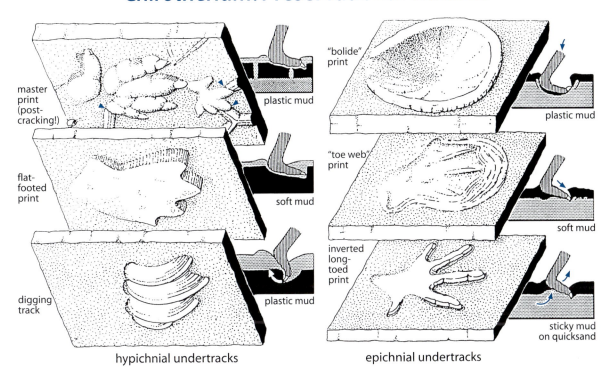

master print (post-cracking!)

plastic mud

flat-footed print

soft mud

digging track

plastic mud

"bolide" print

plastic mud

"toe web" print

soft mud

inverted long-toed print

sticky mud on quicksand

hypichnial undertracks

epichnial undertracks

Conneticut Valley: Undertrack Repetition

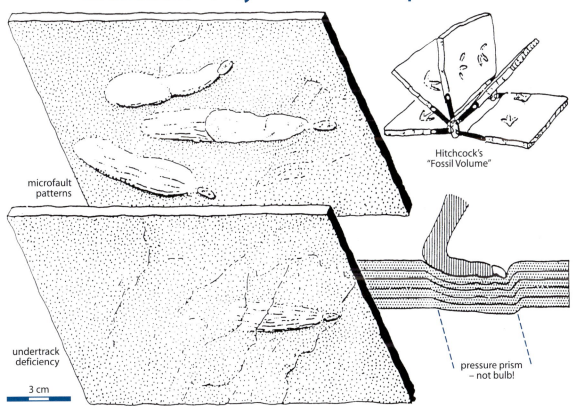

microfault patterns

undertrack deficiency

3 cm

Hitchcock's "Fossil Volume"

pressure prism – not bulb!

3

Plate 3
Tambach Ichnotope

Our third representative ichnotope is a Lower Permian sandstone near Tambach (Thuringia, Germany). By its red color and sedimentary structures (including mud-crack fills) it resembles the Triassic *Chirotherium* Sandstones in the same area. Its trackways were made by reptiles with a wider gait, such as pelycosaurs, but they also differ from typical *Chirotherium* by some interesting preservational features.

Let us begin with a slab (not illustrated) that is exhibited in front of the museum of Gotha (Germany) and as a cast in the travelling exhibit "Fossil Art". It shows the hypichnial casts of a trackway and suncrack fillings. In this case, the cracks intersect the footprints, as expected in the normal sequence: moist mud/desiccation/burial under sand. There are, however, two irritating details. (1) The mud cracks consist of two superimposed systems, one more delicate than the other. (2) There are many raindrop impressions, but only in the upper right corner. As rain does not stop at a sharp line (remember the beduin who insists to leave by the left door of a taxi, because it is raining on the right side!), the rest of the mud surface must have already been covered by sand.

The second example (top figures) is on exhibit in the Tübingen museum (Germany). It also appears to represent a simple desiccation sequence. The large trackway was obviously made while the underlying mud was still very wet, because its footprints are deep, smeared, and have squeezed-out rims. When the mud dried, cracks followed the line of these footprints, whose original outlines have been **de-cracked** in the right picture by cutting-out and refitting the clay tablets. Yet there is a paradox: in the lower part of the trackway, footprints are missing. Instead there is a much smaller footprint belonging to another trackway. It had obviously been made earlier and was covered by a small sand dune when the large animal ran over it. The clear outline of the small footprint (arrow), and the distinct impressions of pads and claws indicate that it was made not in soft, but more plastic mud. Probably this represents different track generations separated by a rainfall that turned the mud wet again.

A third slab, also in front of the Gotha museum (Germany), and as a cast in "Fossil Art", further complicates the preservational history of the Tambach ichnotope. In addition to a vertebrate undertrack, it contains minute whirls of parallel scratches (***Tambia spiralis***). Their origin was a riddle until the solution came with a unique trace fossil from the **Oligocene** *Uintatherium lagerstaette* of Wyoming. It has the shape of an inverted beehive and its thick wall consists of a spirally coiled cylindrical burrow, in which meniscate backfill lamellae indicate downward burrowing. When the right depth was reached, the animal dug increasingly narrower coils and finally turned

upward again to excavate a cavity inside. Most probably this is the pupa chamber of a coleopteran grub that built the beehive for additional protection (perhaps assisted by toxicity) during metamorphosis into some kind of large beetle (see Pl. 20 for more examples). Alternatively, the structure could have been made by a dung beetle to store its crop for the next generation. In any case, *Tambia spiralis* probably corresponds to the bottom of a similar chamber – with the difference that the sand/mud interface preserved only the scratches of the producer. Whereas vertical sections failed to reveal the chamber itself, its former presence can be inferred from a round **collapse caldera** surrounding each scratch spiral.

As *Tambia* is superimposed on the footprints, it was certainly dug at a later stage. The association of the two, however, is not random: many scratch spirals coincide exactly with the toe-tip impressions, as if they had been an integral part of the vertebrate track! We conclude that the grubs dug until reaching the bottom of the sand layer. There they preferred places where depressions in the interface concentrated moisture. As pupa chambers and dung silos penetrate to about double their width, they also tell us that the sand layer was at least 5 cm thick before the mud cracks eventually opened and rifted parts of the footprints apart as in the top figure (see sequence in the plate). Perhaps most of the sharp "suncracks" in the fossil record formed under sand cover without ever having seen the sun.

■ Large footprint (L. Permian, Tambach); *arrow:* small footprint in place of large one (see Pl. 3)

Plate 3 · Tambach Ichnotope   11

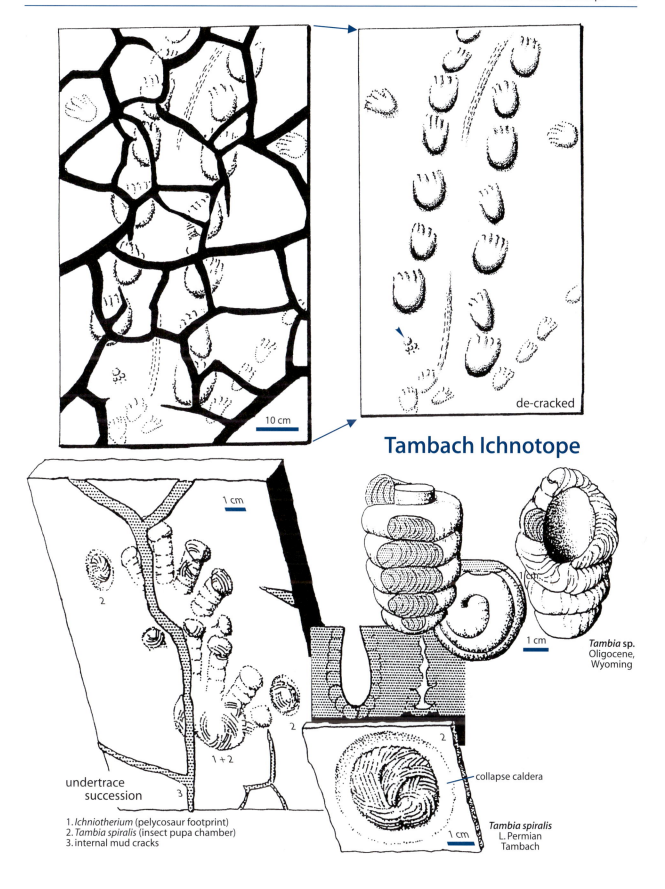

10 cm

de-cracked

Tambach Ichnotope

Tambia sp.
Oligocene,
Wyoming

1 cm

1 cm

2

undertrace
succession

3

1+2

2

collapse caldera

Tambia spiralis
L. Permian
Tambach

1. *Ichniotherium* (pelycosaur footprint)
2. *Tambia spiralis* (insect pupa chamber)
3. internal mud cracks

4

Plate 4
Coconino-Type Ichnotopes

Anyone who hiked down into the Grand Canyon of Arizona knows the Permian Coconino Sandstone and the reptilian tracks exhibited along the trail. Yet, few of the visitors wonder about their preservation. All the vertebrate tracks we have discussed so far were made in wet substrates, because dry mud is unsuited for footprinting. The Coconino Sandstone, however, is a typical dune deposit. Dune sand must be dry to be transported by wind. It is also soft enough for tracking. In the morning light, dune surfaces present a beautiful record of the beetles, millipedes, scorpions, lizards and snakes that had moved over it during the night. But once the dew is gone, how can such delicate impressions survive the deposition of a new sand layer? In dunes, our own tracks usually penetrate deeply and fail to show clear outlines or shoe profiles. In contrast, Coconino-type footprints (and similar ones in the Permian Cornberger Sandstein of Germany) have not only distinctive outlines. They also show details whose cotour is incompatible with a surface origin: push-back hills that are stepped by microfaults rather than having smooth slopes.

The figured trackway (*Chelonichnus* from the Cornberger Sandstone; local museum at Rotenburg near Fulda, Germany) also exemplifies the phenomenon of undertrack deficiency. Only its left file preserves the traces of the hand as well as the foot, the latter overstepping the hand impression and being less deeply impressed. On the right side, however, foot impressions are completely absent. Probably the animal climbed the dune at an angle (see dip lines), so that the uphill foot did not penetrate deep enough to reach the bedding plane now exposed.

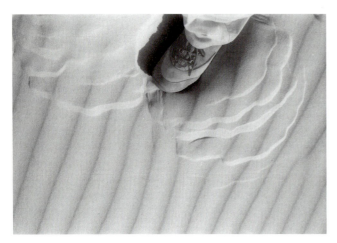

■ Fault deformation around dune footprint

An undertrack origin explains not only this, but also another strange phenomenon. When Eddie McKee, then naturalist in the Grand Canyon, measured cross laminations in the Coconino Sandstone, he found that the vertebrate tracks run almost exclusively uphill! This remained a riddle, until a similar behavior was observed in **trilobite tracks** on a rippled Silurian storm sand. In spite of the differences in scales and environments, the two examples have in common that they are undertracks. As more force is required in climbing, uphill footsteps penetrate deeper and therefore have a higher undertrack potential than those made downhill.

For lay persons, dinosaur footprints have a unique fascination. They are also easy to forge, because no other material than the sediment is involved. Therefore one should be critical of uncontrolled new discoveries. A slab of Keuper (Upper Triassic) sandstone in the Tübingen museum (*Saurischichnus*) exemplifies this. The circumstances of its discovery (it was found by a local doctor on the floor of a farmer's kitchen) suggest that it was not an intented forgery, but the delight of a loving father generations ago. He must have produced the footprints with a chisel, with the result that the centers (rather than the toe tips) appear most deeply impressed – a situation incompatible with the construction of a vertebrate foot.

Human footprints are commonly noted in the current creationist debates as a proof for the coexistence of man with dinosaurs. In such cases it is telling that the claimed evidences are always negative epireliefs (rather than less forgeable casts). Others are casts preserved imperfectly enough to allow any wishful interpretation.

Vertebrate paleoichnology is an important field of its own. Tetrapod tracks are sufficiently "fingerprinted" to be ascribed to certain taxonomic groups, at least at the ordinal level. Nevertheless an experienced vertebrate paleontologist is needed to retrieve all the biohistoric information encoded in such tracks. In the absence of other remains, they may also be used for biostratigraphic correlation. One should bear in mind, however, the modifications imposed on track patterns by different gaits, and on footprint morphology by differences in substrate consistence and modes of preservation. Such features may confuse ichnotaxonomy, but they also allow us to view tetrapod tracks as experiments in paleosol mechanics and to use them as paleoenvironmental indicators. For completeness, one should also mention the *burrows* made by vertebrates. On land, many small mammals with a nocturnal lifestyle use them to spend the day and raise their young. The large corkscrew burrows of prairie dogs (*Daimonelix*) have particularly vexed early paleontologists, but they are clearly distinguished by size and geometry from helicoidal burrows in subaquatic environments, on which this book is focused.

Plate 4 · Coconino-Type Ichnotopes   13

Undertracks Going Uphill

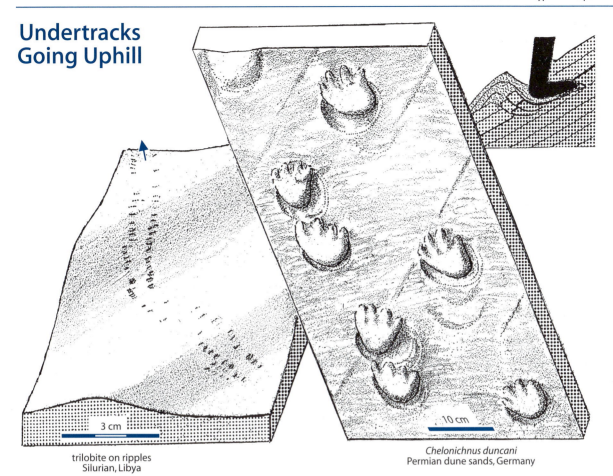

3 cm

trilobite on ripples
Silurian, Libya

Chelonichnus duncani
Permian dune sands, Germany

10 cm

Forgery

10 cm

Saurischichnus primitivus v. Huene

5

Plate 5
Fish Trails

The term "swimming trail" (or "swim trail") appears an oxymoron, a contradiction in itself: swimming is locomotion in a medium that provides buoyancy and low friction. Accordingly, a swimming fish should avoid contact with the bottom. Yet, such trails are rather common in the fossil record, particularly in lake deposits. This has nothing to do with the lower density of freshwater, nor with a particular behavioral trait of its inhabitants. More likely it relates to exceptionally low water levels, in which swimming was impossible without touching the bottom. Another factor may be the laminated character of many lake sediments, which favored the formation of preservable undertraces. So it is probably no coincidence that all examples shown on this plate come from freshwater deposits.

The motion of the fin tips traced from movies of a modern **dogfish** provides the model for the trace fossil *Undichna*. Its pattern reflects the division of labor among fins. The tail fin (in sharks, its trailing lower wing is the shorter one) acts as the propeller. Therefore the undulating trace of the tail fin has the largest amplitude. Other fins act as keels and for active steering. Accordingly they follow the undulating course of the tail fin, but they are offset by the axial distance between the fins plus the motion of the body before the wave reached the pelvic and pectoral fins. Away from the tail fin there is also a decrease in amplitude.

In *Undichna consulca* only the tail (dotted line) and an unpaired anal fin have left trails. The broad central groove (shaded) must have been made by the mouth, because it does not undulate.

Undichna britannica shows no such groove, but the trails of a paired fin alternate in the rhythm of the undulation.

The paired trails of *Undichna radnicensis* undulate with an offset of half a period, so the corresponding fins were farther away from the tail fin. In *Undichna simplicitas* there are two paired fin trails, the front ones with an unusually wide spread. The asymmetry in the figured specimen is probably caused by a side current. *Undichna quina* and *Undichna gosiutensis* differ again by the number, gauge and offsets of the paired fin trails.

Except for the undulation of the elements, the situation is completely different in the large trackway called *Parundichna*. It was found as enforcement on a riverbank and was traced to the original construction site, where a Lower Keuper (Upper Triassic) sandstone cropped out. In contrast to the previous examples it is not preserved in a laminated sediment, but as positive hyporelief on the sole of a sandy event bed. There is no nonpaired trace of a tail or anal fin and the paired fins have left sets of multiple sigmoidal scratches. They correspond to projecting fin rays that actively combed the sediment (**A**), but nevertheless alternate like the footprints in a tetrapod trackway.

At first glance (**B**) the braided pattern looks like the product of one pair of rayed fins acting in alternation, but in fact it is made by two pairs that touched bottom only during the active inward stroke (**C**). Drawing only the front scratches in each set and supplementing the unrecorded outward recovery stroke (**D**), one can reconstruct the sinusoidal courses of the four fins, resulting from the pendulum motion superimposed on the head-on motion of the body (**E**).

The most probable tracemaker was a large coelacanth crossopterygian driven by the unpaired fins in the elevated rear part of the body. The four paired fins acted on the bottom in the rhythm of terrestrial quadrupeds, probably to shy up benthic prey rather than for locomotion.

Without showing examples, it should be mentioned that stingrays leave another kind of trace, which was first noted in the Tertiary of New Zealand. These animals are so flat not only for their unique mode of swimming and for blending with the bottom by changing color patterns. They also use this body for creating underneath a chamber for processing the sediment. In a way, their technique resembles that of trilobites (Pl. 11). In both cases, the chamber is flushed by respiration current, but as rays have no legs to dig, their excavations lack distinctive bioglyphic fingerprints.

Plate 5 · Fish Trails   15

Parundichna: Undulation ~ Gait of Paired Fins

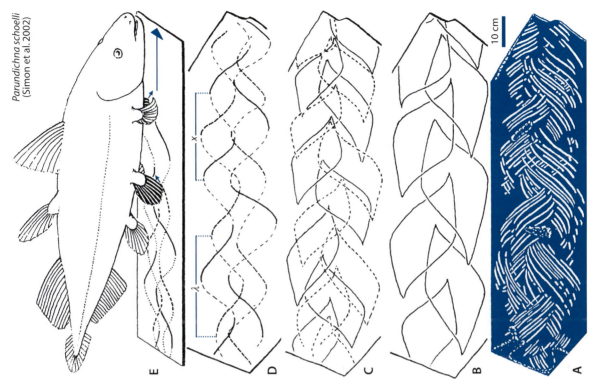

Undichna: Undulation Induced by Tail Fin

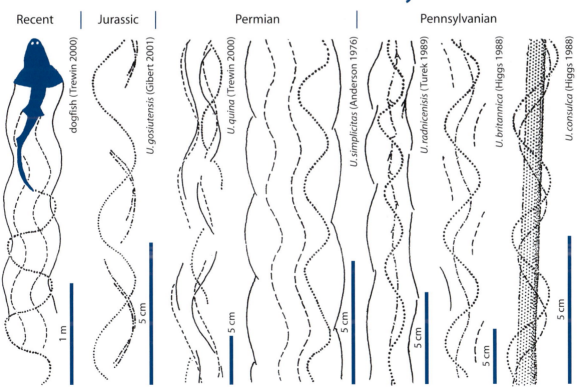

20 cm

■ *Parundichna schoelli*, U. Triassic (Muschelkalk Museum, Ingelfingen)

Arthropod Trackways

Although being supported by an external instead of an internal skeleton, arthropod legs can be functionally compared to those of vertebrates – except that the smaller body size and the larger number of appendages makes their tracks more difficult to interpret. Nevertheless, some groups can be distinguished by characteristic track patterns, as exemplified by the trackways in the upper part of Pl. 7. Except for trilobites and millipedes, their makers used only a small number (≤ 6 pairs) of appendages for walking. Some of them are also heteropodous, which adds to their distinctiveness. But beyond the identification of the tracemakers, arthropod tracks offer many more challenges to the paleodetective.

One difference is the way the legs make contact with the substrate. In land animals (vertebrates; insects), the tips broaden into a foot in order to reduce penetration (snowshoe principle). In contrast, many aquatic arthropods have pointed legs that intentionally pierce into the sediment. This improves not only anchorage against passive displacement by currents or waves. At the same time, sensory setae concentrated near the leg tip allow the animal to probe the substrate for food. Deeper penetration also has an ichnologically important side effect: it increases the likelihood of preservable *undertracks* to be formed. However, unlike the stamped-through footprints of Hitchcock's rock volume (Pl. 2), the impressions of an arthropod undertrack have actually been touched by the tracemaker.

The other requirement for the preservation of continuous arthropod undertracks is a laminated (or biolaminated) sand or silt that splits easily in the fossilized state. As such sediments form mainly below wave base, arthropod tracks tend to be most common in less agitated lake deposits. As we have seen in the vertebrate footprints of the Coconino Sandstone (Pl. 4), laminated dune sands are another lithotope favoring the presentation of pierced-through undertracks. Therefore it is not surprising that eolian sandstones are another favorite ichnotope for tracks including the trackways of large arthropods: scorpions in the Permian (Pl. 7) and eurypterid-like animals in the Upper Cambrian (Pl. 10D).

A very particular ichnotope is represented by the Upper Cambrian Potsdam Sandstone of the Eastern United States and Canada (Pl. 10). It is not laminated, but reflects a narrow taphonomic window in geologic time. In order to understand its nature, we must sidetrack to associated impressions of *jellyfish*. As their body consists up to 98% of water, coelenterate medusae have an extremely low fossilization potential. The majority of what has been described as fossil jellyfish can better be explained as burrow systems (Pl. 47), or pseudofossils (Pl. 62) that happen to have a similar radial symmetry. The experience at modern beaches appears to contradict this statement. As medusae swim primarily for Red Queen filtration, onland winds may wash them ashore in great numbers. Once out of the water, they leave deep and well recognizable impressions in the sand. As the sun reduces the jellyfish body into a thin organic film, this foil will additionally protect the impressions. So, why don't we find them more commonly in ancient beach deposits? The answer is that the impressions became wiped-out by the next tide – unless a resistant microbial film could develop during low tide to produce a kind of death masks. After having disappeared from permanently submerged bottoms in the Cambrian Substrate Revolution (Pl. 65), they persisted on intertidal sand flats during Cambrian times.

What has been said about the jellyfish impressions also applies to the preservation of *surface tracks* of large molluscs (*Climactichnites*, Pl. 63) and arthropods (*Protichnites*, Pl. 10) in the Potsdam Sandstone and its equivalents. Only because they were made on a microbial film rather than loose sediment did the next flood fail to erase them.

Literature

Chapter II

Braddy SJ (1995) The ichnotaxonomy of the invertebrate trackways of the Coconino Sandstone (Lower Permian), northern Arizona. In: Lucas SG, Heckert AB (eds) Early Permian footprints and facies. New Mex Mus Nat Hist Sci Bull 6:219–224 (Taxonomic review of a classic arthropod ichnofauna)

Manton SM (1949) The evolution of arthropodan locomotory mechanisms. Part I. The locomotion of *Peripatus*. Journ Linn Soc London Zool 41:529–570

Manton SM (1952) The evolution of arthropodan locomotory mechanisms. Part 2. General introduction to the locomotory mechanisms of the Arthropoda. Journ Linn Soc London Zool 42(284):93–167

Manton SM (1954) The evolution of arthropodan locomotory mechanisms. Part 4. The structure, habits and evolution of the Diplopoda. Journ Linn Soc London Zool 42(286):300–368

Manton SM (1956) The evolution of arthropodan locomotory mechanisms. Part 5. The structure, habits and evolution of the Pselaphognatha (Diplopoda). J Linn Soc London Zool 43(290):153–187

Manton SM (1977) The arthropoda: Habits, functional morphology, and evolution. Clarendon, Oxford, 527 p (Discussion of experimental work on arthropod trackways)

Schmidtgen O (1927) Tierfährten im oberen Rotliegenden bei Mainz. Paläont Z 9(1):101–107 (Arthropod tracks in Permian redbeds, Germany)

Trewin NH (1994) A draft system for the identification and description of arthropod trackways. Palaeontology 37:811–823. (Proposal for descriptive terminology of arthropod trackways)

Walker EF (1985) Arthropod ichnofauna of the Old Red Sandstone at Dunure and Montrose, Scotland. T Roy Soc Edin-Earth 76:287–297

Plate 6: Limulid Tracks

Bandel K (1967) Isopod and limulid marks and trails in Tonganoxie Sandstone (Upper Pennsylvanian) of Kansas. University of Kansas, Paleontological Contributions 19, pp 1–10 (Description of Xiphosuran trackways from carboniferous marginal marine deposits)

Bromley RG (1996) Trace fossils: Biology, taphonomy and applications, 2nd edn. Chapman & Hall, London, 361 p (Figure 8.14 summarizes successive interpretations of *Kouphichnium*)

Buatois LA, Mangano MG, Maples CG, Lanier WP (1998) Ichnology of an Upper Carboniferous fluvioestuarine paleovalley: The Tonganoxie Sandstone, Buildex Quarry, eastern Kansas, USA. J Paleontol 72:152–180 (Reanalysis of limulid trackways from the Carboniferous site originally documented by Bandel)

Caster KE (1938) A restudy of the tracks of *Paramphibius*. J Paleontol 12:3–60 (This classic paper examined locomotion in *Limulus* and reinterpreted Xiphosuran trackways in the Devonian of New York State)

Caster KE (1939) Were *Micrichnus scotti* Abel and *Artiodactylus sinclairi* Abel of the Newarc Series (Triassic) made by "vertebrates" or limuloids? Am J Sci 237:786–797

Caster KE (1944) Limuloid trails from the Upper Triassic (Chinle) of the petrified forest national monument, Arizona. Am J Sci 242:74–84

Eldredge N (1970) Observations on burrowing behavior in *Limulus polyphemus* (Chelicerata, Merostomata), with implications on the functional anatomy of trilobites. American Museum Novitates 2436, 17 p

Figuier L (1866) La Terre avant la Deluge (1st edn). Hachette, Paris, 435 p (*Kouphichnium* referred to *Pterodactylus*)

Goldring R, Seilacher A (1971) Limulid undertracks and their sedimentological implications. Neues Jahrb Geol P-A 137:422–442 (This paper introduced the undertrack concept to paleoichnologists, although it was already known to trackers; see Brown 1999)

King AF (1965) Demonstration: Xiphosurid trails from the Upper Carboniferous of Bude, north Cornwall. P Geol Soc London 1626: 162–165 (Limulid tracks)

Linck O (1943) Die Buntsandstein-Kleinfährten von Nagold. N Jb Mineral Mh Abt B (1):9–27 (Limulid tracks in Triassic redbed: *Limuludichnulus*)

Linck O (1961) Lebens-Spuren niederer Tiere (Evertebraten) aus dem württembergischen Stubensandstein (Trias, Mittl. Keuper 4) verglichen mit anderen Ichnocoenosen. Stuttgarter Beitr Naturkd, Ser B 66:1–29 (Limulid tracks in fluvial Upper Triassic)

Malz H (1964) *Kouphichnium walchi*, die Geschichte einer Fährte und ihres Tieres. Nat Mus 94(3):81–97

Walther J (1904) Die Fauna der Solnhofener Plattenkalke, bionomisch betrachtet. Naturwiss. Gesellschaft Jena, Med. Denkschriften, Festschrift für E. Haeckel, 1, 135–214 (Referred to *Archaeopterix*)

Plate 7: Other Arthropod Trackways

Anderson AM (1975) Turbidites and arthropod trackways in the Dwyka glacial deposits (early Permian) of South Africa. T Geol Soc S Afr 78:265–273 (Trackways of *Umfolozia* and *Maculichna* with single median drag line)

Braddy SJ, Almond JE (1999) Eurypterid trackways from the Table Mountain Group (Ordovician) of South Africa. J Afr Earth Sci 29:165–177

Brady LF (1947) Invertebrate tracks from the Coconino Sandstone of Northern Arizona. J Paleontol 21(5):466–472, 4 pls.

Brady LF (1961) A new species of *Paleohelcura* Gilmore from the Permian of northern Arizona. J Paleontol 35:201–202 (Additional descriptions of scorpionid trackways from the Coconino Sandstone)

Briggs DEG, Rolfe WDI, Brannan J (1979) A giant myriapod trail from the Namurian of Arran, Scotland. Palaeontology 22(2):273–291, 3 pls.

Briggs DEG, Plint AG, Pickerill RK (1984) *Arthropleura* trails from the Westphalian of eastern Canada. Palaeontology 27:843–855 (Study of the trackway *Diplichnites cuithensis* and its maker, the giant myriapod *Arthropleura*)

Buatois LA, Mángano MG (2003) Caracterización icnológica y paleoambiental de la localidad tipo de *Orchesteropus atavus*, Huerta de Huachi, provincia de San Juán, Argentina: Implicancias en el debate sobre los ambientes de sedimentación en el Carbonífero de Precordillera. Ameghiniana 40:53–70 (Description and paleoecologic interpretation of the ichnofauna associated with *Orchesteropus atavus* at its type locality)

Buatois LA, Mángano MG, Maples CG, Lanier WP (1998) Ichnology of an Upper Carboniferous fluvioestuarine paleovalley: The Tonganoxie Sandstone, Buildex Quarry, eastern Kansas, USA. J Paleontol 72: 152–180 (Analysis of *Diplichnites gouldi* as a myriapod trackway)

Casamiquela RM (1965) Analisis de "*Orchesteropus atavus*" Frenguelli y una Forma afin, del Paleozoico de la Argentina. Estud Icnol 4(24):187–244, 8 pls. (Morphologic and functional analysis of the enigmatic Carboniferous trackway *Orchesteropus atavus*)

Fiege K (1961) Beobachtungen an rezenten Insekten-Fährten und ihre palichnologische Bedeutung. Meyniana 2:1–7 (Actuo-paleontological observations on insect tracks)

Gilmore CW (1929) Fossil footprints from the Grand Canyon. Smithsonian Miscellaneous Collections 77,41 p (Diagnoses of the first known Coconino trackways)

Hanken N-M, Störmer L (1975) The trail of a large Silurian eurypterid. Fossils and Strata 4:255–270, 3 pls (Detailed kinetic analysis of trackway 15 cm wide. Referred to the giant *Mixopterus*)

Heller F (1937) Eine Tiefährtenfundstelle im Rotliegenden Oberhessens. Jber Mitt Oberrh Geol Ver 26:76–78 (Insect traces in Permian redbeds)

Lützner H, Mann M (1988) Arthropodenfährten aus der Phycoden-Folge (Ordovizium) des Schwarzburger Antiklinoriums. Z Geol Wissenschaft 16(6):493–501 (Ordovician arthropod tracks)

Mángano MG, Buatois LA, West RR, Maples CG (2002) Ichnology of an equatorial tidal flat: The Stull Shale Member at Waverly, eastern Kansas. Bulletin of the Kansas Geological Survey 245, 130 p (Large *Diplichnites cuithensis* in Carboniferous fluvial deposits)

Osgood RG (1970) Trace fossils of the Cincinnati area. Paleontographica Americana 6:281–444 (Monograph in which *Allocotichnus* and other Late Ordovician arthropod trackways were analyzed using plastic overlays)

Pirrie D, Feldmann RM, Buatois LA (2004) A new decapod trackway from the Upper Cretaceous, James Ross Island, Antarctica. Palaeontology 47:1–12 (A Cretaceous example of a decapod trackway attributed to brachyurans)

Richter R (1954) Fährte eines "Riesenkrebses" im Rheinischen Schiefergebirge. Nat Mus 84:261–269 (*Palmichnium*)

Savage NM (1971) A varvite ichnocoenosis from the Dwyka Series of Natal. Lethaia 4:217–233 (Arthropod trackways in Permian glacial lake deposits)

Schmidtgen O (1927) Tierfährten im oberen Rotliegenden bei Mainz. Paläont Z 9:101–107 (Description of arthropod tracks in Permian redbeds)

Schmidtgen O (1928) Eine neue Fährtenplatte aus dem Rotliegenden von Nierstein am Rhein. Palaeobiologica 1:245–252 (Permian insect trackways)

Seilacher A, Hemleben C (1966) Beiträge zur Sedimentation und Fossilführung des Hunsrückschiefers, Teil 14, Spurenfauna und Bildungstiefe des Hunsrückschiefers. Hessisches Landesamt für Bodenforschung, Notizblatt 94, pp 40–53 (Description of heteropodous and isopodous trackways)

Seilacher A, Buatois LA, Mángano MG (2005) Trace fossils in the Ediacaran-Cambrian transition: Behavioral diversification, ecological turnover and environmental shift. Palaeogeog Palaeoclim Palaeoecol 227:323–356 (Description of the oldest example of a jumping arthropod, *Tasmanadia cachii,* from lowermost Cambrian strata of the Puncoviscana Formation of Argentina)

Sharpe CFS (1932) Eurypterid trail from the Ordovician. Am J Sci 24:355–361

Taljaard MS (1962) On the palaeogeography of the Table Mountain sandstone series. S Afr Geograph J 25–27, 2 pls (Eurypterid trackways)

Plate 8: Trilobite Tracks

Anderson AM (1975) The "trilobite" trackways in the Table Mountain Group (Ordovician) of South Africa. Palaeont Afr 18:35–45 (*Petalichnus capensis* trackways with intermittent drag marks of paired cerci, associated with resting track *Metaichna rustica*)

Briggs DEG, Rushton AWA (1980) An arthropod trace fossil from the Upper Cambrian Festiniog Beds of North Wales and its bearing on trilobite locomotion. Geologica et Palaeontologica 14:1–8

Osgood RG (1970) Trace fossils of the Cincinnati area. Paleontographica Americana 6:281–444 (With the visual help of transparent overlays, Upper Ordovician trackways were analyzed with regard to possible makers)

Seilacher A (1955) Spuren und Lebensweise der Trilobiten. In: Schindewolf O, Seilacher A (eds) Beiträge zur Kenntnis des Kambriums in der Salt Range (Pakistan). Akademie der Wissenschaften und der Literatur zu Mainz, Abhandlungen der mathematisch-naturwissenschaftliche Klasse, 10, pp 324–327, Pls. 16–21 (Extensive documentation of Cambrian trilobite trackways as *Diplichnites*)

Seilacher A, Hemleben C (1966) Beiträge zur Sedimentation und Fossilführung des Hunsrückschiefers, Teil 14, Spurenfauna und Bildungstiefe des Hunsrückschiefers. Hessisches Landesamt für Bodenforschung, Notizblatt 94, pp 40–53 (Analysis of Devonian trilobite trackways modified by currents)

Plate 9: Adventures of an Early Cambrian Trilobite

Crimes TP (1970) Trilobite tracks and other trace fosils from the Upper Cambrian of North Wales. Geol J 7:47–68 (The ichnogenus *Monomorphichnus* is introduced for undertrace versions of *Dimorphichnus* lacking pusher impressions)

Seilacher A (1955) Spuren und Lebensweise der Trilobiten. In: Schindewolf O, Seilacher A (eds) Beiträge zur Kenntnis des Kambriums in der Salt Range (Pakistan). Akademie der Wissenschaften und der Literatur zu Mainz, Abhandlungen der mathematisch-naturwissenschaftliche Klasse, 10, pp 324–327, Pls. 16–21 (Analysis of an Early Cambrian *Dimorphichnus*)

Seilacher A (1990) Paleozoic trace fossils. In: Said R (ed) The geology of Egypt. AA Balkema, Rotterdam, pp 649–670 (Diagnosis of *Dimorphichnus quadrifidus* and discussion of *Cruziana* and *Dimorphichnus* fingerprints from the Lower Cambrian of Sinai)

Plate 10: Potsdam Sandstone Trackways

Dawson JW (1873) Impressions and footprints of aquatic animals and imitative markings on Carboniferous rocks. Am J Sci Arts (3)5:16–24 (Description of *Protichnites* from eastern Canada)

Hagadorn JW, Dott RH Jr, Damrow D (2002) Stranded on a Late Cambrian shoreline: Medusae from central Wisconsin. Geology 30:147–150 (Spectacular examples of fossil jellyfish in Potsdam Sandstone)

Lyell C (1851) Lower Silurian reptile in Canada. Am J Sci Arts 12(2):120–121 (Despite the misleading title, the subject is *Protichnites*)

MacNaughton RB, Cole JM, Dalrymple RW, Braddy SJ, Briggs DEG, Lukie TD (2002) First steps on land: Arthropod trackways in Cambrian-Ordovician eolian sandstone, southeastern Ontario, Canada. Geology 5:391–394 (Earliest example of a subaerial trackway preserved on Cambrian-Ordovician coastal eolian dunes)

Marsh EO (1869) Description of a new species of *Protichnites,* from the Potsdam sandstone of New York. American Association for the Advancement of Science, Proceedings 17:322–324

Owen R (1852) Description of the impressions and footprints of the *Protichnites* from the Potsdam sandstone of Canada. Q J Geol Soc London 8:214–225 (Diagnosis of *Protichnites*)

Walcott CD (1912) New York Potsdam-Hoyt fauna. Smithsonian Miscellaneous Collections 57:251–304 (Trackways from Potsdam Sandstone)

Walker SE, Holland SM, Gardiner L (2003) *Coenobichnus currani* (new ichnogenus and ichnospecies): Fossil trackway of a land hermit crab, early Holocene, San Salvador, Bahamas. J Paleontol 77:576–582 (A rare example of a Holocene trackway of a land hermit crab preserved in carbonate eolian dunes)

6

Plate 6
Limulid Tracks (*Kouphichnium*)

Modern horseshoe crabs represent a strongly heteropodous mode of arthropod locomotion. Under its broad head-shield, *Limulus* has six pairs of chelate legs, of which the first pair (chelicers) is smaller and used only for eating in a chopstick manner. The next four pairs are larger and serve mainly for locomotion, but as they pierce into the sediment, their distal pincers also probe for potential food.

The sixth pair of legs, in contrast, is modified into strong "*pushers*" that act as antagonists to the frontal legs and must accordingly carry more weight. For this function, setae at the forelast joint have become enlarged into blades that can be either folded together like a fan, or spread out like toes to act as a snowshoe. In the latter function, the last segment, whose pincers are highly reduced, is turned back like the heel of a bird's foot. Because of this similarity, fossil limulid tracks were attributed to birds (*Archaeopteryx*), pterosaurs, or to small tetrapods ("*Paramphibius*") in earlier interpretations.

In order to understand the track *patterns*, however (Pl. 7), one must also take the undertrack concept into account. It is even more relevant here than in vertebrate tracks, because limulid legs pierce relatively deeply into the sediment. The earlier statement, that undertracks have a higher fossilization potential than surface tracks, is even more valid in subaqueous environments: corresponding surface impressions become wiped out by the slightest water movement, while impressions made on internal laminae are immediately cast, even if they are only a few millimeters below the sediment/water interface.

In silts of lake deposits (Upper Devonian, **Pennsylvania**; Upper Carboniferous, **Nova Scotia**), where lamination is commonly accentuated by microbial films, several undertrack copies of the same limulid trackway can commonly be split open. Undertrack preservation, however, also implies the successive disappearance ("**undertrack deficiency**") of less deeply impressed elements of the trackway,

such as the drag impression of the tail spine, or the impressions of pushers and pincers. Even the *shape* of an individual impression may change at different levels. For instance is the **push-back pile** commonly replaced by an anterior **drag pile** in shallow undertrack copies. Another important change results from the folding-together of the pusher blades, as the leg is being pulled out of the sediment. This is why at shallow undertrack levels the pusher impression resembles the trifid footprint of a bird – but one pointing in the wrong direction! In the lithographic limestones of Solnhofen (Upper Jurassic, Germany) a sticky biofilm enhanced this effect.

In an arthropod track from the **Hunsrück Slates**, possibly made by the early limulid *Weinbergina*, several legs had "snow shoes" of delicate setae producing sun-like undertracks. More exactly, the impressions of these setae are twisted like the arms of a spiral nebula. It is this twist that informs us about the attitude of the legs. If the legs had been outspread as in a fly, a left leg would have rotated counterclockwise while the foot rested on the ground, but clockwise in a bandy-legged posture. The observed twist of the setae corresponds to the bandy-legged attitude of limulid legs.

Still more vexing are isolated barrel-like hyporeliefs in **Lower Jurassic** sandstones. Like vertebrate scoopings (Pl. 2), they probably represent **pusher digs** at a deeper undertrack level. Such diggings were made locally and never occur in series, even though the sole faces are flat, but the shapes of the four scratches correspond to the setal blades of a large limulid pusher. Pushers of similar size are recorded by **tiptoe impressions** on bed tops. They do not occur in series, because tops of storm sands get always rippled before they become buried under the muddy tail of the same event. Accordingly, post-event tracks reached the sand/mud interface only in the area of the ripple crests. In connection with the usual undertrack deficiency, this effect adds to the difficulty of reconstructing trackway patterns, even though individual footprints may be perfectly preserved.

■

Limulid undertracks
(U. Carbonif., Nova Scotia);
a upper level, **b** lower level
(see Pl. 6)

Plate 6 · Limulid Tracks (*Kouphichnium*)   21

Limulid Undertracks (*Kouphichnium*)

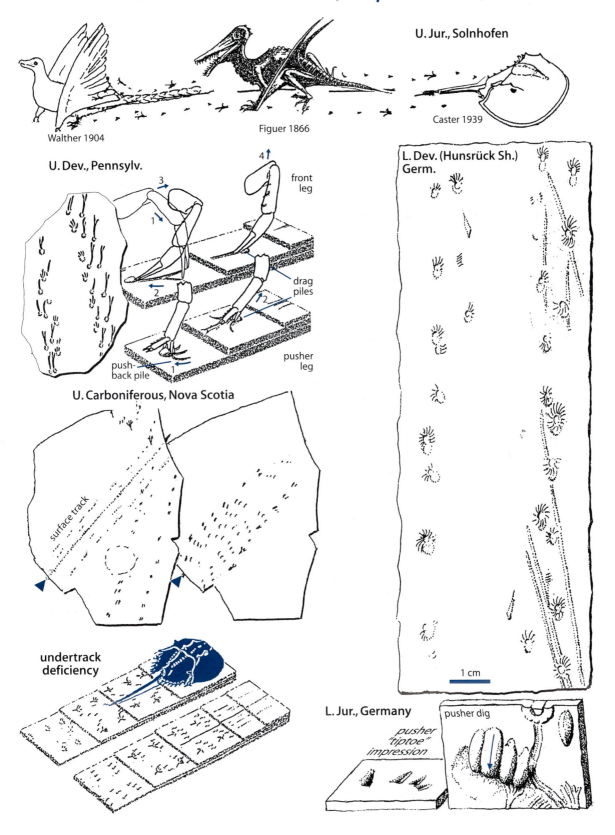

U. Jur., Solnhofen

Walther 1904

Figuer 1866

Caster 1939

U. Dev., Pennsylv.

front leg

drag piles

pusher leg

push-back pile

L. Dev. (Hunsrück Sh.) Germ.

U. Carboniferous, Nova Scotia

surface track

undertrack deficiency

1 cm

L. Jur., Germany

pusher "tiptoe" impression

pusher dig

Plate 7
Other Arthropod Trackways

Another group of chelicerates with a strongly heteropodous mode of locomotion were the Paleozoic *eurypterids*. As in limulids, the sixth pair of legs was the most specialized and acted antagonistically to the walking legs in front. It could also be used for swimming – a task that falls to the gill appendages under the tail shield (opisthosoma) in horseshoe crabs. Tracks referable to eurypterids (**Palmichnium**) are found in finely laminated silts, but the giant size of some forms and the corresponding potential to produce deep undertracks allows their preservation also in coarser-grained sediments. The largest trackway, 77 cm wide, comes from the Carboniferous, a period in which body fossils of these predators also reach gigantic scales and is now in the Pittsburgh Museum.

In the transition from aquatic eurypterids to terrestrial *scorpions*, which took place in the Early Devonian, not only the first pair of legs (chelicerae), but also the second one (the pincers) became modified for feeding purposes. The four pairs left for locomotion are similar in shape (paddles were no longer required) and, instead of being bandy-legged, spread wide enough to produce distinctive, non-overlapping track patterns. Without going into kinematic details, it is important to note that the sets of impressions on either side of a scorpion trackway are no longer symmetrical, but alternate along the midline. So the coordination of leg motions has completely changed compared to aquatic ancestors. It can also be modified into different gaits, as shown by the differences between trackways produced during a *cool* night (*Octopodichnus*) and a **hot** day (*Paleohelcura*). It is probably no coincidence that the figured examples come from the dune deposits of the Coconino Sandstone – the same ichnotope, in which the best dry-land tracks of vertebrates occur (Pl. 4). In scorpion tracks the undertrack origin is not as obvious, but it can be inferred from negative evidence: as in modern dunes, these scorpions were probably associated with beetles and millipedes, but none of their traces have been found. Probably their footprints did not penetrate deep enough to produce preservable undertracks.

In spite of their dominant role in ecosystems, insects, millipedes and other terrestrial arthropods are strongly underrepresented in the trace fossil record. To find their tracks, we rely on wet environments, such as the track of a **water beetle** in Permian lake deposits of Nierstein (Germany), where very fine silt and extremely thin lamination downscaled the threshold for undertrack formation.

But even here, most fossil insect tracks come from aquatic species that would normally swim in the water and produce a bottom track only as the water level becomes very low. Alternatively, giants like the Carboniferous millipede **Arthropleura** crossed the preservational threshold by reaching the size-class of vertebrates.

Hopping Tracks. In normal gait, animals coordinate leg motions in such a way that the body is supported at any moment. The result is a continuous trackway in which corresponding impressions are separated by equal steps. One would think that big leaps, as in grasshoppers, cicadas, or jumping spiders, are infeasible under water. Yet it is not uncommon in laminated subaqueous deposits to find discontinuous series of arthropod track patterns that suggest a hopping rather than walking mode of locomotion.

The oldest hopper is represented by *Tasmanadia* from the lowermost Cambrian Strata of the Puncoviscana Formation, Argentina (Pl. 65). In **Allocotichnus** from the Upper Ordovician of Cincinnati the sets converge towards one end, but their median lines are oblique to the direction of movement. The two sets from the **Lower Devonian** Hunsrück Slates resemble the pattern of eurypterid footprints (except for the doubling of the flipper impressions) and are also laterally offset. *Orchesteropus* from the Carboniferous of Argentina is particularly irritating, because the sets in the long trackways are widely separated and resemble those of limulids, except that there are impressions of two caudal appendages instead of a single telson. In trilobite tracks from the Lower Cambrian of the Grand Canyon (Pl. 8I) and the Upper Cambrian Potsdam Sandstone (Pl. 10A), two long appendages (cerci) intermittently touched bottom as in associated normal trackways, but the sets of footprints are too widely separated for continuous body support. In contrast, the separation of successive sets in the trackway of a Carboniferous centipede (**Diplichnites gouldi**) is not caused by discontinuous movement, but by the combination of metachronal leg motions with a snakelike undulation of the whole body. Accordingly, the sets show the minimum number of legs along the body.

What does this mean? While excluding long jumps, the viscosity of water may actually favor a hopping mode of locomotion, but with the risk to be laterally displaced during the jump. At the same time, hopping tracks would penetrate more deeply and thereby increase the probability of undertrack preservation compared to simple walking tracks. Also, the double pusher impressions in the Hunsrück eurypterid track may reflect an extra beat necessary for jumping.

Plate 7 · Other Arthropod Trackways   23

Arthropod Trackways

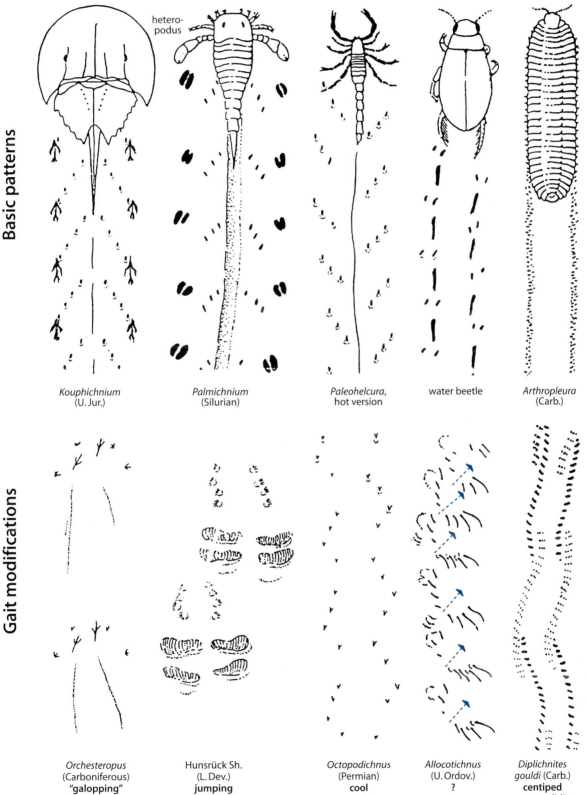

Basic patterns

hetero-podus

Kouphichnium
(U. Jur.)

Palmichnium
(Silurian)

Paleohelcura,
hot version

water beetle

Arthropleura
(Carb.)

Gait modifications

Orchesteropus
(Carboniferous)
"galopping"
xiphosuran?

Hunsrück Sh.
(L. Dev.)
jumping
eurypterid?

Octopodichnus
(Permian)
cool
scorpion

Allocotichnus
(U. Ordov.)
?

*Diplichnites
gouldi* (Carb.)
centiped
or annelid?

8

Plate 8
Trilobite Tracks

We have started our discussion with arthropods, in which a reduced number of legs still allows comparison with the gaits of land vertebrates: in walking, leg motion must follow a program that guarantees continuous support. This can be achieved either by an antagonism between the front and the rear legs (heteropodous modes) or a temporal offset between appendages on the right and left side of the body (isopodous modes). In multilegged arthropods, such as millipedes, centipedes, pill bugs and trilobites, coordination is reduced to metachronal waves of motion passing along the body. For continued support, the distance between successive waves must be smaller than the body length.

While millipedes hold the record with respect to leg numbers, trilobites stand out for lack of leg differentiation. Apart from a pair of flexible antennae in front, and of cerci at the rear end of the body in some Cambrian trilobites, there is only a single kind of appendages. Trilobite legs, however, were functionally differentiated within themselves: the inner branch (endopodite) was used for walking, while the outer one (exopodite) acted primarily in ventilation and the sieving of food particles. Each branch could take on additional functions, for instance in burrowing and swimming. Also, relative proportions of branches may change along the body. While the lack of locomotive differentiation among leg pairs conforms with metachronal coordination, why is there no trilobite with pincers or any other kind of mouthparts? The only reasonable explanation for this paradox seems to be a kind of detritus feeding in which all legs participated. We shall come back to this when talking about trilobite burrowing (Pl. 11).

In general, trackways of multilegged arthropods are difficult to analyze due to the high degree of overstepping. The track of the Carboniferous millipede *Arthropleura* (Pl. 7), for instance, can be identified only by its unusual size. In trilobite tracks, the situation was different for two reasons. First, the gauge decreased, together with leg size, towards the tail end. Second, trilobites had a tendency to sidle, i.e. they commonly moved at an angle to the body axis. Thereby, sets of footprints produced during one metachronal wave became separated, particularly on the trailing side. These sets allow us to retrieve important information:

1. Locomotion was in the direction in which the V-sets open. This criterion is useful, because push-back hills are not always developed and may be ambiguous in undertracks.
2. The direction of the metachronal waves can also be inferred. If they had passed from the head to the tail end, the V-sets would appear too wide and too short compared with leg positions in a reasonable trilobite. As the opposite is the case, we may assume that the waves have passed headwards, as in millipedes.
3. The number of impressions in one set is also a measure for the number of legs involved. However, the smaller legs near the tail end of the trilobite may not have penetrated deep enough to produce an undertrack. Therefore, leg numbers derived from trilobite trackways are generally too low.

The basic trilobite track pattern may be modified by currents. In one Hunsrück Slate specimen (not figured), the tracemaker was suddenly derailed by a lateral current. This tells us which legs were touching bottom at the moment of displacement. In another specimen, a "tail-wind" increased the lengths of the steps, so that the sets are more extended than usual.

In most arthropods, leg impressions hardly deserve the name "footprint", because they lack details such as toe impressions. There are exceptions however, where the tracemaker was very large (Pl. 70) or preservation was exceptional. While the swirl of snowshoe setae in the Hunsrück Slate specimen on Pl. 6 was imposed by rotation of the leg tip during forward movement of the body, a *sigmoidal* variant of footprints, as observed in the trailing sets of oblique trilobite trackways (**G**; **H**), reflects a combination of leg and body motion. If we assume that bandy-legged endopodites scratched by flexing in a medio-posterior direction, leg motion should combine with body motion to produce an oblique scratch; but as leg motion is slower in the beginning and towards the end of each swing, body motion dominates during these phases and causes the scratch to bend at both ends in a sigmoidal fashion.

In Cambrian trilobite trackways, such as the ones figured from the Grand Canyon (**K**), the footprints may be associated with the double trails of the caudal cerci. As the telson of *Limulus* (Pl. 7), they intermittently touched bottom in the rhythm of the metachronal waves. Through this additional support (**I**), the onset of the next metachronal wave could be delayed, so that the trackway pattern resembles that of jumping arthropods (Pl. 7).

In hermit crabs, a similar intermittent support is provided by the gastropod shell used as a domicile. Without modern counterparts, such trackway would be difficult to understand, but in analogy to them, an Upper Cambrian trackway (Pl. 10E) can be interpreted as an early hermit behavior.

Knowing the basic principles of trilobite locomotion, we are now prepared to analyze the complex trackway shown on the next plate.

Plate 8 · Trilobite Tracks   25

Trilobite Trackways (*Diplichnites*)

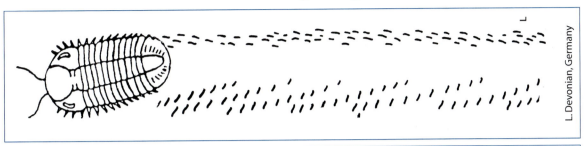

L. Devonian, Germany

galopping

L. Cambr, USA

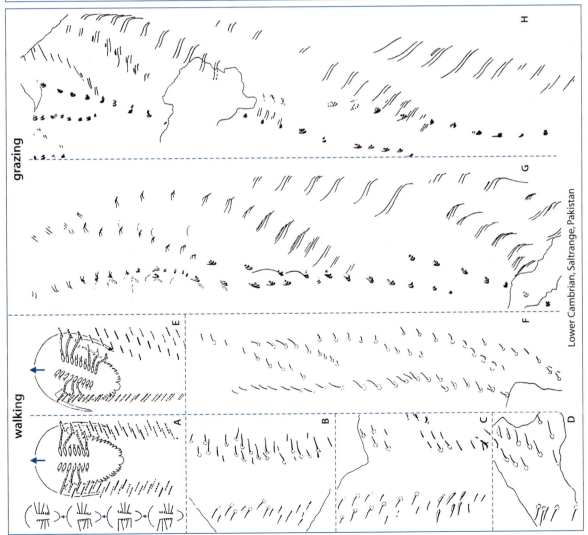

grazing

walking

Lower Cambrian, Saltrange, Pakistan

Plate 9
Adventures of an Early Cambrian Trilobite

At a first look, the figured trackway from the Magnesian Sandstone (Lower Cambrian) near Fort Kussak (Salt Range, Pakistan) is unlike other trilobite trackways, because one cannot distinguish a left and a right series of leg impressions. Instead, sets of sigmoidal scratches *alternate* with similarly oblique sets of more blunt impressions along the same axis. Yet, individual scratches correspond to the pattern of an oblique *Diplichnites* (Pl. 8G,H), into which this trackway actually grades in the upper left corner of the slab. The ***Dimorphichnus*** pattern (see Pl. 70 for a Silurian example) simply reflects the switch from an oblique motion into perfect sidling. So the ichnogenus should be considered as a behavioral variant of *Diplichnites*. Preservational variants have been called ***Monomorphichnus***, deep undertraces, in which only the scratches may be seen.

In an elongate animal, sidling locomotion might appear very inefficient – unless the main purpose were not to get from one place to another, but to screen a maximum surface. After all, this is how we move a broom over the floor. As another analog, brachyuran crabs (which have accordingly become shorter than wide) usually run (and swim) sideways. Yet, when such crabs graze an algal film on a tropical beach, they switch to forward motion.

In a trilobite grazing sideways all appendages on the one side acted in unison and antagonistically to the opposite ones, with legs on the leading side pushing and those on the trailing side scratching. In both actions, legs flexed towards the midline, but only the tips of the pushing legs remained stationary. In reality, however, both sets are oblique relative to the axis of the trackway. The reason is that the legs did not act simultaneously, but in metachronal waves passing from the rear to the front end of the body, as in the walking mode.

With the basic kinematics, the direction of motion, and the head/tail orientation being established, we can now turn to details of this particular trackway.

1. As motions of body and legs fall almost in line in the lateral gait, scratches should be straight. However, some sigmoidality is maintained. This means that legs did not bend transversely, but with a backward component; i.e. leg tips moved *medio-posteriorly* relative to the body axis.
2. Assuming that the legs were similarly flexed in the pushing and the scratching mode, what made the difference between the two functions? Energy input probably had to be higher in the pushing legs, because they had to counteract the scratching moment, as well as push the body ahead. Another difference relates to the *attitude* of the leg tip. From the bifid scratches and pusher impressions it is clear that each leg had two terminal claws. In several instances, however, a pair of much smaller scratch, or pusher, impressions indicates the presence of two smaller setae on the front side of the two main claws.

Because claw configurations are known only in very few trilobites (but not in the chief suspect, *Redlichia*), this fingerprint is criminologically useless. Yet it tells us that friction of the leg tips could be modified in the pushing versus the scratching function.

3. Up to 18 impressions in one step series would fit the leg number of *Redlichia* if one allows for undertrack deficiency.
4. Sediment processing probably conformed to the basic mode of trilobite feeding. After having scraped in medio-posterior direction, each endopodite handed the crop over to the coxae, which acted as a metachronal conveyor belt to forward it to the mouth. As the food string came from the rear, the trilobite mouth opened backwards. It was rather small, but behind it, the gut immediately widened into a "stomach". This organ should more properly be described as a suction pump, because it was suspended by muscles between the exoskeletal glabella on top and the equally dome-shaped hypostome below.

Now we can share the adventure of this particular trilobite more than 500 million years ago. In doing this, we distinguish left and right sides as they appear in the hypichnial cast.

The animal entered the area of the slab from the lower-left corner in a right-lateral gait. Due to a current from the left, it moved rather quickly, so that the sets of push traces made by the right-side legs became relatively stretched. After 64 cm, this motion was interrupted by the encounter with the grazing track of a larger individual (or the trace-maker itself) coming from the opposite side. In response, the animal rotated counterclockwise by 45° and backed off in the opposite direction.

Interestingly, the animal did not stop during this maneuver; rather, the switch of gears became superimposed on the metachronal waves (B_1-B_2). Within the same metachronal wave, the righthand appendages started pushing the body to the right and ended up scratching from right to left, while the legs of the left side first scratched (scratches change direction by the rotation of the body) and pushed by the time the wave had reached the head region.

The rotation (which might have also been imposed due to the current displacing the less supported rear end of the animal) had a purpose: it offset the reverse track (now with the right side scratching) relative to the earlier one, so that double grazing was avoided. When the animal trailed along the previous track, locomotion – now upcurrent – also became slower, reducing the obliquity of the sets.

Eventually, the animal went across its own track (reducing the scratching activity while crossing) and switched from a sidling to an oblique gait in a wide righthand curve. Obviously it had eaten enough.

As paleodetectives we also note a little personal clue: as can be seen from the drag traces produced in the scratching as well as in the pushing function, the left front leg of this particular individual limped.

Plate 9 · Adventures of an Early Cambrian Trilobite   27

Grazing Sideways

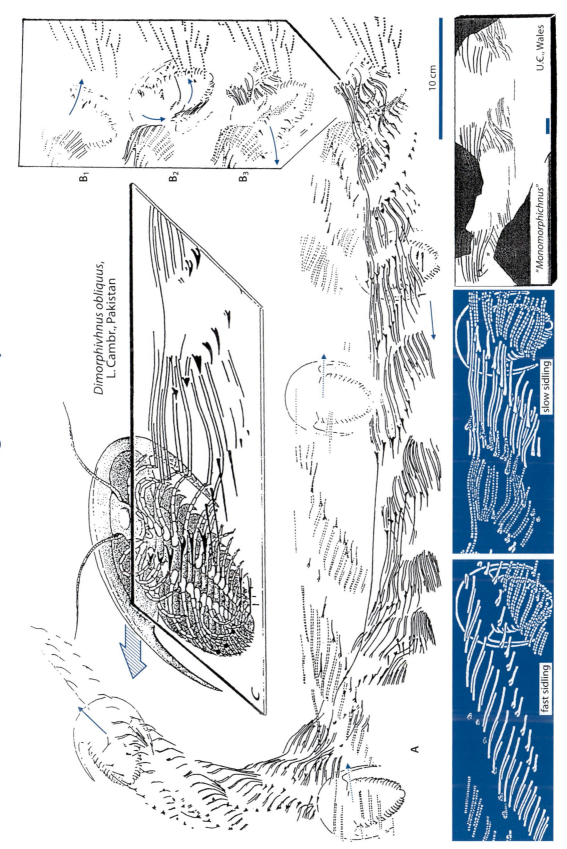

Dimorphivhnus obliquus,
L.Cambr., Pakistan

10 cm

B₁

B₂

B₃

"Monomorphichnus"

U.€., Wales

slow sidling

fast sidling

A

10

Plate 10
Potsdam Sandstone Trackways

Typical eurypterid tracks, as found in Ordovician through Carboniferous nearshore or estuarine sandstones, can be easily recognized by their size, heteropody and impressions of a broad opisthosoma that touched bottom in the rhythm of the leg's metachronal movements (Pl. 8). The Upper Cambrian Potsdam Sandstone of Ontario and New York, as well as equivalent strata in Wisconsin, commonly preserve trackways of similar size with a broad median tail impression, but without recognizable heteropody. This makes it difficult to single out metachronal sets of footprints and even to tell in which direction the animal was moving. Another difficulty is the general coarseness of the bed surfaces. Nevertheless the Potsdam Sandstone provides a unique window into early littoral environments, their biota, and their taphonomy. It is also fortunate that this window has been recently opened again by an extensive commercial quarrying operation in Wisconsin, where impressions of real jellyfish are associated with the tracks. Our plate presents examples of *Protichnites* from three different localities and environments.

One of the tracks figured by Walcott from Essex County (**A**) is almost identical with the trilobite tracks (Pl. 8I,K) from the Grand Canyon, in which each of the widely spaced V-shaped sets of leg impressions has a double drag trace in the middle. As the metachronal wave proceeded towards the head end, the paired cerci of a trilobite could well have intermittently acted as a skid supporting the rear part of the body.

Associated trackways of similar size (**B**) have a median drag trace consisting of only one broad furrow that fits neither trilobite cerci nor the limulid telson. Nor are the footsteps arranged in easily recognizable patterns. Probably we deal here with some kind of early merostomes (true eurypterids are not known before the Ordovician) that used their segmented, but legless, tail part for intermittent support.

A slab from Ontario exhibited in the Redpath Museum (**C**) shows hyporeliefs of two large mollusc trails (*Climactichnites*; see Pl. 63) crossed by two or three trackways of *Protichnites*. Although there is some interaction, the two kinds of tracemakers probably did not actually meet, because the mollusc trails are continuous and do not smear the arthropod footprints. More likely the arthropods were only reacting to preexisting trails. Nevertheless the sharp turns made at the encounter give us a clue to the direction of movement. Just as the course of the rear wheels is backset relative to that of the turning front wheels in a long truck (the reason why in fire engines of old New York they were steered separately), the opisthosomal drag trace should make its turn somewhere behind that of the contemporaneous footprints. On the other hand, the tail would occasionally swing out, producing a discontinuity in the drag trace. According to these criteria, the wide V patterns of the footprints open in the forward direction, as in a scorpion track (Pl. 7). Although metachronal waves cannot be singled out, they are reflected in the rhythmic swelling of the tail drag. On either side there are roughly eight footprints to the rhythm, or four if the legs were bifid.

As evidenced by oscillation ripples and the tracks of large trilobites and molluscs, the previously discussed trackways represent nearshore environments that emerged only during low tides. Not so at a site near Kingston, Ontario, where similar trackways (**D**) climb up the slopes of eolian dunes! As shown in Pl. 4, dunes provide a high fossilization potential for pierced undertracks. But what lured large predators (or scavengers) out of the water at a time long before there were any land plants or animals? My guess is that they lived in estuaries and used the shortcut across the dunes of a barrier island to reach the flotsam of the open shore. This rich food source has always been there, but the barrier of the surf zone makes it inaccessible for smaller sea animals even today.

The intertidal sandstones presently quarried near Mosinee (Wisconsin) have yielded not only *Climactichnites*, regular *Protichnites*, and stranded jellyfish, but also a variant of *Protichnites* that requires a particular explanation (**E**). Its intermittent "tail traces" resemble the ones described above, but instead of forming a string, they are disconnected, oblique to the trail axis, and so closely spaced that they form a regularly shingled pattern. As the sense of imbrication does not change in right and left curves (the whole sequence of tail impressions does swing out), it can only be explained by a permanently oblique attitude of the tail part relative to the body axis. Even more vexing is the fact that the tail was always bent to the *left* side.

In the best preserved trackway (probably the freshest one), each tail impression is also subdivided into four to five segments. But instead of being perpendicular to the axis of the tail, the "segment boundaries" are also oblique, as if one dealt with the impression of a high-spired, dextrally coiled gastropod shell. The same trail also preserves two rows of synchronous footprints reminiscent of the pusher impressions of *Limulus* (Pl. 7).

Two explanations come to mind. (1) The tail was anatomically connected to the front part of the body at a lateral angle. Nothing like this is known in modern arthropods. (2) As in modern hermit crabs, the tail was hidden in a helicoidal gastropod shell. This would explain the uniform obliquity, except that a dextral shell would automatically angle to the *right* side. In a scorpion-like animal that tends to bend its tail dorsally, however, the shell would have been carried on the *left* side. If it only served to keep the gills moist, it could also be much smaller than the hermit.

After all, Ediacaran conditions continued in Cambrian tidal flats not only with respect to the persistence of microbial films (which allowed surface tracks to be preserved), but also by the absence of predators.

Plate 10 · Potsdam Sandstone Trackways   29

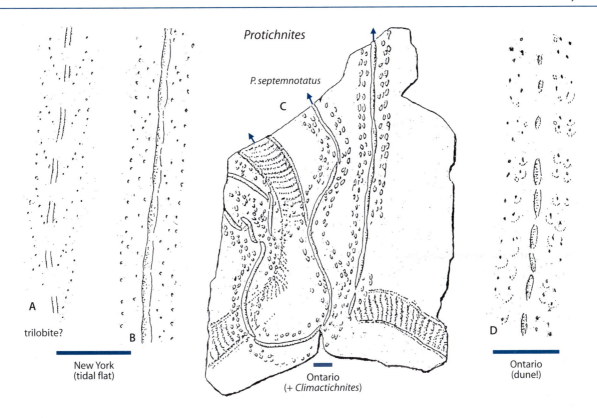

Protichnites

P. septemnotatus

A

trilobite?

B

New York
(tidal flat)

C

Ontario
(+ *Climactichnites*)

D

Ontario
(dune!)

Potsdam SS. (U.Є.)

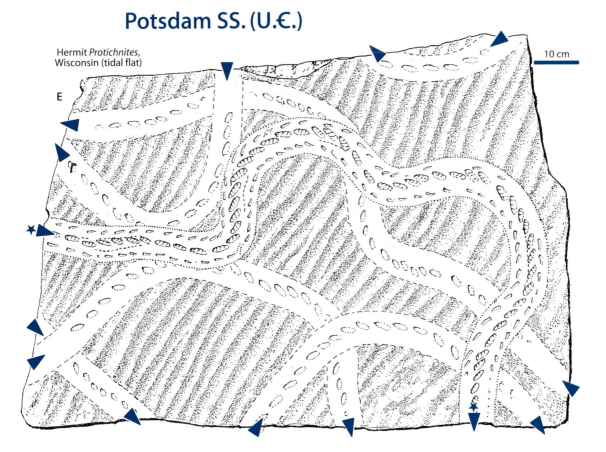

Hermit *Protichnites*,
Wisconsin (tidal flat)

E

10 cm

Protichnites climbing dune sand
(U. Cambrian, Ontario; cast in
FOSSIL ART, 115 cm wide)

Trilobite Burrows

In the last chapter we have become aware of the incongruence between modern arthropod tracks and their fossil record: rather than the familiar impressions seen at the surface, it is the invisible undertracks that are much more likely to be found as fossils. By their deeper penetration, *burrows* have a still higher fossilization potential. This situation is particularly clear in the case of trilobites, which probably produced many more trackways than burrows during their lifetimes. Yet the latter are dominant in the fossil record. Their undertrace nature also implies that they preserve details that would have never survived at the sediment/water interface. Due to the high morphological resolution, an unusually large number of ichnospecies can be distinguished and be used for stratigraphic correlation (see Chap. XIV). In the present context, however, we are mainly concerned with the biological significance of trilobite burrows.

Originally, the reliefs found at the bases of Ordovician sandstone beds (mostly tempestites) were known as *Bilobites*. Because this name (referring to the bilobed profile of the casts) turned out to be preoccupied for a brachiopod, the currently valid name (here used for all bilobed trilobite burrows) is *Cruziana*. It was given by the French explorer Alcide D'Orbigny (1835–1847) in honor of his friend, the general Andres de Santa Cruz, who united Bolivia and Peru after liberation from Spanish rule.

Like most other trace fossils, *Cruziana* was originally interpreted as an algal impression. Today its trace fossil interpretation is well established, but there have been claims that the non-mineralized legs of the trilobites were too weak for active burrowing. Therefore it is useful to repeat the main arguments for trilobite production of *Cruziana*.

1. Environmental (shallow marine) and time ranges (Cambrian to Devonian, with rare later examples) correspond to those of trilobites. So do the size ranges, with giants in the Ordovician. The fact that skeletal remains and burrows rarely co-occur in the same rock is explained by different requirements for their preservation (but see *Flexicalymene*, Pl. 11).
2. In addition to the scratches made by endo- and exopodites, one occasionally observes impressions of pleural spines, as well as furrows made by the edges of head and tail shields. Otherwise the two sources of information remain separated, because trilobite taxonomy is based on the morphology of the dorsal skeleton, whereas differences between ichnospecies relate to the structure and activities of ventral appendages.

Only where burrows co-occur in the same beds, can a correlation be made between the two taxonomies (Pl. 14). This does not mean that all ichnospecies presently assigned to *Cruziana* were necessarily made by trilobites. Other arthropods of similar size, mode of life, and construction (e.g., limulids and aglaspids; Pl. 11) may well have produced similar burrows, but only in the case of the (generally smaller) phyllopod crustaceans (*Isopodichnus*, Pl. 23) does this uncertainty infringe on the stratigraphic usefulness of *Cruziana*. Before discussing their burrowing behavior, we must consider the ways trilobites got their food.

Trilobite Feeding

The strangeness of trilobite design starts with the construction of the segmented *carapace*: instead of surrounding the body like a segmented tube, its duplicatures spread laterally in a wing-like fashion (Pl. 11). Thus the non-mineralized ventral surface, with the vital food groove, gills and legs, remained unprotected, unless the animal hugged the bottom or rolled up like a litter bug, with the tail shield (if there was one) closing against the head shield. Due to this design, body flexure was restricted to the vertical plane, which reduced the ability to move in curves.

Second, there was no differentiation of the appendages into mouth parts, pereiopods, and uropodia, because the paired antennae at the front and cerci at the rear end are probably not modified legs. Otherwise, variation of the legs along the body is restricted to size and the proportion of their two branches. Of these, the lower one (endopodite) served mainly for locomotion. The more flexible exopodite, in contrast, was held aloft and never left an impression in simple trackways.

This outfit made trilobites ill suited for a carnivorous life style (see Pl. 74). They were largely restricted to a kind of particle feeding, in which all legs were equally involved. We dealt with this process already in the grazing behavior of the *Dimorphichnus* maker (Pl. 9). But how did the catch of a rear leg reach the mouth? The clue is encoded in another strange anatomical feature: the trilobite mouth. It was a small opening at the rear end of the *hypostome*, the only element of the ventral integument (other than the doublure) that was stiffened by mineralization. It links with the cephalon by an arched suture, which is reinforced by a thickened rim, so that it could hardly act as a hinge. As the hypostome forms a ventral counterpart to the dorsal glabella by its vaulting and position, the two probably provided the rigid frame for a chamber of the intestinal tract. This organ is usually referred to as the "stomach", but more likely its main function was that of a suction ball – except that its expansion was not effected by energy stored in a rubber wall, but by radial muscles and ligaments attached to the inner surfaces of the glabella and the hypostome. Coxae were tethered to processes corresponding to ex-ternal grooves that trace the boundaries between pleurae and rhachis. Thus the setate inner edge of each coxa could act as a gnatho-base that swung forward during the active stroke of the leg. If particles handed by the sieving expodites to the median food groove were bound by mucus, the metachronal action of the coxae (from rear to front!) could transport the food packages to the mouth, where they could be sucked into the stomach.

Literature

Chapter III

Bergström J (1973) Organization, life and systematics of trilobites. Fossils and Strata 2:69

Fortey RA (2000a) Olenid trilobites: The oldest known chemoautotrophic symbionts? Proceedings of the National Academy of Sciences of the United States of America 97, pp 6574–6578 (Olenid trilobites interpreted as chemoautotrophic symbionts in oxygen-poor sea floors)

Fortey RA (2000b) Trilobite! Eyewitness to evolution. Alfred Knopf, New York, 284 p (Popular book on trilobites)

Fortey RA, Owens RM (1999) Feeding habits in trilobites. Palaeontology 42(3):429–465 (Comprehensive review of the different feeding habits of trilobites)

Goldring R (1985) The formation of the trace fossil *Cruziana*. Geol Mag 122(1):65–72 (Undertrace origin)

Orbigny A d' (1835–1847) Voyage dans l'Amérique méridionale le Brésil, la République orientale de l'Uruguay, la République Argentine, la Patagonie, la République du Chili, la République de Bolivia, la République du Pérou exécuté pendant les années 1826, 1827, 1828, 1829, 1830, 1831, 1832 et 1833. Pitois-Leverault, Paris & Leverault, Strasbourg, 3(4) (Paléontologie), 188 p (Diagnosis of ichnogenus *Cruziana*, then interpreted as an algal impression)

Seilacher A (1959) Vom Leben der Trilobiten. Naturwissenschaften 46:389–393 (Using trace fossils for trilobite Paleobiology)

Seilacher A (1970) *Cruziana* stratigraphy of non-fossiliferous Palaeozoic sandstones. In: Crimes TP, Harper JW (eds) Trace fossils. Geol J, Special Issue 3, pp 447–476 (Preservational and paleobiologic analysis of trilobite trace fossils and proposal of a *Cruziana* ichnostratigraphy)

Seilacher A (1985) Trilobite palaeobiology and substrate relationships. T Roy Soc Edin-Earth 76:231–237 (Analysis of trilobite paleobiology with respect to functional morphology, burrowing behavior, feeding habits and the origin of *Cruziana*)

Seilacher A (1992) An updated *Cruziana* stratigraphy of Gondwanan Paleozoic sandstones. In: Salem MJ, Hammuda OS, Eliagoubi BA (eds) The geology of Libya, 4, Elsevier, Amsterdam, pp 1565–1581 (Revised version of *Cruziana* ichnostratigraphy)

Whittington HB (1992) Trilobites. Boydell Press, Rochester, 145 p, 120 pls. (The interpretation of *Cruziana* as a trilobite trace fossil is questioned)

Plate 11: Trilobite Biology and *Cruziana* Authorship

Baldwin CT (1977) *Rusophycus morgati*: an asaphid produces trace fossil from the Cambro-Ordovician of Brittany and northwest Spain. J Paleontol 51(2):411–413 (Molting burrows)

Bergström J (1976) Lower Palaeozoic trace fossils from eastern Newfoundland. Can J Earth Sci 13(11):1613–1633 (Diagnosis of *Cruziana leiferikssoni*. Interpreted as an opisthocline burrow produced by the trilobite *Stenopilus*)

Bureau ME (1886) Sur la formation de Bilobites à l'epoque actuelle. 1–4 (Seaweed debate of *Cruziana*)

Carrington Da Costa J (1935) O problema das bilobites. Anais da Facudade de Ciencias do Porto 19(3):3–27, 2 pls. (*Cruziana* debate)

Crimes TP (1975) The production and preservation of trilobite resting and burrowing traces. Lethaia 8:35–48 (Interpretation of *Cruziana* as furrows formed at the sediment-water interface)

Dahmer G (1937) Lebensspuren aus dem Taunusquarzit und den Siegener Schichten (Unterdevon). Jahrbuch der Preussisches Geologisch Landesanstalt für 1936 57:523–539, 5 pls. (Rusophyciform burrows referred to *Homalonotus*)

Delgado JFN (1885) Estudo sobre os bilobites e outros fosseis das quartzites da base do systema Silurico de Portugal. Terrenos Paleozoicos de Portugal, 113 p, 42 pls. Academia Real das Sciencias, Lisboa (A superbly illustrated monograph, in Portuguese with French translation, on Lower Ordovician "bilobites", i.e., *Cruziana*)

Mángano MG, Buatois LA (2003) *Rusophycus leiferikssoni* en la Formación Campanario: Implicancias paleobiológicas y paleoambientales. In: Buatois LA, Mángano MG (eds) Icnología: Hacia una convergencia entre geología y biología. Publicación Especial de la Asociación Paleontológica Argentina 9, pp 65–84 (Detailed description of *Cruziana leiferikssoni*. Interpreted as a prosocline burrow produced in intertidal settings)

Mángano MG, Buatois LA (2004) Reconstructing early Phanerozoic intertidal ecosystems: Ichnology of the Cambrian Campanario Formation in northwest Argentina. In: Webby BD, Mángano MG, Buatois LA (eds) Trace fossils in evolutionary palaeoecology. Fossils and Strata 51:17–38 (Paleoenvironmental setting of *Cruziana leifeirikssoni* and its paleoecologic significance)

Nathorst AG (1886) Nouvelles observations sur les traces d'animaux et autres phénomènes d'origine purement mécanique décrits comme "Algues fossiles". Konglinga Svenska Vetenskapsakademien, Handlingar 21(14), 58 p, 5 pls. (Pioneering interpretation of *Cruziana* and others as trace fossils rather than seaweeds)

Radwanski A, Roniewicz P (1963) Upper Cambrian trilobite ichnocoenosis from Wielka Wisniówka (Holy Cross Mountains, Poland). Acta Palaeontol Pol 8(2):259–280 (Pl. 2: Several *Cruziana polonica* with coxal impressions, probably melting burrows)

Schmalfuss H (1981) Structure, pattern and function of cuticular terraces in trilobites. Lethaia 14:331–341 (Detailed study of trilobite terrace lines)

Seilacher A (1985) Trilobite palaeobiology and substrate relationships. T Roy Soc Edin-Earth 76:231–237 (Analysis of trilobite paleobiology with respect to functional morphology, burrowing behavior, feeding habits and the origin of *Cruziana*)

Shone RW (1979) Giant *Cruziana* from the Beaufort Group. T Geol Soc S Afr 82:371–375 (*Cruziana* homeomorph in Triassic fluvial sandstones)

Zonneveld JP, Pemberton SG, Saunders TDA, Pickerill R (2002) Large, robust *Cruziana* from the Middle Triassic of northeastern British Columbia: Ethologic, biostratigraphic, and paleobiologic significance. Palaios 17:435–448 (Detailed documentation of a large Triassic *Cruziana* not made by trilobites)

Plate 12: Trilobite Fingerprints

Seilacher A (1962) Form und Funktion des Trilobiten-Daktylus. Paläont Z, H. Schmidt-Festband, pp 218–227 (Functional morphology of trilobite appendages)

Plate 13: *Cruziana* Modifications

Crimes TP (1973) The production and preservation of trilobite resting and furrowing traces. Lethaia 8:35–48 (Denies undertrace origin)

Fenton CL, Fenton MA (1937) Trilobite "nests" and feeding burrows. Am Midl Nat 18:446–451 (Diagnosis of *Cruziana jenningsi*)

Jensen S, Bergström J (2000) *Cheiichnus gothicus* igen. et isp. n., a new *Bergaueria*-like arthropod trace fossil from the Lower Cambrian of Västergötland, Sweden. Geol Foren Stock For 122:293–296 (Introduction of ichnogenus *Cheiichnus*)

Lessertisseur J (1956) Sur un bilobite nouveau du Gotlandien de L'Ennedi (Tchad, AEF.), *Cruziana ancora*. B Soc Geol Fr 6:43–47 (Silurian *Cruziana ancora*)

Orłowski S, Radwański A, Roniewicz P (1971) Ichnospecific variability of the Upper Cambrian *Rusophycus* from the Holy Cross Mts. Acta Geol Pol 21:341–348 (Detailed description and interpretation of *Cruziana polonica*)

Radwański A, Roniewicz P (1972) A long trilobite-trackway, *Cruziana semiplicata* Salter, from the Upper Cambrian of the Holy Cross Mts. Acta Geol Pol 22:439–447 (This paper documents the association of *Cruziana semiplicata* and *C. polonica*, proposing an olenid maker)

Seilacher A (1970) *Cruziana* stratigraphy of non-fossiliferous Palaeozoic sandstones. In: Crimes TP, Harper JW (eds) Trace fossils. Geol J, Special Issue 3, pp 447–476 (Analysis of several *Cruziana* ichnospecies)

Plate 14: *Cruziana semiplicata*

Colchen MM (1964a) Sur une coupe a travers les formations paléozoiques de la Sierra de la Demanda (Burgos-Logrono, Espagne). CR Soc Geol Fr 10:422

Colchen MM (1964b) Successions lithologiques et niveaux repères dans le Paléozoique Antécarbonifère de la Sierra de la Demanda (Burgos-Logrono, Espagne). CR Acad Sci 259(9):4758–4761

Crimes TP (1970) Trilobite tracks and other trace fosils from the Upper Cambrian of North Wales. Geol J 7:47–68 (Analysis of *Cruziana semiplicata* from the Upper Cambrian of Wales)

Färber A, Jaritz W (1964) Die Geologie des westasturischen Küstengebietes zwischen San Esteban de Pravia und Ribadeo (NW-Spanien). Geol Jb 81:679–738, Pls. 42 und 43 (Pl. 41, Fig. 3: pirouetting *Cruziana semiplicata*)

Fortey RA, Seilacher A (1997) The trace fossil *Cruziana semiplicata* and the trilobite that made it. Lethaia 30:105–112 (U. Cambrian, Oman)

Nathorst AG (1888) Herrn Lebesconte's neueste Bemerkungen über *Cruziana*. N Jb Mineral 1:205–207 (Seaweed debate)

Neto de Carvalho C (2006) Roller coaster behaviour in the *Cruziana rugosa* group from Penha Garcia (Portugal): Implications for the feeding program of trilobites. Ichnos 13(4):255–265

Radwanski A, Roniewicz P (1972) A long trilobite-trackway, *Cruziana semiplicata* Salter, from the Upper Cambrian of the Holy Cross Mts. Acta Geol Pol 22(3):439–447 (150 cm long smoothly curved burrow referred to olenids; perfect preservation)

Seilacher A (1997) Fossil art. An exhibition of the Geologisches Institut Tübingen University. The Royal Tyrell Museum of Palaeontology, Drumheller, Alberta, Canada, 64 p (Illustration of more scribbling *C. semiplicata* from Spain)

Plate 15: Burrowing Behavior of Silurian Trilobites

Bottjer DJ, Droser ML, Savrda CE (1987) New concepts in the use of biogenic sedimentary structures for paleoenvironmental interpretation. SEPM Pacific Section, pp 1–65 (Ichnofabrics)

Frey RW, Pemberton SG (1987) The *Psilonichnus* Ichnocoenose, and its relationship to adjacent marine and nonmarine Ichnocoenoses along the Georgia Coast. B Can Petrol Geol 35(3):333–357

Frey RW, Seilacher A (1980) Uniformity in marine invertebrate ichnology. Lethaia 13:183–207

Lessertisseur J (1956) Sur un bilobite nouveau du Gotlandien de L'Ennedi (Tchad, AEF.), *Cruziana ancora*. B Soc Geol Fr 6:43–47 (Diagnosis of *Cruziana ancora*)

Seilacher A (1996) Evolution of burrowing behavior in Silurian trilobites: Ichnosubspecies of *Cruziana acacensis*. In: Salem MJ, Busrewil MT, Misallati AA, Sola M (eds) The geology of Sirt Basin, 3, Elsevier, Amsterdam, pp 523–530 (Definition of *Cruziana acacensis* ichnosubspecies and their biostratigraphic significance)

Seilacher A, Cingolani C, Varela C (2003) Ichnostratigraphic correlation of early Paleozoic quartzites in central Argentina. In: Salem MJ, Oun KM, Seddig HM (eds) The geology of Northwest Libya. Earth Science Society of Libya 1, Tripoli, pp 275–292 (Correlation between Silurian of Argentina and North Africa using *Cruziana acacencis* and *C. ancora* as index fossils)

Plate 11
Trilobite Biology and *Cruziana* Authorship

Even though trilobite authorship may be questioned in some cases, *Cruziana* burrows as a whole enrich our understanding of trilobite paleobiology. Together with the functional morphology of the dorsal skeleton, they tell us that these rulers of the early Paleozoic biosphere differed from other arthropods not only in feeding (see introduction), but also in the way they processed the sediment when they burrowed.

Trilobite Biology. Ordinary trackways (*Diplichnites*; Pl. 8) told us about the direction of metachronal waves and *Dimorphichnus* added the information that the active stroke of the trilobite leg was normally in medio-posterior direction. Kinematics also led to the conclusion that trilobite endopodites were bandy-legged. Such an attitude is also observed in trilobite carcasses (*Phacops*) of the Hunsrück Slates, but it may be largely a taphonomic effect comparable to the inwardly coiled legs of a dead spider. Nevertheless, the taphonomic signal may be revealing. Unlike stiffer arthropod legs operated by antagonistic muscle systems, spider legs have only flexor muscles, while extension of the legs is hydraulic. As muscle shrinkage is responsible for the postmortem curling of the spider legs, we may assume a similar mechanism in trilobites.

If this kinetic model is translated into stationary burrowing, the medio-posterior active stroke creates a problem, because it accumulates sediment instead of removing it from under the body. Yet, *Cruziana* casts clearly show that the medio-posterior stroke direction was also maintained in burrowing: they are never surrounded by a groove corresponding to a rim of pushed-out sediment.

The important conclusion is that trilobites dug primarily for feeding, not to hide. In this process, the vaulted pleural carapace played an essential part by providing a hood, under which sediment processing could go on. This function is also expressed by **terrace lines** on the ventral *doublures* and on the hypostome of the mineralized carapace. By being ratcheted towards the periphery, they kept the sediment from slumping into the *filter chamber*. This chamber was not completely closed: in contrast to *Limulus*, water could enter under the arched front of the cephalon and leave at the tail end. This model agrees with the tendency of stationary (rusophyciform) burrows to face ambient currents recorded by sedimentary structures. Thus the concerted action of the exopodites could flush the filter chamber and export processed sediment and faecal pellets at the rear side. At the same time, these feather-like structures could strain food particles from the sediment excavated by the endopodites, but their distal ends could also assist in digging.

Accordingly, a trilobite would normally lie horizontally in its burrow, with the head and tail ends bent up by dorsal flexure of the whole animal – just like the Ordovician *Flexicalymene* found in its burrow. However, this basic behavior was modified in some groups. The maker of *Cruziana leiferikssoni* from the uppermost Cambrian of Newfoundland, for example, may have dug its unusually deep burrows tail-on, an attitude in which carapaces of *Stenopilus* are preserved in penecontemporaneous limestones. This trilobite has a highly vaulted tail shield, the upper surface of which bears terrace lines ratcheted in reverse (anterior) direction. Tail-on burrowing, however, required also a modified leg activity: scratching had to be medio-anterior and the flush stream two-way, backwards below, and headwards above the canopy of the exopodites. Alternatively, *C. leiferikssoni* may have been dug head-down, like *C. acacensis* in Pl. 15.

C. morgati, as well as *C. radialis* (Pl. 13) and *C. polonica* (Pl. 67), suggest yet another reason for a trilobite to get dug in the sediment, **molting**. In this critical stage it would need protection. But at the same time it took advantage of an effect we know from walking with rubber boots in sticky mud – only that the trilobite *wanted* to get rid of its old hull. In this case there was no filtration chamber left open below the carapace. When pulling its legs out of their ecdysial boots, the animal rather pressed its body down for purchase, so that coxae and proximal podomeres of the endopodites left sharp impressions.

***Cruziana* Authorship.** Apart from the smoking gun of **Flexicalymene**, there are cases in which parts of the trilobite's dorsal skeleton have left an impression. Thus, impressions of the pleural spines can be seen in the figured specimens of *Cruziana dispar*; *C. goldfussi* and *C. pudica*. The impression of the headshield in *C. rugosa* shows that it had no genal spine, while the angularity of the endopodal scratches leaves no doubt that it was not a tail shield.

A head shield is also impressed in the three illustrated burrows from the **Upper Ordovician** (see also Pl. 68), but the difference between the scratches in the front and rear part of the burrow points to a leg differentiation that is unknown in trilobites.

Finally, the large bilobed burrow from the Rhaetic (**Upper Triassic**) sandstone near Tübingen, Germany, would certainly have been called *Cruziana* had it been found in the Paleozoic. Yet, associated trackways leave no doubt that the tracemaker was a horseshoe crab. Another Triassic homeomorph of *Cruziana*, but on the small side, is shown on Pl. 23 under the name *Isopodichnus*. It was made by a phyllopod brine shrimp, a kind of crustacean that even today employs the trilobite mode of sediment processing.

Such kinds of homeomorphy may pose problems in nomenclature. Yet they should not keep us from interpreting clearer Paleozoic examples in terms of trilobite biology.

Plate 11 · Trilobite Biology and *Cruziana* Authorship   35

Cruziana Authorship

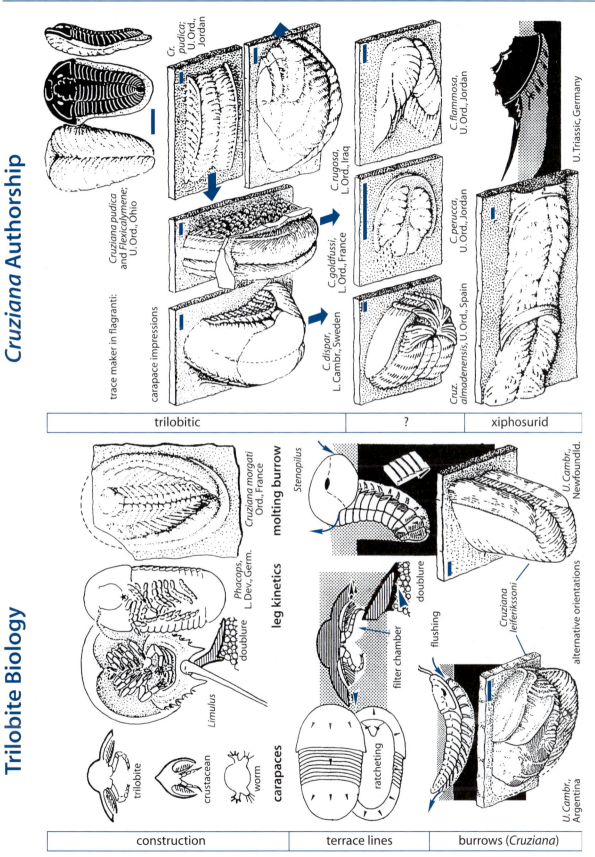

trace maker in flagranti:

carapace impressions

Cruziana pudica and *Flexicalymene;* U.Ord., Ohio

Cr. *pudica;* U.Ord., Jordan

C.rugosa, L.Ord., Iraq

C.goldfussi, L.Ord., France

C.dispar, L.Cambr., Sweden

C. flammosa, U.Ord., Jordan

C.perucca, U.Ord., Jordan

Cruz. *almadenensis,* U.Ord., Spain

U.Triassic, Germany

| trilobitic | ? | xiphosurid |

Trilobite Biology

Cruziana morgati Ord., France

Phacops, L.Dev., Germ.

Limulus

trilobite

crustacean

worm

carapaces

leg kinetics

molting burrow

Stenopilus

doublure

filter chamber

ratcheting

flushing

doublure

U.Cambr., Newfoundld.

Cruziana leiferikssoni

alternative orientations

U.Cambr., Argentina

| construction | terrace lines | burrows (*Cruziana*) |

Plate 12
Trilobite Fingerprints

At this point, a few remarks are in place about the muscles that activated the various motions of trilobite legs. Normally, each motion requires a pair of antagonistic muscles at each joint, for instance one behind the coxa for tilting the whole leg backward in the active walking stroke and an antagonist for the forward relaxation swing. This rule does not necessarily apply to the flexing motion. As mentioned before, spiders use the pressure of the body fluid to extend their legs after they have been actively flexed. As fluid pressure dissipates after death, dry spider carcasses always have the legs flexed together. This is also the case in trilobite carcasses (*Phacops*, Pl. 11). There are other reasons why the spider model might apply to trilobite legs. (1) They were also unmineralized; (2) as the active phase was always in the flexing mode, an undifferentiated hydraulic process was sufficient for reextension.

As tracemaking is restricted to the active phases, such considerations are rather irrelevant for ichnologists. All the more important is the differentiation of the leg tips for various purposes.

Endopodite Claw Traces. So far, we have treated the endopodite as an articulated stick. In reality, its function was enhanced by *setae* – not only at the proximal gnathobase, but also near the distal tip. In body fossils, their knowledge is restricted to the few cases in which the non-mineralized appendages have been preserved by bacterial pyritization. As such preservation is restricted to low-oxygen environments, this sample is necessarily biased. Ichnological evidence (mostly from well aerated sandy facies) complements this knowledge, even though the connection to established trilobite taxa remains usually unknown. Scratch morphology has nevertheless become an important database for the distinction of *Cruziana* ichnospecies, because due to their undertrace nature these fingerprints are usually well recorded. To single out the scratch set made by an individual leg in arrays of almost parallel scratches is a more difficult task. Our table of typical claw patterns gives an idea about the diversification of endopodites for their burrowing function, culminating in mole-like multiclawed shovels that were never preserved as body fossils.

As shown by *Cruziana acacensis* (Pls. 15 and 70), the setae producing "combed" scratch sets must not necessarily have been placed on a single podomere, as is assumed in the present diagram.

Exopodal Brushings. In contrast to the endopodites, the exopodites usually had a featherlike structure, fitting their function as gills, strainers, and ventilation fans. Nevertheless they often left delicate "brushings", which are always close to the lateral margins of the burrow and become more dominant towards the tail end of the burrow. Such brushings record a rearward active stroke, but their curvature is not the same in all ichnospecies. In the stationary (rusophyciform) burrows of **Cruziana aegyptica** (Lower Cambrian of Sinai) they are convex to the outside, as should be expected in an appendage swinging around the base. In all other forms (including the associated **Cruziana salomonis**), the brushings are convex to the center. This shape can only be explained by a bandy-legged posture of the exopodite.

For ventilating, straining and swimming functions, the exopodites had to be feathery with many thin setae. Their scratches, or brushings, are shown in the figures. In more fossorial species, however, setae near the tips may be sturdier in order to participate in excavation. Sets that had originally been interpreted as exopodal in *Cruziana acacensis* (Pl. 15) because they are rather delicate and maintain marginal positions in the burrow, were more probably made by smaller setae on a more proximal podomere of the endite. This is suggested by their congruence with endopodal scratch sets. In burrows from the Lower Cambrian of Laurentia (Pl. 66), whole exopodites were modified into comblike structures. Their scratches extend over most of the two lobes and mimic burrows from the Lower Ordovician of Gondwana (Pl. 68), in which the endopodite was modified in a similar way.

■ 3-clawed endopodal scratches: **a** *Cruziana semiplicata* (U. Cambr., Spain), **b** *Cruziana omanica* (Tremadoc., Oman), **c** *Cruziana petraea* (Caradoc., Jordan)

Plate 12 · Trilobite Fingerprints   37

Trilobite Fingerprints

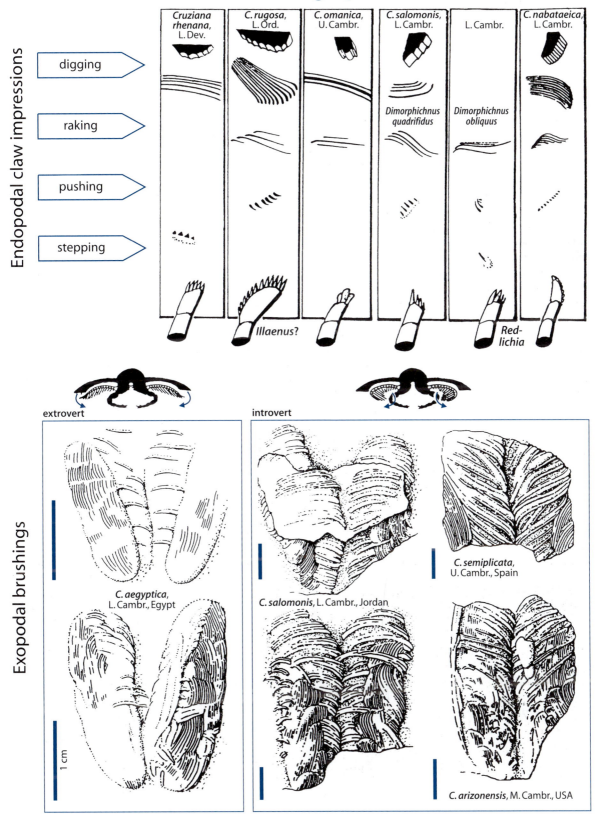

Endopodal claw impressions

digging

raking

pushing

stepping

Cruziana rhenana, L. Dev.

C. rugosa, L. Ord.

C. omanica, U. Cambr.

C. salomonis, L. Cambr.

L. Cambr.

C. nabataeica, L. Cambr.

Dimorphichnus quadrifidus

Dimorphichnus obliquus

Illaenus?

Red-lichia

Exopodal brushings

extrovert

introvert

C. aegyptica, L. Cambr., Egypt

C. salomonis, L. Cambr., Jordan

C. semiplicata, U. Cambr., Spain

C. arizonensis, M. Cambr., USA

1 cm

Plate 13
Cruziana Modifications

Apart from the fingerprints, there are other features to be considered in *Cruziana* ichnotaxonomy.

They relate to preservation, burrowing techniques and behavioral programs.

Undertrace Deficiency. Differences due to undertrack effects should be treated as preservational noise without taxonomic value. Nevertheless they must be singled out. In stationary (i.e. rusophyciform) trilobite burrows, such differences are restricted to size (mainly the length). In *Cruziana barbata* (Middle Cambrian), however, there is also a change in scratch patterns at different undertrack levels. In the deepest level (which is most common due to the high preservation potential), one sees only the scratches of the front legs (**a**). Because they dug proversely (i.e. in medio-anterior direction), this variant looks like a Santa-Claus moustache. Where the covering sand layer was thinner, the medio-posterior scratches of the rear endopodites add a goat beard (**b**). Yet in only a single specimen (**c**; Göttingen, Germany, collection) was the sand/mud interface shallow enough to depict the complete burrow with additional exopodite brushings and pleural grooves.

Instrumental Variation. *Cruziana semiplicata* (Upper Cambrian) is one of the forms in which stationary burrowing was consistently replaced by continuous plowing (see Pl. 14). Thereby the short "resting" traces transform into long cruzianaeform ribbons, along which rusophyciform resting places can nevertheless be recognized. We shall learn about the identity of the maker and its regular scribbling behavior in the following plate. In the non-scribbling Welsh occurrence, the profile of the plowings may change from two-lobed with only endopodal scratches (**a**), to ones with additional pleural grooves (**b**), and to the most typical variant with endopodal as well as exopodal lobes (**c**). Associated molting burrows (**d**; *Cruziana polonica*) show coxal impressions in the center.

Burrowing Techniques. *Cruziana jenningsi* (Lower Cambrian, USA, Canada) shows combed exopodal scratches (see other Laurentian forms, Pl. 66). In addition, there are impressions suggesting that the head shield helped to shovel-out sediment from the unusually deep burrow dug in a head-down attitude.

In *Cruziana dispar* (Lower Cambrian, Sweden) we observe a similar discordance between scratches in the front and rear parts of the body as in *C. barbata*. In addition, the steep flanks of unusually deep burrows may show impressions of the head shield and of pleural spines that trace the successive deepening of the burrow.

The giant *Cruziana radialis* (Ordovician, Australia) differs from other rusophyciform trilobite burrows in that the angle between opposite scratches does not decrease towards the rear end (see *C. jenningsi*). Instead, scratches gradually change to a medio-anterior direction. In addition, the central part of the burrow shows the impressions of the coxae, as typical for molting burrows (Pl. 11).

Burrowing Attitude. As scratch patterns made in the front and the rear part of the body may differ, cruzianaeform burrows will look different if the animal plowed in a head-down (prosocline) or in a tail-down (opisthocline) position.

Behavioral Variants. Some trilobites evolved particular behavior patterns that resulted in more effective foraging.

In the Lower Cambrian Mickwitzia Sandstone of Sweden, one often observes depressions with faint scratches radiating from the center (**Cheiichnus**). They can be explained by rotation of the body around the front end. Their trilobite origin, however, is uncertain, because bilobedness has been eliminated by the rotation.

Small multiclawed trilobites from the Lower Ordovician of northern **Iraq** managed to move in continuous scribbles, although not as regularly as in the Cambrian example shown in Pl. 14. Recently, "pirouettes" have been also found in a variant of *C. rugosa* in the Lower Ordovician of Portugal. In some multiclawed plowings from the Grès Armoricain of **France** (Nantes collection) the same group of trilobites shaved a deep burrow with a smoothly curved bottom. Because this burrow is much too long to have been made by stationary digging and as the lateral walls are too steep to have stood freely, it may have been a protrusive spreite burrow that maintained a U-shaped open tunnel at the bottom.

Palmate feeding tunnels (*C. ancora*) will be discussed in Pl. 15. Interestingly, trilobites appear not to have employed the meandering mode of grazing, possibly because they could not perform the sharp lateral turns needed for such a performance.

Plate 13 · *Cruziana* Modifications   39

Cruziana Modifications

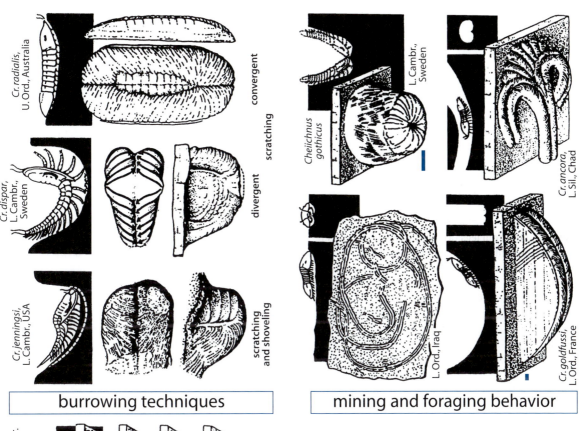

burrowing techniques

mining and foraging behavior

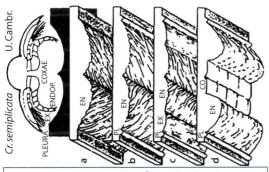

instrumental variation

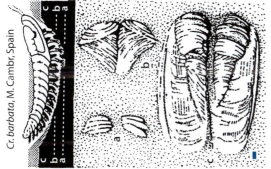

undertrace deficiency

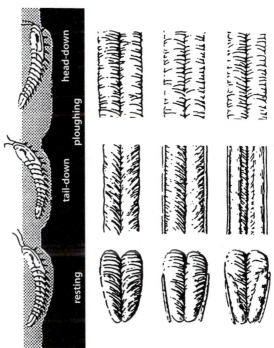

burrowing attitude

14

Plate 14
Cruziana semiplicata

The name of this ichnospecies refers to the fact that only the median halves of the two lobes bear the typical endopodal scratches, while the outer halves appear to be smooth. In Pl. 12 we learned that the latter are actually exopodal brushings. In stationary burrows, exopodal lobes are prominent only in the posterior part, but in tail-down plowings (Pl. 13), the exopodal lobes become a continuous feature. Originally having been described from the uppermost Cambrian of Wales, this ichnospecies was later found at the same level in northern Spain, eastern Newfoundland, northwest Argentina, Germany, Poland and Oman. All these areas were at that time situated at the northern margin of the Gondwana paleocontinent.

Authorship. From the presence of three-clawed endopodites, bandy-legged (introvert) exopodites and downturned pleural spines alone one could not guess what kind of trilobite species might have been responsible for this very characteristic trace fossil. But there was another possibility: in the deserts of Oman, sandy storm beds contain skeletal remains together with the burrows. This rare situation allowed, in cooperation with a trilobite expert (Richard Fortey), to test the co-occurrence of skeletal fragments and traces bed by bed. Remains of the genus *Maladioidella* were actually found, but only in the trace-bearing beds, their size varying with the width of the associated burrows. With this link, it is now possible to expand the known geographic range of this trilobite genus.

Search Behavior. In Wales, *Cruziana semiplicata* shows a variety of preservation styles, but otherwise it resembles other trilobite burrows: furrows differ in length, but remain essentially straight. In the Sierra de la Demanda of Spain, however, a steeply dipping bedding plane exposes regular "pir-ouettes", in which each furrow, many meters in length, covers a given area like a coiled garden hose. Deeper rusophyciform depressions mark resting stations along the furrow without a change in course. This tells us that the animal followed a fixed program over a considerable time. In order to create the circles, it had to laterally tilt its body either to the left or right side. As lateral flexure was limited, trilobites could not perform tight meanders. Circular scribbling, or looping, was probably the best they could do to systematically forage a given surface.

Thirty years after the first discovery, we went back to the same locality and uncovered a large adjacent area with more pirouettes (left slab). They were made by four different-sized individuals, of which two turned clockwise, two others counterclockwise, but always in the same sense within each system.

Circular scribbling also occur in the Oman population, although the outcrop situation prevents excavation of complete systems. Once this behavior has been checked in other occurrences, it might be reasonable to distinguish between linear and scribbling ichnosubspecies of *Cruziana semiplicata* and thus bring ichnotaxonomic resolution closer to the level of species in the body fossil record.

More recently, similar "trilobite pirouettes" were also discovered in the Lower Ordovician of Penha Garcia (Portugal). Their maker, however, was related to that of the larger *Cruziana rugosa* (Pl. 66), because its endopodites had about twelve comb-like claws. So the scribbling program evolved independently in at least two groups of trilobites.

Casts of both occurrences are now contained in the traveling exhibit Fossil Art. In the Spanish locality (Sierra de la Demanda) we also had a depressing experience: Only a few weeks after the additional surface had been excavated, cleaned and cast with the help of colleagues from the University of Zaragoza, it was vandalized. Sometimes it pays to take a container of latex right to the field!

5 cm

Cruziana semiplicata
(U. Cambrian, Spain)

Plate 14 · *Cruziana semiplicata*   41

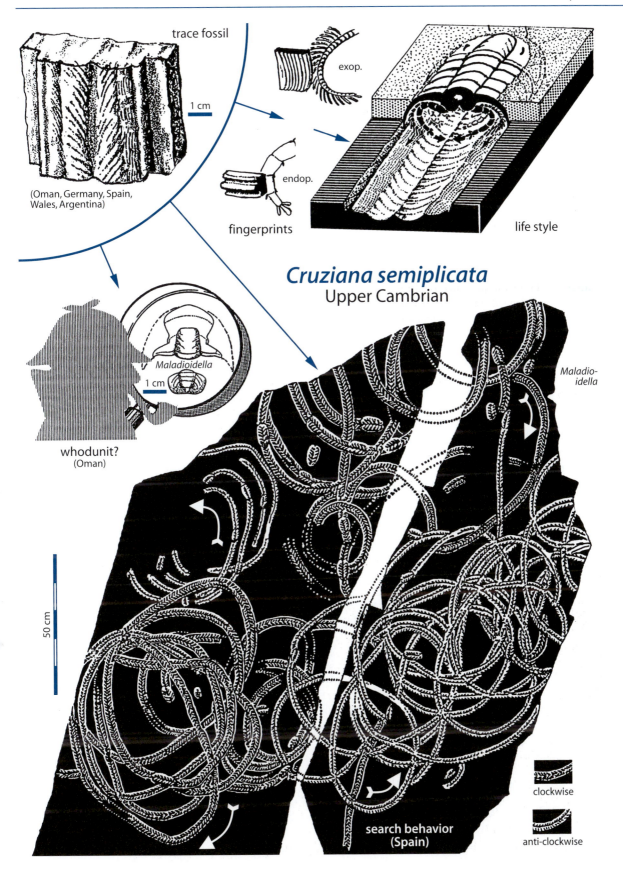

trace fossil

exop.

1 cm

(Oman, Germany, Spain,
Wales, Argentina)

endop.

fingerprints

life style

Cruziana semiplicata
Upper Cambrian

Maladioidella

1 cm

whodunit?
(Oman)

Maladio-idella

50 cm

clockwise

anti-clockwise

**search behavior
(Spain)**

Plate 15
Burrowing Behavior of Silurian Trilobites

By Silurian times, trilobites were already in decline – at least this is the impression one gets from diversity curves based on skeletal remains from the limestones and shales of Europe. Such remains are hardly ever found in the "Nubian Facies" of North Africa. Here, continued cratonic sedimentation and resedimentation resulted in thick sandstone series in which calcareous body fossils had little chance to be preserved. Still the clastic facies favored the preservation of trilobite burrows and it is from these that a remarkable case of behavioral evolution can be derived.

Although there are comparable occurrences in Chad, Saudi Arabia, Algeria, Benin, Brazil, and Argentina (Pl. 70), the best exposures are in the escarpments of the Akakus Range on the western side of the Murzuq Basin in southwest Libya. Here, the sandstones prograde northward into a euxinic basin, whose blackshales are not only the source of petroleum, but also of graptolites. From these we know that the trace-bearing Akakus Sandstone is Middle to Late Silurian in age.

Claw formulas reflect particular body features. Therefore they are the main criteria for the distinction of ichnotaxonomic subgroups, one of which is characterized by *Cruziana acacensis*. Here, the two distal podomeres of the endopodites were equipped with five blunt claws of nearly equal size. In larger members (*Cruziana bonariensis*, Pl. 70) one can also recognize a median furrow in each scratch, corresponding to a notch in the claw. In the marginal zone of the burrow there are scratch sets that might be referred to exopodites. While being finer than the endopodal scratches, they have a similar pattern and number of scratches (5 to 6). So they could also be referred to a more proximal podomere of the endopodite. In any case these were trilobites highly adapted to burrowing. Other characteristics are an outline slightly tapering towards the front end and the tendency to burrow in a head-down attitude (prosocline, Pl. 13). So far, members of this group are restricted to the Silurian of Gondwana (Pls. 69, 70).

Cruziana acacensis. In the Akakus section, some horizons are covered with spectacular arthrophycid worm burrows (Pls. 42 and 43), but otherwise the dominant trace fossil is *Cruziana acacensis*. It occurs in rusophyciform as well as short cruzianaeform versions, both showing the characteristic endopodal scratches. In some cases, impressions of the head shield and of the antennae are also preserved (*C. acacensis acacensis*). This feature and the orientation against the ambient current (cross bedding) show that the front end is actually tapering.

Most interesting, however, are deeper modifications (treated as ichnosubspecies), which show areas smoothened by the head shield in addition to the characteristic leg scratches. In *C. acacensis sandalina*, the surface smoothed by the cephalon is overhanging, so that the cast looks like an Oriental shoe with the bilobed scratch pattern on the sole. In *C. acacensis retroversa*, it is the scratch-bearing surface that has become overhanging, as if the body axis had been tilted into a headstand position during the burrowing process. On the flanks one can also distinguish "exopodal" scratch sets, which are relatively coarse, but still finer than the five endopodal distal scratches. Only the former can be seen on the flanks of *C. acacensis laevigata*, because on the sole face of these elongate burrows leg scratches have been completely obliterated by the scraping edge of the shield. Burrowing in a headstand would have been impossible in an open pit; it required the support of sand sinking down behind the animal. Because these variants are represented by many specimens from various horizons, they may in the future become useful for a high-resolution stratigraphy based on the evolutionary change of a basic burrowing program and on bed-by-bed sampling.

Cruziana ancora. *Cruziana acacensis sandalina* also occurs farther east in northern Chad, where it is associated with the smaller *Cruziana ancora*. Without the typical bilobed scratch pattern, this form would be interpreted as a worm burrow, because it not only has a smaller diameter; it also consists of a U-shaped tunnel with a palmate foraging structure at one end. The curves implied in the palmate branches were executed in a typical trilobite fashion: leaning to the side like a banking airplane, the animal turned by bending dorsally. In this fashion, the lateral branches have been plowed repeatedly, as shown by their high profile and protrusive backfill structure.

While the basic burrowing program of *Cruziana ancora* is clear (dig a U-shaped base tunnel and add increasingly curved side branches on both sides of the distal exit), the body orientation is somewhat problematic because the faint endopodal scratches converge towards the openings of the branches.

In this dilemma, an occurrence from the other side of the Atlantic provides complementary evidence. In the Balcarce Quartzites of Argentina, several complete burrows were excavated from one large rock surface. Thereby it was possible not only to confirm the U-shape of the base tunnel, but also to establish the orientation: the palmate opening always points upcurrent.

This orientation and the directions of the scratches on the floor of the U-tunnel indicate that the trilobite headed towards the palmate end of the tunnel. In the Argentinian form (*Cruziana ancora angusta*), the lateral branches are less developed. There are also two sharp ridges that could not have been excavated by legs. They probably correspond to the more numerous ridges at the front end of the Chad specimens and could be explained as diggings of frontal spines that were operated by dorsal flexure of the whole body. Alternatively they could correspond to antennae extending towards the surface.

Even though interpretations may change in future studies, these trace fossils suggest that some trilobites responded to increasing predator pressures by evolving complex burrowing programs that allowed them to become more or less infaunal – in spite of their constructional handicaps.

Plate 15 · Burrowing Behavior of Silurian Trilobites   43

Silurian Ichno-subspecies
based on burrowing behavior

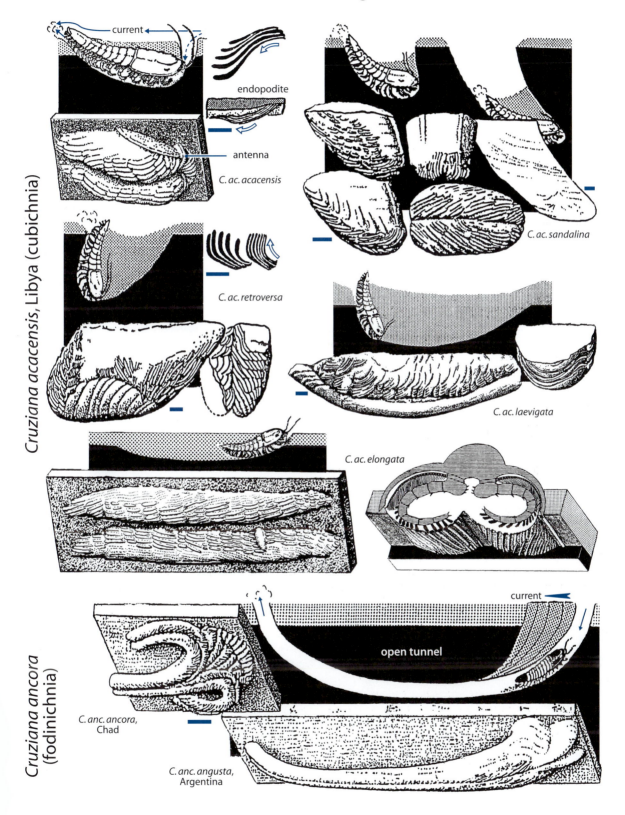

current

endopodite

antenna

C. ac. acacensis

C. ac. sandalina

C. ac. retroversa

C. ac. laevigata

C. ac. elongata

Cruziana acacensis, Libya (cubichnia)

current

open tunnel

C. anc. ancora, Chad

C. anc. angusta, Argentina

Cruziana ancora (fodinichnia)

1 cm

■ *Cruziana acacensis retroversa* (Akakus Ss., SW Libya)

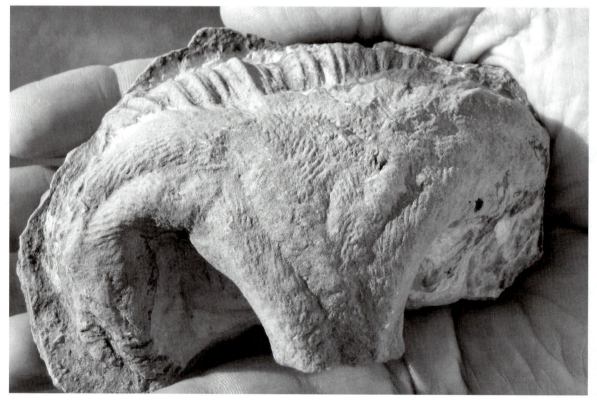

■ *Cruziana ancora ancora* (Silurian, Chad)

Arthropod Tunnel Systems

Due to their relatively weak appendages, trilobites were ill-suited for deep burrowing. Only few of them produced permanent tunnels, in contrast to decapod *crustaceans* such as crabs and shrimps, whose legs are not only stiffer, but also have more freedom to bend in various directions. The differentiation of only five pairs of legs into pereiopods also allowed them to be operated individually, rather than collectively, in locomotion and burrowing. So it is not surprising that crustaceans had less difficulty than trilobites to excavate more compact mud and to produce open tunnels as permanent domiciles. This ability also gave them a lead in the competitive trend to penetrate the sediment to deeper and deeper levels (infaunal tiering). In addition, this trend (and the ability to actively reinforce tunnel walls in looser sediments; Pl. 18) increased the preservation potential of the crustacean burrows, because they (1) remained open long enough to be passively filled with different (mostly coarser) sediment; (2) they were unlikely to become eroded and (3) they were not erased by deeper burrows during the upward shift of tiers following sedimentation. Whereas trilobite burrows are commonly penetrated by deeper-tier worm burrows (often misinterpreted as prey; Pl. 74), bioturbational overprints on crustacean tunnels are restricted to exploitation of their sedimentary fill by *Chondrites* (Pl. 74). As another preservational advantage, the relatively lose fill was a preferred site for concretionary prefossilization. The resulting calcareous, sideritic, or chertified casts are resistant enough to become secondarily reworked, or to weather out, without losing the details of scratch patterns (Pl. 16). Open tunnels also acted as traps, in which delicate microfossils are preferentially preserved in three dimensions.

The tunnel systems of terrestrial *insects* have a much lower preservation potential. An exception are domiciles whose walls became actively solidified by an organic cement. This is the case in the nests of soil bees, in the depositories of dung beetles, and in pupa chambers (Pl. 3). The latter are so resistant that they may survive reworking and limited transport, just as body fossils.

There is also the question, whether insect nests should be classified as trace fossils? It becomes critical in objects like bee or wasp nests constructed above ground, or in the minings made by larvae in leaves or bark. Coprolites and cololites are a similar case. Such objects are fossilized and their interpretation poses the same problems as ordinary trace fossils. Still they are not covered by the present text, which only deals with traces in the sense of biogenic *sedimentary* structures.

Literature

Chapter IV

Bromley RG (1996) Trace fossils: Biology, taphonomy and applications, 2nd edn. Chapman & Hall, London, 361 p (Pages 85–105 offer a thorough review of crustacean burrow architecture)

Frey RW, Curran HA, Pemberton SG (1984) Tracemaking activities of crabs and their environmental significance: The Ichnogenus *Psilonichnus*. J Paleontol 58(2):333–350 (Simple crab tunnels)

Genise JF (2000) The ichnofamily Celliformidae for *Celliforma* and allied ichnogenera. Ichnos 7:267–284 (Review of hymenopteran trace fossils and their terminology)

Genise JF (2004) Ichnotaxonomy and ichnostratigraphy of chambered trace fossils in palaeosols attributed to coleopterans, ants and termites. In: McIlroy D (ed) The application of ichnology to palaeoenvironmental and stratigraphic analysis. Geological Society of London, Special Publication 228, pp 419–453 (Extensive review of coleopteran, ant and termite trace fossils, including a discussion of their stratigraphic distribution in paleosols)

Pervesler P, Dworschak PC (1985) Burrows of *Jaxea nocturna* Nardo in the Gulf of Trieste. Senck Marit 17(1/3):33–53 (Modern thalassinid tunnel systems consisting of spiral U-tube with attached chambers)

Rindsberg AK (1994) Ichnology of the Upper Mississippian Hartselle Sandstone of Alabama, with notes on other carboniferous formations. Geological Survey of Alabama, Bulletin 158, 107 p, 22 pls. (Interpretation of *Rusophycus hartselleanus* as indicating ventilation tubes)

Rindsberg AK, Kopaska-Merkel DC (2005) *Treptichnus* and *Arenicolites* from the Steven C. Minkin Paleozoic Footprint Site (Lang-settian, Alabama, USA). In: Buta RJ, Rindsberg AK, Kopaska-Merkel DC (eds) Paleozoic footprints in the Black Warrior Basin of Alabama. Alabama Paleontological Society Monograph 1, pp 121–141 (*Treptichnus* and related ichnogenera are reconsidered based on examination of the Carboniferous types; a longitudinally striate form from Alabama is named as *T. apsorum*)

Sakagami SH, Michener CD (1962) The nest architecture of the sweat bees (Halictinae): A comparative study of behavior. University of Kansas Press, Lawrence, Kansas, 135 p (The evolution of sweat bees can be deduced in part from their burrow architecture)

Sellwood BW (1971) A *Thalassionides* burrow containing the crustacean *Glyphaea udressieri* (Meyer) from the Bathonian of Oxfordshire. Palaeontology 14(4):589–591

Sheehan PM, Schiefelbein DRJ (1984) The trace fossil *Thalassionides* from the Upper Ordovician of the Eastern Great Basin: Deep burrowing in the Early Paleozoic. J Paleontol 58(2):440–447 (Ophiomorphid burrows of this age must have been produced by non-callianassid shrimp-like crustaceans)

Uchman A (1995) *Treptichnus*-like traces made by insect larvae (Diptera: Chironomidae, Tipulidae). In: Buta RJ, Rindsberg AK, Kopaska-Merkel DC (eds) Paleozoic footprints in the Black Warrior Basin of Alabama. Alabama Paleontological Society Monograph 1, pp 143–146 (Carboniferous *Treptichnus* is compared to the burrows of modern dipteran insect larvae)

Utashiro T, Horii Y (1965a) Ecology and burrows of *Scopimera globosa* and *Ilyoplax pusillus*: Biological studies in "Lebensspuren" Part VII, (12):110–143

Utashiro T, Horii Y (1965b) Some knowledges on the ecology of *Ocypode stimpsoni* Ortmann and on its burrows. Biological Study of "Lebensspuren" Part 6, (3):121–141

Plate 16: Crab and Shrimp Burrows

Basan PB, Frey RW (1977) Actuo-palaeontology and neoichnology of salt marshes near Sapelo Island, Georgia. In: Crimes TP, Harper JC (eds) Trace fossils 2. Geol J, Special Issue 9, pp 41–70 (Resin casts of modern crustacean burrows)

Carmona NB, Buatois LA, Mángano MG (2004) The trace fossil record of burrowing decapod crustaceans: Evaluating evolutionary radiations and behavioral convergence. In: Webby BD, Mángano MG, Buatois LA (eds) Trace fossils in evolutionary palaeoecology. Fossils and Strata 51:141–153 (Evaluation of the fossil record of shrimp and shrimp-like burrows through the Phanerozoic. Paleozoic examples are attributted to behavioral convergence)

Dworschak PC (1981) The pumping rates of the burrowing shrimp *Upogebia pusilla* (Petagna) (Decapoda: Thalassinidae). J Exp Mar Biol Ecol 52:52–35

Dworschak PC (1983) The biology of *Upogebia pusilla* (Petagna) (Decapoda, Thalassinidea). I. The burrows P.S.Z.N.I.: *Marine Ecology* 4(1):19–43 (Y-shaped burrows recorded by epoxy casts)

Frey RW, Pemberton SG (1987) The *Psilonichnus* ichnocoenose, and its relationship to adjacent marine and nonmarine ichnocoenoses along the Georgia coast. B Can Petrol Geol 35:333–357 (Definition of the *Psilonichnus* ichnofacies, including a characterization of various supralitoral crab burrows)

Frey RW, Curran HA, Pemberton SG (1984) Tracemaking activities of crabs and their environmental significance: The ichnogenus *Psilonichnus*. J Paleontol 58:333–350 (Description of crab burrows and introduction of new ichnospecies of *Psilonichnus*)

Miller MF, Curran HA (2001) Behavioral plasticity of modern and Cenozoic burrowing thalassinidean shrimp. Palaeogeog Palaeoclim Palaeoecol 166:219 236 (Analysis of burrowing strategies in callianassid shrimp)

Muñiz F, Mayoral E (2001) El icnogénero *Spongeliomorpha* en el Neógeno Superior de la Cuenca del Guadalquivir (área de Lepe-Ayamonte, Huelva, España). Rev Esp Paleontol 16:115–130 (General review of the ichnogenus *Spongeliomorpha* with illustration of specimens from the Miocene of Spain)

Myrow PM (1995) *Thalassinoides* and the enigma of early Paleozoic open-framework burrow systems. Palaios 10:58–74 (Detailed documentation of a Paleozoic *Thalassinoides*)

Pemberton SG, Frey RW, Walker RG (1984) Probable lobster burrows in the Cardium formation (Upper Cretaceous) of Southern Alberta, Canada, and comments on modern burrowing decapods. J Paleontol 58(6):1422–1435

Radwanski A (1977) Burrows attributable to the ghost crab *Ocypode* from the Korytnica basin (Middle Miocene; Holy Cross Mountains, Poland). Acta Geol Pol 27(2):217–225 (Large burrow casts with diagenetic overprint)

Plate 17: Other Arthropod Tunnels and Nests

Ausich WI (1979) *Hondichnus monroensis* n.gen.sp. a new early Mississippian trace fossil. J Paleontol 53:1155–1159 (Junior synonym of *Margaritichnus*)

Bandel K (1967) Trace fossils from two upper Pennsylvanian sandstones in Kansas. Paleontological Contributions (Paper 18):1–13 (*Margaritichnus*)

Bergström J (1972) Appendage morphology of the trilobite *Cryptolithus* and its implications. Lethaia 5:85–94

Brown RW (1934) *Celliforma spirifer*, the fossil larval chambers of mining bees. J Wash Acad Sci 24(12):532–539 (Club-shaped pupa chambers from continental Eocene resembling *Tambia* lack spiral structure inside apical wall)

Brown RW (1935) Further notes on fossil larval chambers of mining bees. J Wash Acad Sci 25(12):527–528

Buckman JO (1997) An unusual new trace fossil from the Lower Carboniferous of Ireland: *Intexalvichnus magnus*. J Paleontol 71:316–324 (similar to *Margaritichnus*)

Buckman JO (2001) *Parataenidium*, a new *Taenidium*-like ichnogenus from the Carboniferous of Ireland. Ichnos 8:83–97 (Diagnosis of *Parataenidium*)

Eberth DA, Berman DS, Sumida SS, Hopf H (2000) Lower Permian terrestrial paleoenvironments and vertebrate paleoecology of the Tambach Basin (Thuringia, central Germany): The upland Holy Grail. Palaios 15:293–313 (*Tambia* associated with terrestrial fauna)

Frey RW, Pemberton SG, Fagerstrom JA (1984) Morphological, ethological and environmental significance of the ichnogenera *Scoyenia* and *Ancorichnus*. J Paleontol 58:511–528 (Reevaluation of some meniscate burrows)

Genise JF, Sciutto JC, Laza JH, Gonzalez MG, Bellosi E (2002) Fossil bee nests, coleopteran pupation chambers and tuffaceous palaeosols from the Late Cretaceous Laguna Palacios Formation, central Patagonia (Argentina). Palaeogeog Palaeoclim Palaeoecol 177:215–235

Hasiotis ST (2003) Complex ichnofossils of solitary and social soil organisms: understanding their evolution and roles in terrestrial paleoecosystems. Palaeogeog Palaeoclim Palaeoecol 192:259–320 (Review of modern insect burrows and Jurassic termite nests)

Houck K, Lockley M (1986) Pennsylvanian biofacies of the Central Colorado Trough. Geology Department Magazine 2:1–64 (Mention *Margaritichnus* and *Curvolithus*)

Lockley MG, Rindsberg AK, Zeiler RM (1987) The paleoenvironmental significance of the nearshore *Curvolithus* ichnofacies. Palaios 2:255–262 ("*Margaritichnus*" [*Parataenidium*] as an indicator of high-energy nearshore environments)

Mángano MG, Buatois LA (2003) Trace fossils. In: Benedetto JL (ed) Ordovician fossils of Argentina. Universidad Nacional de Córdoba, Secretaría de Ciencia y Tecnología, pp 507–553 (Description of Lower Ordovician *Trichophycus*)

Mángano MG, Buatois LA, West RR, Maples CG (2000) A new ichnospecies of *Nereites* from Carboniferous tidal-flat facies of eastern Kansas, USA: Implications for the *Nereites-Neonereites* debate. J Paleontol 74:149–157 (Discussion of the *Eione-Margaritichnus* problem)

Osgood RG (1970) Trace fossils of the Cincinnati area. Paleontographica Americana 6:281–444 (Description of Upper Ordovician *Trichophycus* in its type area)

Seilacher A (1990) Paleozoic trace fossils. In: Said R (ed) The geology of Egypt. AA Balkema, Rotterdam, pp 649–670 (*Trichophycus* of proposed trilobite origin)

Seilacher A, Hemleben C (1966) Beiträge zur Sedimentation und Fossilführung des Hunsrückschiefers, Teil 14, Spurenfauna und Bildungstiefe des Hunsrückschiefers. Hessisches Landesamt für Bodenforschung, Notizblatt 94, pp 40–53 (Description of *Ctenopholeus*)

Seilacher A, Meischner D (1965) Fazies-Analyse im Paläozoikum des Oslogebietes. Geol Rundsch 54:596–619 (Ordovician *Trichophycus* as a trilobite burrow in the Oslo Graben of Norway)

Tschinkel WR (2003) Subterranean ant nests: Trace fossils past and future? Palaeogeog Palaeoclim Palaeoecol 192:321–333 (Modern ant nests consist of flat horizontal chambers connected by vertical shafts. They reach up to 4 m deep)

Volk M (1968) *Trichophycus thuringicum*, eine Lebensspur aus den Phycoden-Schichten (Ordovizium) Thüringens. Senck Leth 49(5/6):581–585, 1 pl. (Longitudinal scratches)

Plate 18: Ophiomorphids

Berger W (1957) Eine spiralförmige Lebensspur aus dem Rupel der bayrischen Beckenmolasse. Neues Jb. Geol. Paläontol. Mh. 12:538–540 (*Gyrolithes*)

Bromley RG (1967) Some observations on burrows of thalassinidean Crustacea in chalk hardgrounds. Q J Geol Soc London 123:157–182 (One of the first studies to document in detail the architecture of thalassinidean burrows in firmgrounds)

Bromley RG, Ekdale AA (1984) Trace fossil preservation in flint in the European chalk. J Paleontol 58:298–311 (Preservational analysis of Cretaceous chalk trace fossils, including "paramoudras")

Bromley RG, Ekdale AA (1998) *Ophiomorpha irregulaire* (trace fossil): Redescription from the Cretaceous of the Book Cliffs and Wasatch Plateau, Utah. J Paleontol 72:773–778 (Redescription of *Ophiomorpha irregulaire*)

Bromley RG, Frey RW (1974) Redescription of the trace fossil *Gyrolithes* and taxonomic evaluation of *Thalassinoides, Ophiomorpha* and *Spongeliomorpha*. Bull Geol Soc Denmark 23:311–335 (Comprehensive review of crustacean tunnel systems)

Bromley RG, Schulz M-G, Peake NB (1975) Paramoudras: Giant flints, long burrows and the early diagenesis of chalks. Biol Skr Dan Vid Sel 20(10):3–31, 5 pls. (Description of paramoudras as *Bathichnus paramoudrae*)

Brönnimann P, Masse JP (1969) Thalassinid (Anomura) coprolites from Barremian-Aptian passage beds, Basse Provence, France. Rev Micropalaeontol 11(3):153–160

Buckland W (1817) Description of the paramoudra, a singular fossil body that is found in the Chalk of the north of Ireland, with some general observations on flints in chalk. Trans Geol Soc Lond (1)4:413–423 (First description of paramoudras)

Cheng Y-M (1972) On some Lebensspuren from Taiwan. Acta Geol Taiwan (15):13–22 (*Ophiomorpha* with dendritic branching)

Curran HA (1976) A trace fossil brood structure of probable callianassid origin. J Paleontol 50(2):249–259 (Brood chambers in Pleistocene *Ophiomorpha* with radiating larval tunnels)

Curran HA, Frey RW (1977) Pleistocene trace fossils from North Carolina (U.S.A.), and their Holocene analogues. In: Crimes TP, Harper JC (eds) Trace fossils 2. Geol J, Special Issue (9):139–162

Dworschak PC, Rodrigues S de A (1997) A modern analogue for the trace fossil *Gyrolithes*: burrows of the thalassinidean shrimp *Axianassa australis*. Lethaia 30:41–52 (Description of corkscrew-shaped shrimp burrows in modern environments)

Farrow GE (1971) Back-reef and lagoonal environments of Aldabra Atoll distinguished by their crustacean burrows. Symp Zool Soc Lond (28):455–500 (Spiral burrows of male *Ocypod* are dextrally or sinistrally coiled depending on whether the right or the left claw is larger)

Fiege K (1944) Lebensspuren aus dem Muschelkalk Nordwestdeutschlands. Neues Jahrb Geol P-A 88(3):401–426 (Diagnosis of *Pholeus abomasoformis*)

Frey RW, Howard JD (1975) Endobenthic adaptations of juvenile thalassinidean shrimp. Bull Geol Soc Denmark 24:283–297

Frey RW, Howard JD, Pryor WA (1978) *Ophiomorpha*: Its morphologic, taxonomic and environmental significance. Palaeogeog Palaeoclim Palaeoecol 23:199–229

Fürsich FT (1973) A revision of the fossils *Spongeliomorpha, Ophiomorpha* and *Thalassinoides*. Neues Jahrb Geol P M 1973:719–735 (Review of fossil crustacean tunnel systems; the three ichnogenera are synonymized under *Spongeliomorpha*)

Garcia-Ramos JC Garcia Domingo A, Gonzalez Lastra JA, Hernaiz P, Ruiz P, Valenzuela M (1984) Significado ecologico y aplicaciones sedimentologicas de la traza fosil *Tubotomaculum* del Eoceno-Mioceno inferior de Andalucia. Homenaje a Luis Sanchez de la Torre 20:373–378 (Identical to figured burrow from Borneo)

Häntzschel W (1934) Schraubenförmige und spiralige Grabgänge in turonen Sandsteinen des Zittauer Gebirges. Sternspuren, erzeugt von einer Muschel: *Scrobicularia plana* (da Costa). Senckenberg 16(4/6):313–330

Häntzschel W (1935) *Xenohelix saxonica* n.sp. und ihre Deutung. Senckenberg 17(1/2):105–108 (*Gyrolithes*)

Häntzschel W (1952) Die Lebensspur *Ophiomorpha* Lundgren im Miozän bei Hamburg, ihre weltweite Verbreitung und Synonymie. Mitt Geol Staatsinst Hamburg (21):142–153, 2 pls (Synonymy and global distribution)

Howard JD (1978) Sedimentology and trace fossils. In: Basan PR (ed) Trace fossil concepts. Society of Economic Paleontologists & Mineralogists (SEPM). Short Course 5, 13–47 (Reaction of *Ophiomorpha* makers to erosion and deposition)

Jordan R (1981) Sind submarine Gas- und Schlammvulkane in der Schreibkreide-Fazies Nordwesteuropas Anlaß für die Genese der Paramoudras? Neues Jahrb Geol P M (7):419–424 (Nonbiogenic origin of Paramoudra structures)

Keij AJ (1965) Miocene trace fossils from Borneo. Paläont Z 39:220–228 (Description of sideritized ophiomorphids from the Miocene of Borneo)

Kennedy WJ, Sellwood BW (1970) *Ophiomorpha nodosa* Lundgren, A marine indicator from the Sparnacian of South-East England. P Geologist Assoc 81(1):99–110

Kennedy WJ, Jakobson ME, Johnson RT (1969) A *Favreina-Thalassinoides* association from the Great Oolite of Oxfordshire. Palaeontology 12(4):549–554, 1 pl.

Kilpper K (1962) *Xenohelix* Mansfield 1927 aus der miozänen Niederrheinischen Braunkohlenformation. Paläont Z 36(1/2):55–58 (Tertiary *Gyrolithes*)

Knaust D (2002) Ichnogenus *Pholeus* Fiege, 1944, revisited. J Paleontol 76:882–891

Macsotay O (1967) Huellas problematicas y su valor paleoecologico en Venezuela. Universidad Central de Venezuela, Escuela de Geo-logia, Minas y Metalurgia, Geos 16:7–79 (Spectacular *Gyrolithes*)

Meiburg P, Speetzen E (1970) Ein Problematikum aus dem Turon von Lengerich (Westfalen). Neues Jahrb Geol P M (1):10–17 (Pyritized version of *Ophiomorpha*)

Perkins BF, Stewart CL (1971) Amon Carter Park. In: Perkins BF (ed) Trace fossils: A field guide to selected localities in Pennsylvanian, Permian, Cretaceous, and Tertiary rocks of Texas and related papers. Louisiana State University School of Geoscience Miscellaneous Publication 71-1, pp 80–88 (*Thalassinoides* with "median furrows" (draft fill) in the Cretaceous of Texas)

Pollard JE, Goldring R, Buck SG (1993) Ichnofabrics containing *Ophiomorpha*: Significance in shallow-water facies interpretation. J Geol Soc London 150:149–164 (Analysis of *Ophiomorpha* ichnofabrics with reference to colonization windows)

Seilacher A (1968) Sedimentationprozesse in Ammonitengehäusen. Akademie der Wissenschaften und der Literatur in Mainz, mathematisch-naturwissenschaftliche Klasse 1967, pp 191–203 (Mentions fill channels in crustacean burrows)

Sellwood BW (1971) A *Thalassionides* burrow containing the crustacean *Glyphaea udressieri* (Meyer) from the Bathonian of Oxfordshire. Palaeontology 14(4):589–591

Uchman A (1995) Taxonomy and paleoecology of flysch trace fossils: The Marnoso-arenacea formation and associated facies (Miocene, Northern Apennines, Italy). Beringeria 15:1–115 (Discussion of deep marine crustacean burrows)

Plate 19: Rhizocoralliids

Arkell WJ (1939) U-shaped burrows in the Corallian Beds of Dorset. Geol Mag 76(904):455–458 (*Diplocraterion*)

Bather FA (1909) Fossil representatives of the lithodomous worm *Polydora*. Geol Mag 6(537):108–110

Bromley RG, D'Alessandro A (1983) Bioerosion in the Pleistocene of southern Italy: Ichnogenera *Caulostrepsis* and *Maeandropolydora*. Riv Ital Paleontol S 89(2):283–309 (Shell borers)

Fürsich FT (1974a) Ichnogenus *Rhizocorallium*. Paläont Z 48(1–2):16–28

Fürsich FT (1974b) On *Diplocraterion* Torell 1870 and the significance of morphological features in vertical spreite-bearing, U-shaped trace fossils. J Paleontol 48:952–962

Fürsich FT, Mayr H (1981) Non-marine *Rhizocorallium* (trace fossil) from the Upper Freshwater Molasse (upper Miocene) of southern Germany. Neues Jahrb Geol P M 1981:321–333 (Occurrence of *Rhizocorallium* in continental firmgrounds)

Goldring R (1962) The trace fossils of the Baggy Beds (Upper Devonian) of north Devon, England. Paläont Z 36:232–251 (Analysis of equilibrichnia including *Diplocraterion yoyo*)

Hempel C (I960) Über das Festsetzen der Larven und die Bohrtätigkeit der Jugendstadien von *Polydora ciliata* (Polychaeta sedentaria). Helgoland Wiss Meer 7(2):80–92 (Burrowing mechanism)

Hertweck G (1971) *Polydora ciliata* auf lebenden Herzmuscheln. Nat Mus 101(11):458–466 (Shell borers)

Howell BF (1945) *Skolithos*, *Diplocraterion*, and *Sabellidites* in the Cambrian Antietam Sandstone of Maryland. Bull Wagner Free Institute of Science 20:4, 2 pls

Schäfer W (1972) Ecology and palaeoecology of marine environments. University of Chicago Press, Chicago, 568 p

Plate 20: Rhizocoralliid Modifications

Andree K (1926) Bedeutung und zeitliche Verbreitung von *Arenicoloides* Blanckenhorn und verwandten Formen. Palaeontologische Zeitschrift 8(1):120–128 (Epireliefs of *Diplocraterion*)

Bentz A (1928) Fossile Röhrenbauten im Unterneocom des Isterbergs bei Bentheim. Jb. Preuss. Geol. L.-A. 49:1173–1183, 1 pl. (*Rhizocorallium hohendahli*)

Bromley RG (1996) Trace fossils: Biology, taphonomy and applications, 2nd edn. Chapman & Hall, London, 361 p (Introduction of equilibrichnia as illustrated by *Diplocraterion*)

Chaplin JR (1996) Ichnology of transgressive-regressive surfaces in mixed carbonate-siliciclastic sequences, Early Permian Chase Group, Oklahoma. In: Witzke B, Ludvigson GA, Day J (eds) Paleozoic sequence stratigraphy: Views from the North American Craton. Geological Society of America Special Paper 306: 399–418 (Firmground surfaces with *Glossifungites*)

Cole SL III, McDowell RR (n.d.) Implications of *Bifungites* from the Upper Devonian of West Virginia, USA. (Interpretation of arrowlike projections of *Bifungites* as an adaptation enabling the inhabitant to wedge itself in the burrow against predators)

Damborenea S (1981) *Bifungites* sp. (trace fossils) in Lower Ordovician beds of Perchel Region, Quebrada de Humahuaca, Jujuy Province. Rev. Asoc. Paléont. Argent., pp 2–20

Desio A (1940) Vestigia problematiche paleozoiche della Libia. Pubblicazioni dell'Istituto di Geologia, Paleontologia e Geografia Fisica della R. Università di Milano, Serie P, Pubblicazione 2, 47–92 (Diagnosis of *Bifungites*)

Dubois P, Lessertisseur J (1964) Note sur *Bifungites*, trace problématique du Dévonien du Sahara. B Soc Geol Fr 6(7):626–634

Farrow GE (1966) Bathymetric zonation of Jurassic trace fossils from the coast of Yorkshire, England. Palaeogeog Palaeoclim Palaeoecol 2:103–151 (Slipperlike *Rhizocorallium*)

Fillion D, Pickerill RK (1984) On *Arthraria antiquata* Billings, 1872 and its relationship to *Diplocraterion* Torell, 1870 and *Bifungites* Desio, 1940. J Paleontol 58:683–696 (Taxonomic review of *Bifungites* and *Arthraria*)

Fürsich FT (1981) Invertebrate trace fossils from the Upper Jurassic of Portugal. Comun Serv Geol Portugal 67(2):153–168, 6 pls (Rhizocoralliids)

Fürsich FT, Mayr H (1981) Non-marine *Rhizocorallium* (trace fossil) from the Upper Freshwater Molasse (Upper Miocene) of southern Germany. Neues Jahrb Geol P M 6:321–333

Goldring R (1962) The trace fossils of the Baggy Beds (Upper Devonian) of north Devon, England. Paläont Z 36:232–251 (A key paper including analysis of the equilibrichnion *Diplocraterion yoyo*)

Hamm F (1929) Über Rhizocoralliden im Kreidesandstein der Umgegend von Bentheim. Mitt Provinzialstelle Naturdenkmalpflege Hannover 2:101–107 (cf. *Rhizocorallium habichi*)

Hosius A (1893) Über marine Schichten im Wälderthon von Gronau (Westfalen) und die mit denselben vorkommenden Bildungen (*Rhizocorallium Hohendali*, sog. Dreibeine). Z Dtsch Geol Ges 45:34–53 (Triple rhizocoralliids in Cretaceous firmground)

Karaszewski W (1974) *Rhizocorallium*, *Gyrochorte* and other trace fossils from the Middle Jurassic of the Inowłódz Region, Middle Poland. Bull Acad Pol Sci 21(3–4):199–204, 4 pls.

Lange W (1932) Über spirale Wohngänge, *Lapispira bispiralis* ng. et n.sp., ein Leitfossil aus der Schlotheimien-Stufe des Lias Norddeutschlands. Z Dtsch Geol Ges 84:537–543 (Diagnosis of *Lapispira*)

Mansfield WC (1927) Some peculiar fossil forms from Maryland. US National Museum, Proceedings 71(16), 9 p (Description of *Xenohelix* (*Gyrolithes*) *marylandica* from the Miocene of Maryland, USA)

Mayer G (1952) Bisher bekannte und neue Vorkommen der Trias-Lebensspur *Rhizocorallium jenense* Zenker. Beitr Natkd Forsch Südwestdtschl 11(2):111–115

Mayer G (1953) Über ein *Rhizocorallium*-Vorkommen im Jura der Langenbrücker Senke. Jber Mitt Oberrh Geol Ver 35:22–25

Mayer G (1954) Ein neues *Rhizocorallium* aus dem Mittleren Hauptmuschelkalk von Bruchsal. Beitr Natkd Forsch Südwestdtschl 13(2):80–83

Osgood RG (1970) Trace fossils of the Cincinnati area. Paleontographica Americana 6:281–444 (Redescription of *Diplocraterion*, *Bifungites*, and *Arthraria*)

Pemberton SG, MacEachern JA, Saunders T (2004) Stratigraphic applications of substrate-specific ichnofacies: Delineating discontinuities in the rock record. In: McIlroy D (ed) The application of ichnology to palaeoenvironmental and stratigraphic analysis. Geological Society, London, Special Publication 228, pp 157–178 (*Glossifungites* ichnofacies)

Rieth A (1932) Über ein rhizocoralliumartiges Problematicum aus dem Arietenkalk (Lias alpha Betzingen). Zbl Mineral, Abt B 10: 518–524

Rodriguez J, Gutschick RC (1970) Late Devonian – Early Mississippian ichnofossils from western Montana and northern Utah. In: Crimes TP, Harper JC (eds) Trace fossils. Geol J, Special Issue 3, pp 407–438 (Arrowlike and other *Bifungites*)

Schlirf M (2000) Upper Jurassic trace fossils from the Boulonnais (northern France). Geologica et Palaeontologica 34:145–213 (Slipperlike *Rhizocorallium*)

Schloz W (1968) Über Beobachtungen zur Ichnofazies und über umgelagerte Rhizocorallien im Lias alpha Schwabens. N Jb Palaeont Mh 11:691–698 (*Rhizocorallium* diagenesis)

Uchman A, Buniak I, Bubniak A (2000) The *Glossifungites* ichnofacies in the area of its nomenclatural archetype. Ichnos 7: 183–193 (*Glossifungites* at its type locality resembles *Rhizocorallium*)

Valenzuela M, García-Ramos JC, Gonzalez Lastra J, Suarez De Centi C (1985) Sedimentación ciclica margo-calcárea de plataforma en el Lías de Asturias. Trabajos de Geología, Univ. de Oviedo 15, pp 45–52 (Fig. 5: *Glossifungites* in Jurassic firmground)

Veevers JJ (1962) *Rhizocorallium* in the Lower Cretaceous rocks of Australia. BMR Bull 62(2):3–14, 3 pls. (Pl. 1: *Glossifungites* preservation)

■ *Thalassinoides* (U. Triassic, Italy)

16

Plate 16
Crab and Shrimp Burrows

Crab Burrows. On tropical beaches today, the dominant tunnel builders are brachyuran crabs. Their tail is folded under the body, so that the animals are wider than long. They can run *sideways*, and reverse gears, at considerable speed. In tube-dwelling species (e.g., fiddler crabs), body length has become even further reduced, so that the now cigar-shaped animal fits perfectly into the cylindrical tunnel, as shown in the diagram. These crabs are active during low tide, grazing the algae that develop at the sediment surface. While scraping the sediment with the pincers in a medio-posterior direction, they employ a *forward* gear in order to scan a broader surface (like *Dimorphichnus*, Pl. 9), but upon the slightest disturbance they switch to sideward running and retreat into the burrow. Other activities also take place at the surface, such as defense of the territory against neighbors, or courtship. For this purpose male fiddler crabs (*Uca*) have enlarged one of their two claws into a colorful flag waved according to specific codes. For a patient observer willing to remain motionless in spite of mosquito bites, these activities provide a fascinating spectacle. They also leave a record in the form of tracks and scratchways radiating from burrow entrances, but this protocol does not survive the next flood and never enters the fossil record, because the algal skin does not resist erosion. What we find are the vertical burrows. Only in exceptional cases – as in the figured chertified cast from the **Cretaceous chalk** of northern Germany – are leg scratches preserved on their surfaces. Their pattern registered by rolling the cast over clay like a cylinder seal reflects a brachyuran tracemaker. Otherwise burrow morphologies show little variation, because most brachyuran galleries serve the single purpose of shelter against enemies, high tides and storms.

Shrimp Burrows. The great majority of post-Paleozoic crustacean burrows can be ascribed to *ghost shrimps*, as represented by modern species of *Callianassa* and *Upogebia* (Pl. 18). These are relatives of the hermit crabs whose exoskeleton is weakly mineralized (and hence transparent: "ghost" shrimps) except for the claws, which are the only body parts likely to be fossilized. In modern environments, ghost shrimp are rarely seen because of their truly troglodytic existence; like moles, they spend all their lives in the underground tunnel systems. Only the expert recognizes their presence from the volcano-like cones of flushed-out sediment (Pl. 18). On these mounds, one may also find the characteristic faecal pellets, whose complex internal structure makes them stratigraphically useful in the fossil record (*Favreina*). More recently, paleontologists working in modern mud flats managed to make epoxy casts of complete burrow systems and dig them out by underwater "vacuum cleaners". By their size and complexity, such casts would deserve a place in a gallery of modern art, but for the biologist they express complex behavioral programs and reveal functions that different parts of the tunnel system had in the life of the inhabitant shrimp family.

Fossil examples show similar morphologies, but they also provide additional information. In the figured specimen from the Miocene of Spain, for instance, pointed ends and claw traces running at a slight angle to the burrow axis reflect a certain digging technique, in which the pincers probably made headway, while the other pereiopods widened the tunnel by scraping the wall in tailward direction. As the scratch pattern is the same all around the cast, the animal must have also rotated along its axis while excavating (a trilobite would have had difficulties in doing this). There are also finer "brushings" on both sides of the branching points. They were probably made by the setate tail appendages, which serve as gills and also ventilate and flush the tunnel system. Such details cannot be seen in a modern cast, which depicts the wall after it has been lined with mucus. Behavioral modifications will be discussed in Pl. 18.

Neglecting that through Earth history shrimplike crustaceans other than *Callianassa* probably produced similar burrows (the earliest known examples are from the Paleozoic), one may informally group all branching crustacean tunnels into one ichnofamily, Ophiomorphidae (name derived from the most familiar representative, *Ophiomorpha*). Even if they are taxonomically heterogeneous, they share a number of characters:

1. They maintain a deeper tier than other burrows and are therefore preferentially preserved.
2. Vertical shafts tend to branch at depth into a boxwork of largely horizontal tunnels.
3. At the dichotomous branching points, tunnel diameter increases as in elk antlers, in order to provide space for the animal to turn around by somersaulting.
4. If scratches are preserved, they run at a slight angle in two directions, so that their overcrossing produces a rhombic network.

Plate 16 · Crab and Ghost-Shrimp Burrows   51

Scratch Patterns

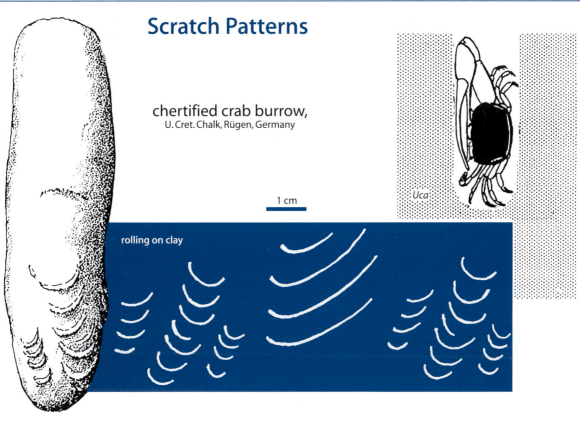

chertified crab burrow,
U. Cret. Chalk, Rügen, Germany

Uca

1 cm

rolling on clay

shrimp burrow,
Miocene, Spain (firmground)

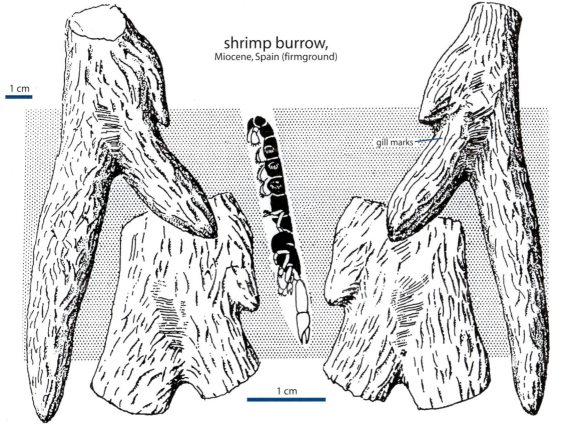

gill marks

1 cm

1 cm

17

Plate 17
Other Arthropod Tunnels and Nests

The previous examples made it clear that the architecture of a particular tunnel system is not sufficient to identify the owner, because different groups of animals may have independently evolved the same burrowing techniques and behavioral programs. Thus additional clues are necessary to single out rhizocoralliids made by crustaceans (Pts. 19–20), but due to preservational constraints only few of the potential criteria are available in any single occurrence. This dilemma increases with regard to less known, or less distinctive, burrow morphologies. This chapter presents cases, in which at least an arthropod origin may be assumed.

Teichichnoid Forms. One may question whether this group of burrows should be referred to *Teichichnus* (Pl. 41), whose arthropod affiliation rests on the weakest of all criteria: size. Yet the Cambrian *Teichichnus rectus* well represents the basic architectural principle. Even though only the retrusive spreite is preserved, there was clearly a generating tunnel. While functioning as a U-tube, it had no vertical shafts and more closely resembled an inverted arch. Unfortunately no scratches or faecal pellets are preserved in *T. rectus*; nevertheless its thumb-like diameter (as in all following examples) is the only argument for a crustacean, rather than a wormlike, tracemaker.

A form from the Middle Cambrian sandstones of Öland, Sweden (not figured) looks very much like *Teichichnus* in vertical outcrops, but as the retrusive spreite lamellae are bilobed, it may represent trilobite-made tunnels. It would be worth the effort to search for a specimen penetrating into an underlying shale in order to corroborate such an interpretation by *Cruziana*-like scratches.

In the architecturally similar *Trichophycus* from Ordovician limestones, such scratches are preserved (hence the name, meaning "hairy seaweed") and sometimes occur in sets of up to six parallel scratches. They suggest a trilobite maker (e.g., *Cryptolithus*); but scratch patterns in other occurrences are more likely made by crustaceans.

Serial Teichichnids. Besides providing only limited protection, shallow teichichnid burrow systems have the disadvantage that retrusive spreite production must end when the tunnel gets too close to the sediment surface. Therefore it is not surprising that some forms tend to expand their mines by intermittent dislocation of one tunnel exit. This results in a more complex spreite structure: while backfills are retrusive at every station, expansion must proceed in a protrusive mode. New fields may be opened also in *Trichophycus*, but in the following forms this was done in a regular fashion.

The earliest example is *Ctenopholeus* from the Lower Devonian Hunsrück Slates. In *Margaritichnus* from the Lower Carboniferous (Colorado, Kansas, and Morocco), the shafts follow in line at short distances, but instead of opening to the surface, they appear to end blindly like upside-down elephant feet, but still this may result from only the spreite being preserved. In material from Morocco, openings proceed protrusively and in one case alternate, as in the Irish *Intexalvichnus* of the same age. All these forms must be studied by serial sectioning before the underlying programs can be fully understood.

Pholeus abomasoformis from the Muschelkalk of northern Germany (Pl. 18) appears to stem from a crustacean unrelated to ghost shrimps. Probably the inhabitant used the wider chamber as shelter, but added a narrower vertical shaft for easier ventilation. Otherwise the fossil appears to be a simple internal cast of the cavity, without a backfill body.

Insect Burrows. In terrestrial habitats, insects make domiciles in various substrates, some of which also provide food (wood or leaves). In the present context we focus on wet-sediment burrows made by insect larvae. They are usually backstuffed behind the animal, rather than forming transversal backfills as in rhizocoralliid and teichichnid burrows. As one would expect, insect burrows are generally restricted to nonmarine sediments.

Scoyenia (Pl. 32) occurs in redbeds of Permian and younger ages, where it forms straight ridges on the soles of thin sand layers. In this mode it may preserve delicate longitudinal scratches which are arranged in groups and give the tunnel cast a somewhat "segmented" appearance. In stretches where the animal happened to backfill mud instead of sand, the hypichnial ridges may also switch into sharp-edged grooves. Potential makers are insect larvae that bulldoze below microbial mats in the style of "undermat miners" (Pl. 45).

In mud puddles, insect burrows appear as elevated ridges tracing the course of the tunnel underneath. The figured *modern* spiral trace was observed in such a puddle. In a fabricational sense, it reflects a program, in which probings are made only to one side and stop short before hitting a previous tunnel. When there is no more space left, the larva turns into a pupa, from which the fly emerges after metamorphosis. So the function of this complex pattern is probably that of a protective fence around the most vulnerable stage of the insect's life cycle. The beehive-shaped wall of *Tambia* from the Oligocene titanothere beds of Wyoming, in another preservation and associated with Permian vertebrate tracks (Pl. 3), reflects a similar strategy. Similar pupal chambers are known from the Tertiary of South America and Australia.

Plate 17 · Other Arthropod Tunnels and Nests   53

Backfilled Arthropod Tunnels

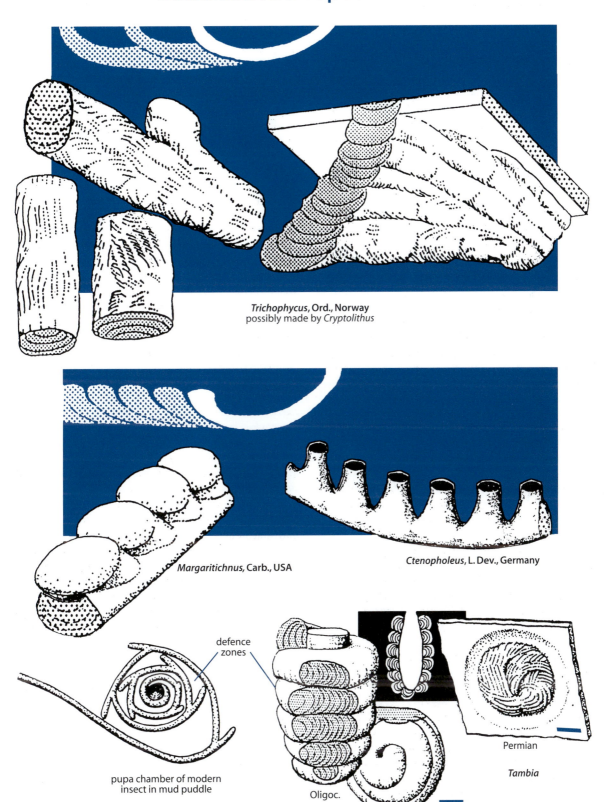

Trichophycus, Ord., Norway
possibly made by *Cryptolithus*

Margaritichnus, Carb., USA

Ctenopholeus, L. Dev., Germany

defence
zones

pupa chamber of modern
insect in mud puddle

Oligoc.

Permian

Tambia

18

Plate 18
Ophiomorphids

Tunnel Systems. We have already talked about ghost shrimp burrows in the text to Pl. 16. Basically there are several *shafts* connected at depth by a horizontal **gallery system.** The tunnels may reach 10 cm in diameter and anastomose into hexagonal meshes (the figured *Thalassinoides suevicus* comes from the Lower Jurassic). There may also be bulb-shaped "turnarounds" between branchings, as well as a **corkscrew** extension to lower levels (*Gyrolithes*), whose function will be discussed below.

At first glance, *Granularia* from Late Cretaceous (and younger) deepsea turbidites (flysch facies) looks very different. It reaches only the diameter of a pencil and the branchings are sparse and form angles smaller than 120°. Yet, the presence of turnarounds, scratches and pelletoidal linings (from which the name is derived) suggest an ophiomorphid relationship. Therefore this may be another example for the general onshore → offshore trend. In the deepsea environment the shrimp responsible became not only miniaturized, but also penetrated more deeply (possibly several meters; see Pl. 72). This behavior allowed them to reach the nutrient-rich bases of new sandy turbidites. At the same time, the original boxwork changed into a more centralized system with a reduced number of shafts and long probes along the turbidite sole.

Preservational Modifications. On Pl. 16 we discussed scratch patterns (*Spongeliomorpha* preservation). They are preserved in the cast if the tunnel was dug in stiff mud or as it crossed the interface between sand and an underlying mud layer, but due to secondary wall linings they may not be seen on the inside of the actual tunnel (*Thalassinoides* preservation). In clean sand, however, the shrimp has to protect the wall against collapse and against erosion by its own ventilatory current. This is done with distinctive *mud pellets*. As they are globular and much larger than thalassinid fecal pellets, they probably consist of material sorted out and shaped by the mouth parts like the sand balls of modern ghost crabs (*Ocypode*). Because these balls tend to become diagenetically mineralized, specimens weathered out from loose sands show a typical cobblestone pattern (*Ophiomorpha* preservation). In other cases the modification of the fossil burrow is strictly diagenetic: shafts served as a conduit, so that a concretionary halo formed in the surrounding sediment. In less consolidated matrix, such as the Cretaceous Chalk, a weathered-out "*Paramoudra*" may be several meters high and by far exceed the diameter of the burrow nucleus.

In consolidated storm sands, horizontal gallery systems are exposed on bedding planes. On top surfaces, they form positive **epireliefs** with smooth surfaces (casts of inner tunnel) and rims of weathered-out mud pellets. On sole surfaces they occur either as three-dimensional *Thalassinoides*, or as **washed** out hyporeliefs without sharp margins, i.e. tunnels dug in stiff mud became uncovered and buried again during a storm.

In micritic limestones, one commonly observes a cylindrical tube running along the crest of the shrimp burrow, as if a worm had been creeping along the already filled tunnel. In reality it is an artifact related to the filling process itself. Comparable **draft fill** channels are known in ceratite steinkerns, in which the phragmocone chambers became gradually mud-filled by draft currents. Eventually only a channel with the diameter of the narrow siphuncular passages is left on top of the fill. This principle can be applied to *Callianassa* burrows, because their openings are always narrower than the tunnel and its inhabitant, who never leaves its burrow voluntarily.

Functional Modifications. The micritic *Krebsscheren-Kalke* in the Upper Jurassic of southern Germany must have been a paradise for ghost shrimps: their pincers (the only well-calcified parts of their exoskeleton) are so common that the formation was named after them and that contemporary tube-worms used them for constructing their walls. Ophiomorphids show not only draft fills, but also modifications that are clearly biological. One variant looking like the base of a candle holder is difficult to explain. It may have served for food *storage* (some modern *Callianassa* species store plant material for fermentation), as a brood chamber, or simply as a terminal turnaround.

A typical *Thalassinoides* in the Cretaceous of **Texas** has a stack of teichichnoid lamellae (Pl. 41) below the horizontal tunnels. Such backfill structures are actually more common, but have escaped attention in other occurrences.

In an occurrence in the Miocene of **Borneo**, burrows are selectively sideritized. During this process the internal structure got lost, but the whole backfill bodies weather out with perfectly preserved surface patterns. What resembles the turnaround swellings in ophiomorphid burrows was in reality a *sanitary dump* for faecal pellets, whose ellipsoidal shape (enlarged picture) suggests that the maker was not a ghost shrimp.

A last modification are the corkscrew tunnels (*Gyrolithes*). Their connection with *Ophiomorpha* is shown by the specimen from **Switzerland.** Vertical sections in Miocene sands of **New Zealand** look like puppet faces, because the lining of mud pellets is restricted to the roofs, where it was most essential. Another cast from Tertiary limestones in **Venezuela** (Univ. of Caracas coll.) has a draft-**fill channel.** It indicates that the spiral section was not dead-ended, but connected with the surface at both ends, in spite of being at the deepest level of the boxwork.

Consequently, *Gyrolithes* may be a farming burrow. Being actively flushed by oxygenated surface water from above, and supplied with reduced pore water from below, the floor of the corkscrew tunnel would have been an ideal place to farm sulfuric bacteria. Regarding the functional significance, comparison with the much larger *Daimonelix* in Miocene paleosols of Nebraska and the Permian of South Africa is pointless: these were made by tetrapods (rodents and therapsids, respectively), for whom a spiral staircase is more convenient than climbing up and down in a chimney. Ghost shrimps, in contrast, are able to bend their legs to the dorsal side, and have no problem moving in a vertical shaft.

Plate 18 · Ophiomorphids   55

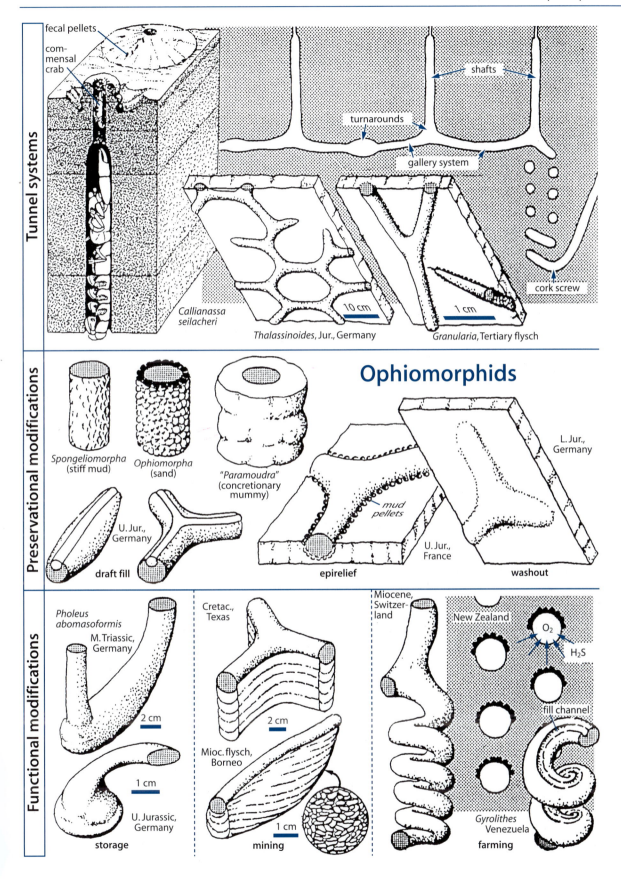

Ophiomorphids

Tunnel systems

fecal pellets
com-
mensal
crab

shafts
turnarounds
gallery system
cork screw

*Callianassa
seilacheri*

Thalassinoides, Jur., Germany   10 cm

Granularia, Tertiary flysch   1 cm

Preservational modifications

Spongeliomorpha
(stiff mud)

Ophiomorpha
(sand)

"*Paramoudra*"
(concretionary
mummy)

U. Jur.,
Germany

draft fill

mud
pellets

U. Jur.,
France

epirelief

L. Jur.,
Germany

washout

Functional modifications

*Pholeus
abomasoformis*
M. Triassic,
Germany

2 cm

1 cm

U. Jurassic,
Germany

storage

Cretac.,
Texas

2 cm

Mioc. flysch,
Borneo

1 cm

mining

Miocene,
Switzer-
land

New Zealand

O_2

H_2S

fill channel

Gyrolithes
Venezuela

farming

19

Plate 19
Rhizocoralliids

So far we have been able to relate trace fossils to certain groups of animals – at least at the level of classes and phyla. In the group of fossils that may informally be called rhizocoralliids, such distinction cannot be consistently made, because their main character is a particular technique of burrow construction rather than kinship. Today it is practiced by unrelated aquatic animals, such as worms, crustaceans and insect larvae (Pl. 17); so the producer can only be inferred in fossil forms.

Polydora. The polychaete worm *Polydora* certainly does not fit the heading of this chapter. Yet its borings in hard substrates are well suited as a model for the much larger rhizocoralliids, the majority of which was probably made by shrimplike crustaceans.

Polydora burrows are most familiar as bioerosional shell borings. As fossils, they are known as *Caulostrepsis* and may also be found in calcareous rock grounds. Essentially they are U-tubes in the shape of an old-fashioned hairpin. What makes them distinctive is that the area between the two limbs is filled with weakly cemented sediment grains. After the inhabitant has died, this backfill becomes readily washed out, so that the burrow transforms into a slit with a dumbbell-shaped cross section. By breaking it open, or making a resin cast and freeing it with hydrochloric acid, one can also see traces of former U-tubes between the two shafts. They record the gradual deepening of the U (Pl. 36) required by the growth of the tenant.

In order to maximize safety against predators and erosion, penetration should be perpendicular to the surface of the substrate, but in mollusk shells the limited thickness of the substrate forces the borer to deviate parallel to the surface. The figured example from a Red Sea **pearl oyster** shows some interesting modifications.

1. The original hairpin first turned flat and then developed two separate lobes at a deeper level. Did lengthening only serve to accommodate the growing length of the worm? Alternatively, it could be related to foraging on organic components or on microscopic borers, such as algae, within the shell.
2. Between two lobes there is a backfill structure ("spreite") on the wrong side (asterisk). This suggests that the worm mistook the sharp bend as a cue and penetrated its own backfill.
3. One of the apertures followed the growth of the host shell by extending into a shallow radial groove that ends at the shell margin. The animal probably parasitized on the inhaling current of the host for its own ventilation and possibly for suspended food. The other opening stayed behind and functioned as an exhaustor.

Polydora (perhaps even the same species!) may also dig into **stiff mud** exposed along tidal channels. In this case,

the hairpin tubes are always perpendicular to the surface and never become lobate. The block diagram shows the animal at the base of such a tunnel. Note that the parapodial setae of the fifth body segment are modified into a nearly horizontal shovel operated by muscles. As shown by steeply oblique scratches in the burrow wall, they remove sediment from the floor of the tunnel and transport it to the ceiling, where it is plastered with mucus secreted by the body wall.

How can this technique transform into shell boring? The answer is probably that the mucus is acid enough to chemically soften the substrate for subsequent mechanical removal in the described fashion.

The third picture shows the behavior of *Polydora* after it was placed into a petri dish filled with a layer of very **soft mud**. In this experimental situation, the animal could not burrow vertically; instead it dug along the glass wall, so that the burrowing process could be directly observed over several hours.

Fossil Rhizocoralliids. Being larger (tube diameters reach more than a centimeter) and lithified, fossil examples are much more suitable for the study of backfill structures than modern ones. The term "*spreite*" used for them (as well as the name *Rhizocorallium*) comes from the time when fossil burrows were considered as seaweeds: for German botanists, **spreite** is the part spreading between the veins of a plant leaf. Sections reveal that the rhizocoralliid spreite consists of stacked lamellae of reworked sediment whose shapes correspond to the ceiling of the U-bend, i.e. they resemble the rim of a bicycle wheel, whose curvature is opposite in longitudinal and cross sections. Such a structure is called **protrusive**, indicating that the U became deeper at every stage. This applies to most occurrences of the vertical *Diplocraterion*, but in the Rhaetic Sandstone (Upper Triassic, southern Germany), the spreite is always **retrusive**, with the lamellae looking like the fender of a bicycle and the terminal tunnel being on top. In a Devonian sandstone, the late Roland Goldring observed a combination of the two structures. The name he gave, *Diplocraterion yoyo*, well describes the down and up motion of the U-tube. As the switch to the retrusive mode (and vice versa) implies reworking of the former backfill, the up and down was probably a response to erosion and sedimentation. In other words, *Diplocraterion* was not a feeding burrow, but the easily flushable domicile of a suspension feeder. *Rhizocorallium* itself follows the same principle, but the plane of its spreite is inclined or bends into the bedding plane at depth. It also is never retrusive and may become excessively long without gaining increased security. The conclusion that its maker was a sediment feeder is corroborated by rod-like fecal **pellets** lining the tunnel wall and making up most of the spreite. As the ratio between pellet and tunnel diameters (the latter corresponding to the circle in drawings) is the same in large and small *Rhizocorallium*, they must be the products of the owner, whose narrow anus matches a crustacean better than a worm.

Plate 19 · Rhizocoralliids   57

Rhizocoralliids

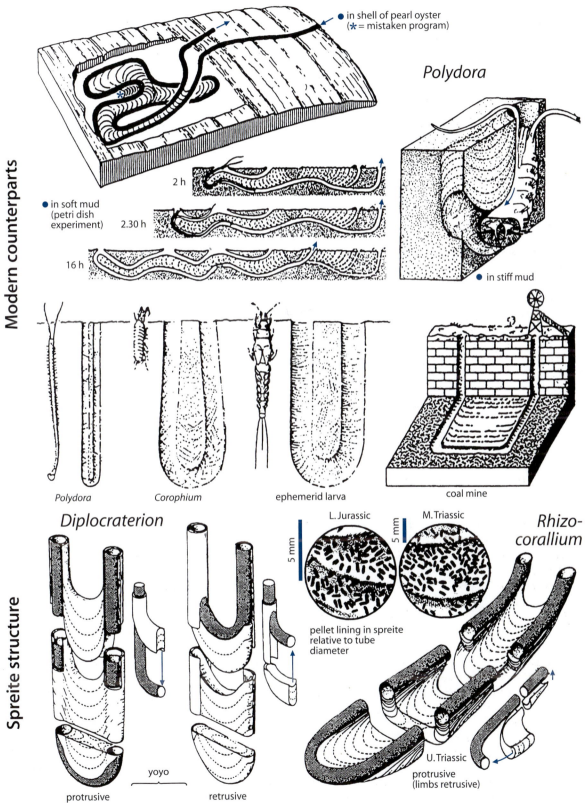

Modern counterparts

● in shell of pearl oyster
(✱ = mistaken program)

Polydora

● in soft mud
(petri dish
experiment)

2 h

2.30 h

16 h

● in stiff mud

Polydora   *Corophium*   ephemerid larva   coal mine

Spreite structure

Diplocraterion

L. Jurassic   M. Triassic

*Rhizo-
corallium*

pellet lining in spreite
relative to tube
diameter

5 mm   5 mm

yoyo

protrusive   retrusive

U. Triassic
protrusive
(limbs retrusive)

20

Plate 20
Rhizocoralliid Modifications

Through the Phanerozoic, the basic rhizocoralliid program has been modified in ways that can be explained by changes in function (dwelling versus feeding burrows) and fabrication (behavioral programs), while other differences are due to preservation. Yet, it often remains uncertain whether the makers were arthropods or worms.

Functional Modifications in Softgrounds. Rhizocoralliid burrows are primarily flushable domiciles that can be adapted to the growth of the inhabitant without it having to leave, but the same technique can also be used to mine the sediment for food. Modifications indicate whether or not this additional function was important.

Rhizocorallium. Excessive length of the tube (with more energy required for flushing), inclined or horizontal burrowing (with little gain in security), and faecal pellets suggest that *Rhizocorallium* was a feeding burrow, whose effectiveness could be improved by modified programs.

A large form found in highly bioturbated sands of the Upper Jurassic (Boulogne, France) is **slipper-shaped**. In contrast to the diagram in Pl. 19 (based on specimens from the Upper Triassic of the German Alps), the retrusive teichichnoid spreite structures below the terminal tunnel are not the accidental product of sediment falling from the roof of the U-tunnel. Rather, the slipper shape reflects a fixed two-stage program: the animal first increased tube length by constructing an inclined *protrusive* spreite and then switched to an upward *retrusive* mode. In this phase the tube became again shorter, but without reworking parts of the earlier spreite. However, this process could not be continued indefinitely; it is a kind of count-down program.

A corkscrew version (**Lapispira bispiralis**, Lower Jurassic; spreite hypothetical) was possible because it maintains the inclination of the spreite. In contrast to spiral worm burrows (*Zoophycos*, Pl. 38; *Daedalus*, Pl. 44) its central shaft is not straight, but forms a steeper screw. Wider horizontal circles occur in the **Permian** (*Bellerophon* Limestone) of Austria.

On the other hand an irregularly winding course within the bedding plane (Triassic and Jurassic) often leads to a *lobate* spreite: without a gravitational compass, the animal relied on the signal of its own body flexure to induce spreite construction. Thereby it interpreted accidental bends in the primary limb tunnels as a signal to produce secondary lobes. The gain of new exploration fields evidently outweighed the disadvantage of a longer ventilation tunnel.

All these variants support the view that *Rhizocorallium* was basically a feeding burrow; but its irregular stratigraphic distribution does not (yet) allow to establish a behavioral genealogy.

Diplocraterion. The occurrence in high-energy sands, vertical orientation, and the response to sedimentation (*Diplocraterion yoyo*, Pl. 19) fit the paradigm of simple domiciles. Yet *Diplocraterion* is also found in silts and muds deposited in quiet waters. One ichnospecies (***Diplocraterion cincinnatiensis***) occurs in finely laminated silt beds and its tunnel resembles the outline of an elephant's foot rather than a U. As these burrows usually end at the base of the silt bed, it seemed reasonable to assume that the encounter with the underlying mud was responsible for the deformation. Occasional specimens, however, end in the same fashion at a higher level. This supports a *chemosymbiotic* function: the two lower corners were the pumping stations for H_2S water from the mud and their interference was reduced by distancing them beyond the regular width of the spreite. This view is corroborated by the fact that the only associated trace fossil is *Chondrites* (Pl. 48), another suspect for chemosymbiosis. Additional information comes from a different kind of preservation: on the soles of tempestites, *Diplocraterion* may be expressed by casts in the shape of a dumb bell (**Bifungites**). They formed when a previous silt layer became stripped away to its mud base and reburied during the same event. In the Ordovician, however, the swollen ends are trifoliate, rather than simple globes resulting from erosion of a marginal tunnel. In conclusion, there were probably three probes radiating from each corner of *Bifungites biclavatum*.

In Cretaceous shallowmarine sandstones (**Wyoming**; Germany) one commonly observes small spreite burrows that could well have been made by *Corophium* (Pl. 19). However there is an additional "escape hatch" rising obliquely from the base of the U-tube. More likely it was made by a suspension feeder for distancing the two openings in the final state, i.e. to place the sewage outlet further away from the eating table. This also tells us that the animal had a three-stage burrowing program: (1) head-on piercing used to make the initial U-tube; (2) vertical spreite construction to accommodate growth; (3) head-on construction of the terminal **ventilation shaft**.

Modifications in Firmgrounds. Firmgrounds result from erosion of muddy sediment to a level, at which it had already become sufficiently stiffened by compaction; i.e. they trace a stratigraphic gap (diastem) on the order of hundred years (Pl. 73). Burrowing in stiff mud poses, first, a fabricational problem: the animal has to work hard and be equipped with strong claws or setae (trilobite limbs would not do). Accordingly, firmground burrows functioned as mere domiciles, because their construction is overly expensive relative to the nutrient content of the substrate. Second, a stiff substrate provides better protection than soft sediment against predators and erosion, so that burrows need not be as deep and not perfectly vertical. Firmground burrows also have preservational ▶

Plate 20 · Rhizocoralliid Modifications   59

Rhizocoralliid Modifications

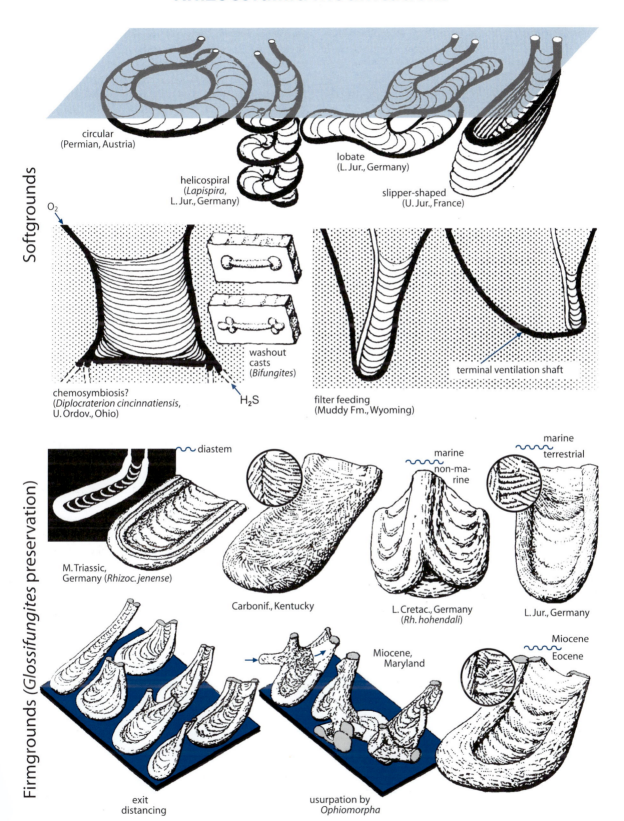

Softgrounds

circular
(Permian, Austria)

helicospiral
(*Lapispira*,
L. Jur., Germany)

lobate
(L. Jur., Germany)

slipper-shaped
(U. Jur., France)

O_2

chemosymbiosis?
(*Diplocraterion cincinnatiensis*,
U. Ordov., Ohio)

H_2S

washout
casts
(*Bifungites*)

terminal ventilation shaft

filter feeding
(Muddy Fm., Wyoming)

Firmgrounds (*Glossifungites* preservation)

diastem

M. Triassic,
Germany (*Rhizoc. jenense*)

Carbonif., Kentucky

marine
non-ma-
rine

L. Cretac., Germany
(*Rh. hohendali*)

marine
terrestrial

L. Jur., Germany

Miocene
Eocene

Miocene,
Maryland

exit
distancing

usurpation by
Ophiomorpha

advantages. As they do not readily collapse, the open tunnels become passively filled with looser sediment that becomes preferentially cemented during early diagenesis.

Our first example is a horizontal *Rhizocorallium jenense* from a low-grade firmground in the Middle **Triassic** limestone of Germany. Along the crest of its marginal tunnel commonly runs a fill channel, as in ophiomorphids from a similar facies (Pl. 18). So we may conclude that – at least in softer mud – the openings of a *Rhizocorallium* burrow were also constricted and that the inhabitant did not normally emerge at the surface.

The other figured specimens from the Carboniferous of **Kentucky** and the Miocene of **Maryland** differ from softground *Rhizocorallium* in various respects. (1) Their outline resembles a rabbit ear rather than a U with parallel limbs. This is to be expected in domichnial rhizocoralliids: if flexure of the body was the signal for burrowing, the spreite became automatically wider as the animal grew larger. (2) As concretionary cementation of the fill sediment stopped at the interface with the dense host mud, the surface of the casts preserves scratches in considerable detail. Their pattern suggests a crustacean maker. (3) Despite of being dwelling burrows like *Diplocraterion*, they are inclined like the feeding burrow *Rhizocorallium*. Inclination, however, facilitates not only the exploitation of nutrient-rich horizons, but also climbing up and down the tube. As the inhabitant allowed itself such comfort, its burrow became similar to *Rhizocorallium* in spite of not being a feeding burrow. For this

reason, one should maintain the old name *Glossifungites* for firmground versions of rhizocoralliid burrows, whether vertical or inclined. This applies also to the specimens from the **Lower Cretaceous** and the **lowermost Jurassic** of Germany. Both were produced by marine crustaceans penetrating into muds deposited in a different regime. The former one regularly produced three *Diplocraterion*-like spreite bodies linking three vertical shafts ("tripods"). The simpler burrow from the boundary between a Triassic red-bed and a Jurassic limestone bed is much smaller (only a few millimeters wide) and could well have been made by *Corophium*. Still it preserves the typical scratch pattern observed in modern examples (Pl. 19).

The lower row shows specimens collected at a locality (Susquehanna River, Maryland, USA), where Eocene shales are disconformably overlain by bioclastic sands of Miocene age. But while the Eocene bioturbation of the shale is hardly recognizable, the burrows dug into the same bed during the Miocene transgression can be easily collected, because they weather out as ear-shaped casts. Firmground conditions are also expressed by associated burrows of pholadid bivalves and of ghost shrimps, which are other characteristic members of the *Glossifungites* Ichnofacies (Pl. 71). Belonging to a deeper tier, the *Spongeliomorpha* tunnels were probably made after *Glossifungites*, so the ghost shrimps could reduce their burrowing effort by partly using the shafts of their rhizocoralliid predecessors.

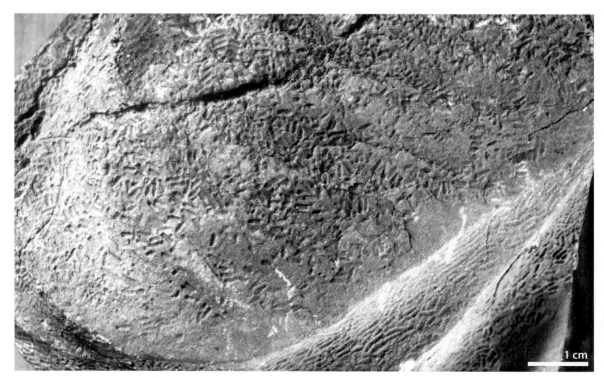

■ Fecal pellets of *Rhizocorallium* (U. Muschelkalk, Germany)

Resting Traces

V

The term "resting traces" (cubichnia) is in a way a misnomer, because it relates to the *purpose* of these burrows, rather than to the activities by which they are made. Their shared character is a kind of behavior that one can best observe when swimming over a sandy sea bottom. As one scrapes the surface with spread fingers, shrimp and fish emerge from it, swim a few meters and disappear again in the sand. All this happens with incredible speed in order to escape potential predators. Camouflage color patterns provide additional protection in an environment that has no natural hiding places. Traces produced in this connection are shallow and therefore have a relatively low fossilization potential. On the other hand, corresponding undertraces provide more clues to the body shape and burrowing technique of the tracemaker than diggings intentionally made for other purposes. Accordingly, *cubichnia* (Pl. 31) are primarily an ethologic category, in which authorship is distinguished at the level of ichnogenera and ichnospecies. Tracemakers come from different phyla (coelenterates, mollusks, echinoderms, arthropods); significantly, the only distinctive resting trace of a worm (*Aphrodite*, Pl. 23) is produced by setate parapodia rather than by peristalsis.

As functions may intergrade, this group has no clear-cut boundaries. Rusophyciform trilobite burrows (Pl. 11), for instance, were resting traces in the sense that they temporarily hid the maker, but they served probably more for sediment processing. On the other hand, a sea anemone or sea pen (Pl. 25) may use the same burrow for most of its lifespan without attempting to turn it into a more sophisticated and durable domicile. Before dealing with particular cases, we shall discuss the burrowing techniques used in different phyla.

Literature

Chapter V

Lessertisseur J (1955) Traces fossiles d'activitié animale et leur significance paléobiologique. Soc Geol Fr Mem New Ser 74:1–150

Nathorst AG (1881) Om spår af nagra evertebrerade djur m. m. och deras palæontologiska betydelse (Mémoire sur quelques traces d'animaux sans vertèbres etc. et de leur portée paléontologique) Konglinga Svenska Vetenskapsakademien, Handlingar (2) 18 (1880), 1–60 (Swedish), 61–104 (abridged, French)

Schäfer W (1972) Ecology and palaeoecology of marine environments. University of Chicago Press, Chicago, 568 p

Seilacher A (1953b) Studien zur Palichnologie. II. Die fossilen Ruhespuren (Cubichnia). Neues Jahrb Geol P-A 98, pp 87–124 (Discussion of different kinds of resting traces)

Plate 21: Burrowing Techniques

Ansell AD, Trueman ER (1967) Burrowing in Mercenaria mercenaria (L.) (Bilvalvia, Veneridae). J. Exper. Biology 46(1):105–115 (Physiology of burrowing)

Dahmer G (1937) Lebensspuren aus dem Taunusquarzit und den Siegener Schichten (Unterdevon). Jahrbuch der Preussisches Geologisch Landesanstalt für 1936 57:523–539, 5 pls. (Devonian Imbrichnus interpreted as faeces)

Dorgan KM, Jumars PA, Johnson B, Boudreau BP (2005) Burrow extension by crack propagation. Nature 433:475 (Energy-saving locomotion in cohesive sediments)

Dworschak PC (1987) Burrows of Solecurtus strigillatus (Linné) and S. multistriatus (Scacchi). Senck Marit 19(3/4)131–147

Hallam A (1970) Gyrochorte and other trace fossils in the Forest Marble (Bathonian) of Dorset, England. In: Crimes TP, Harper JC (eds) Trace fossils. Geol J, Special Issue 3, pp 189–200 (Diagnosis of Imbrichnus)

Howard JD (1969) X-ray radiography for examination of burrowing in sediments by nearshore invertebrate organisms. Sedimentology 11:249–258 (Bioturbation by modern shallow-marine invertebrates revealed by X-radiography)

Howard JD, Elders CA (1970) Burrowing patterns of haustoriid amphipods from Sapelo Island, Georgia. In: Crimes TP, Harper JC (eds) Trace fossils. Geol J, Special Issue 3, pp 243–262, 9 pls. (Modern crustacean burrow)

Mángano MG, Buatois LA, West RR, Maples CG (1998) Contrasting behavioral and feeding strategies recorded by tidal-flat bivalve trace fossils from the Upper Carboniferous of eastern Kansas. Palaios 13:335–351 (A detailed analysis of bivalve burrows and their ethologic significance)

Maples CG, West RR (1989) Lockeia, not Pelecypodichnus. J Paleontol 63:694–696 (Lockeia proposed as a senior synonym of Pelecypodichnus)

Savazzi E (1994) Functional morphology of boring and burrowing invertebrates. In: Donovan SK (ed) The palaeobiology of trace fossils. Johns Hopkins University Press, Baltimore, pp 43–82

Schäfer W (1972) Ecology and palaeoecology of marine environments. University of Chicago Press, Chicago, 568 p (Very useful review of burrowing mechanisms of infaunal invertebrates)

Seilacher A (1961) Krebse im Brandungssand. Nat Volk 91:257–264 (Analysis of the mole crab Emerita)

Seilacher A, Seilacher-Drexler E (1994) Bivalvian trace fossils: A lesson from actuopaleontology. Courier Forschungsinstitut Senckenberg 169, pp 5–15 (Analysis of bivalve trace fossils. Protovirgularia is interpreted as protobranch undertrack)

Trueman ER (1966) Bivalve molluscs: Fluid dynamics of burrowing. Science 152:523–525 (Excellent summary of bivalve burrowing mechanisms)

Trueman ER (1975) The Locomotion of soft-bodied animals. Edward Arnold, London, 200 p (A comprehensive account of the way in which bivalve molluscs and worms move through soft sediments)

Plate 22: Undertrace Experiments

Schäfer W (1972) Ecology and palaeoecology of marine environments. University of Chicago Press, Chicago, 568 p (Fish resting traces)

Seilacher A (1953) Studien zur Palichnologie. I. Über die Methoden der Palichnologie. Neues Jahrb Geol P-A 96:421–452 (Discussion of present figures)

Plate 23: Bilateral Resting Traces

Bandel K (1967) Isopod and limulid marks and trails in Tonganoxie Sandstone (Upper Pennsylvanian) of Kansas. University of Kansas, Paleontological Contributions 19, pp 1–10 (Description of Xiphosuran trackways from carboniferous marginal marine deposits)

Braddy SJ, Briggs DEG (2002) New Lower Permian nonmarine arthropod trace fossils from New Mexico and South Africa. J Paleontol 76:546–557 (Description of the arthropod resting traces Tonganoxichnus, Hedriumichnus, Rotterodichnium and Quadrispinichnia)

Brady LF, Haas O (1949) Oniscoidichnus, new name name for Isopodichnus Brady 1947 not Bornemann 1889 – Possibility of synonymous homonyms. J Paleontol 23(5):1

Bromley RG, Asgaard U (1972) Notes on Greenland trace fossils. Rapport Grønlands Geologiske Undersøgelse 49:5–30 (Isopodichnus synonymized with Cruziana)

Eagar RMC, Baines JG, Collinson JD, Hrady PG, Okolo SA, Pollard JE (1985) Trace fossil assemblages and their occurrence in Silesian (Mid-Carboniferous) deltaic sediments of the Central Pennine Basin, England. pp 99–149 (Limulicubichnus made by Belinurns)

Feldmann RM, Osgood RG Jr, Szmuc EJ, Meinke DW (1978) Chagrinichnites brooksi, a new trace fossil of arthropod origin. J Paleontol 52:287–294 (Introduction of Chagrinichnites brooksi as resting trace of a primitive decapod)

Fortey RA (1999) Olenid trilobites as chemoautotrophic symbionts. Acta Univ Carol Geol 43:355–356 (Special adaptation in oxygen-poor sea floors)

Frey RW (1968) The Lebensspuren of some common marine invertebrates near Beaufort, North Carolina. I. Pelecypod Burrows. J Paleontol 42(2):570–574

Fürst M (1954) Der Creußener Fährtenhorizont sowie Bemerkungen zu Cruziana didyma Salter (=Isopodichnus). Geol Bl NO-Bayern 4(4):113–119, 1 pl. (Triassic Isopodichnus together with Rhizocorallium)

Greiner H (1972) Arthropod trace fossils in the Lower Devonian Jacquet River Formation of New Brunswick. Can J Earth Sci 9(12):1772–1777 (Arthropod resting traces together with trackways)

Hannibal JT, Feldmann RM (1983) Arthropod trace fossils, interpreted as echinocarid escape burrows, from the Chagrin Shale (Late Devonian) of Ohio. J Paleontol 57:705–716 (Chagrinichnites osgoodi as escape trace of a phyllocarid)

Hesselbo SP (1988) Trace fossils of Cambrian aglaspid arthropods. Lethaia 21:139–146 (The ichnogenus Raaschichnus is introduced)

Holm G (1887) Om förekomsten af en Cruziana i öfversta Olenidskiffern vid Knifvinge i Vreta Kloster socken i Östergötland. Aftryck nr Geol Fören i Stockholm Förhandl 9(6):411–418, 1 pl. (Cruziana sp. possibly made by chemosymbiotic trilobites)

Hoppe W (1965) Die Fossilien im Buntsandstein Thüringens sowie ihre stratigraphische und ökologische Bedeutung. Geologie 14(3):272–323 (Isopodichnus in L. Triassic redbeds)

Kennedy WJ (1967) Burrows and surface traces from the Lower Chalk of southern England. Bull Br Mus (Nat Hist) Geol 15:125–167 (Pseudobilobites as crustacean resting trace)

Linck O (1942) Die Spur Isopodichnus. Senckenberg 25:232–255 (Review of the ichnogenus Isopodichnus)

Mángano MG, Buatois LA, Maples CG, Lanier WP (1997) Tunganoxichnus, a new insect trace from the Upper Carboniferous of eastern Kansas. Lethaia 30:113–125 (Diagnosis of Tonganoxichnus)

Mángano MG, Labandeira CC, Kvale EP, Buatois LA (2001) The insect trace fossil *Tonganoxinus* from the middle Pennsylvanian of Indiana: Paleobiologic and paleoenvironmental implications. Ichnos 8:165–175 (Analysis of an additional occurrence)

Minter NJ, Braddy SJ (2006) Walking and jumping with Paleozpic Apterygote insects. Palaeontology 49:872–835 (Model analysis showing that *Tanganoxichnas* is part of a jumping mode of locomotion)

Müller AH (1955) Über die Lebensspur *Isopodichnus* aus dem Oberen Buntsandstein (Unt. Röt) von Göschwitz bei Jena und Abdrücke ihres mutmaßlichen Erzeugers. Geologie 4(5):481–489 (Figured hyporeliefs belong to *Lockeia*; associated epireliefs of pseudofossil *Aristophycus* are interpreted as gill impressions)

Müller AH (1956) Weitere Beiträge zur Ichnologie, Stratinomie und Ökologie der germanischen Trias. Geologie 5(4/5):405–423 (*Isopodichnus* and "helminthoid" pseudofossils from L. Triassic, Germany)

Pollard JE (1985) *Isopodichnus*, related arthropod trace fossils and notostracans from Triassic fluvial sediments. T Roy Soc Edin-Earth 76:273–285 (Detailed study of *Isopodichnus* in Triassic redbeds)

Radwanski A, Roniewicz R (1967) Trace fossil *Aglaspidichnus sanctacrucensis* n.gen., n.sp., a probable resting place of an aglaspid. Acta Palaeont Pol 12(4):545–554 (Large burrow from the Upper Cambrian of the Holy Cross Mountains referred to aglaspids)

Schindewolf OH (1923) Über Spuren mariner Würmer im Mittleren Buntsandstein Oberhessens. Zbl Mineral (21):662–670 (Introduction of *Isopodichnus*)

Speck J (1945) Fährtenfunde aus dem subalpinen Burdigalien und ihre Bedeutung für Fazies und Paläogeographie der oberen Meeresmolasse. Eclogae Geol Helv 38:411–416 (Description of "*Isopodichnus*" tugiensis)

Szmuc EJ, Osgood RG, Meinke DW (1976) *Lingulichnites*, a new trace fossil genus for lingulid brachiodop burrows. Lethaia 9:163–167 (Slit-like burrows in Upper Devonian siltstones)

Trewin N (1976) *Isopodichnus* in a trace fossil assemblage from the Old Red Sandstone. Lethaia 9:29–37 (Description of *Isopodichnus* from Devonian redbeds)

Trusheim F (1931) Aktuo-paläontologische Beobachtungen an *Triops cancriformis* Schaeffer (Crust. Phyll.). Senckenberg 13(5/6):234–243

Vallon LH, Röper M (2006) *Tripartichnus* n.igen. A new trace fossil from the Buntsandstein (Lower Triassic) and from the Solnhofen Lithographic Limestones (Upper Jurassic), Germany. Paläont Z 80:156–166 (Two ichnospecies of small resting traces, referred to euthycarcinid and palinurid crustaceans, respectively)

Wisshak M, Volohonsky E, Seilacher A, Freiwald A (2004) A trace fossil assemblage from fluvial Old Red deposits (Wood Bay Formation; Lower to Middle Devoniana) of NW-Spitsbergen, Svalbard. Lethaia 37:149–163 (Diagnosis of *Cruziana polaris* and *Svalbardichnus*)

Plate 24: Asterozoan Resting Traces

Chamberlain CK (1971) Morphology and ethology of trace fossils from the Ouachita Mountains, southeast Oklahoma. J Paleontol 45:212–246 (Carboniferous *Asteriacites*)

Da Costa G (1979) Novos Ichnofósseis Devonianos da Formacao Inajá, no Estado de Pernambuco. An Acad Bras Ciênc 51(1):121–132 (*Asteriacites*)

Goldring R (1964) Trace-fossils and the sedimentary surface in shallow-water marine sediments. In: van Straaten LMJU (ed) Deltaic and shallow marine deposits (Sixth International Sedimentological Congress, Proceedings). Elsevier, Amsterdam, pp 136–143

Jones DJ (1935) Some Asteriaform Fossils from the Francis Formation of Oklahoma. Am Midl Nat 16:427–428

Jörg E (1957) Ein Erstfund von *Asteriacites* Schlotheim (Cubichnia, Asterozoa) aus der burdigalen Meeresmolasse des Bodenseegebietes. Beitr Natkd Forsch Südwestdtschl 16(1):34–36, 1 pl.

Karaszewski W (1973) A star-like trace fossil in the Jurassic of the Holy Cross Mts. Bull Acad Pol Sci 21(2):157–160, 2 pls. (*Asteriacites*)

Knorr GW, Walch JEI (1755) Sammlung von Merkwürdigkeiten der Natur und Alterthümern des Erdbodens, welche petrificirte Cörper enthält. Nürnberg (p 50, Figs. 1–5 as earliest illustr. of *Asteriacites lumbricalis*)

Maerz RH Jr, Kaesler RL, Hakes WG (1976) Trace fossils from the Rock Bluff Limestone (Pennsylvanian, Kansas). University of Kansas Paleontological Contributions 80, pp 1–6 (Definition of the ichnogenus *Pentichnus* for ophiuroid dwelling burrows)

Malaroda R (1952) Asteroidi werfeniani della Regione Dolomitica. Nature 43:63–68 (L. Triassic *Asteriacites*)

Mángano MG, Buatois LA, West RR, Maples CG (1999) The origin and paleoecologic significance of the trace fossil *Asteriacites* in the Pennsylvanian of Kansas and Missouri. Lethaia 32:17–30 (Study of superbly preserved specimens of *Asteriacites*)

Mundlos R (1966) Ruhespuren von Schlangensternen und ihre mutmaßlichen Erzeuger im Lias alpha von Emmerstedt. Der Aufschluss (10):257–263 (*Asteriacites*, L. Jurassic, Germany)

Osgood RG (1970) Trace fossils of the Cincinnati area. Paleontographica Americana 6:281–444 (Ordovician *Asteriacites stelliforme*)

Rindsberg AK (1994) Ichnology of the Upper Mississippian Hartselle Sandstone of Alabama, with notes on other carboniferous formations. Geological Survey of Alabama, Bulletin 158, 107 p, 22 pls. (Carboniferous *Asteriacites* and *Pentichnus*, the latter attributed to stalkless crinoids)

Seilacher A (1953) Studien zur Palichnologie. II. Die fossilen Ruhespuren (Cubichnia). Neues Jahrb Geol P-A 98:87–124 (Detailed analysis of the origin of *Asteriacites*)

Seilacher A (1990) Paleozoic trace fossils. In: Said R (ed) The geology of Egypt. AA Balkema, Rotterdam, pp 649–670 (Diagnosis of *Asteriacites gugelhupf*)

Plate 25: Coelenterate Resting Burrows

Alpert SP (1973) *Bergaueria* Prantl (Cambrian and Ordovician), a probable actinian trace fossil. J Paleontol 47:919–924 (Interpretation as actinian burrows)

Frey RW (1970) The Lebensspuren of some common marine invertebrates near Beaufort, North Carolina, II. Anemone Burrows. J Paleontol 44(2):308–311

Hallam A (1960) *Kulindrichnus langi*, a new trace-fossil from the Lias. Palaeontology 3:64–68 (Introduction of *Kulindrichnus*)

Howell BF, Hutchinson RM (1958) New Lower Paleozoic coelenterate from Washington. Bull Wagner Free Institute of Science 33(2):13–15, 2 pls. (*Bergaueria magna* made by actinians or pennatulids)

Narbonne GM, Hofmann H (1987) Ediacaran biota of the Wernecke Mountains, Yukon, Canada. Palaeontology 30:647–676 (Ediacaran *Bergaueria*)

Pemberton SG, Frey RW, Bromley RG (1988) The ichnotaxonomy of *Conostichus* and other plug-shaped ichnofossils. Can J Earth Sci 25:886–892 (Taxonomic review of coelenterate resting burrows)

Prantl F (1945) Two new problematic trails from the Ordovician of Bohemia. Bull Int Acad Sci 46(3):1–11 (*Bergaueria perata*)

Prantl F (1946) Two new problematic trails from the Ordovician of Bohemia. Bull Int Acad Sci 46:49–59 (*Bergaueria perata*)

Radwanski A, Roniewicz P (1963) Upper Cambrian trilobite ichnocoenosis from Wielka Wisniówka (Holy Cross Mountains, Poland). Acta Palaeontol Pol 8(2):259–280 (*Bergaueria* in Upper Cambrium of Poland)

Seilacher-Drexler E, Seilacher A (1999) Undertraces of sea pens and moon snails and possible fossil counterparts. Neues Jahrb Geol P-A 214:195–210 (Analysis of various coelenterate resting burrows)

Van Der Meer Mohr CG, Okulitch VJ (1967) On the occurence of Scyphomedusa in the Cambrian of the Cantabrian Mountains (NW Spain). Geologie en Mijnbouw 46:361–362 (*Astropolichnus*)

21

Plate 21
Burrowing Techniques

Crustacean-type Burrowing. The mole crab *Emerita* lives in beach sands, where it filters the backwash with its feathery antennae. It has to change locations several times during each tide and must get reburied at the next location with the right orientation, before the wave is gone and before being located by a hungry seabird. In contrast to shrimps, mole crabs burrow tail-first by means of the pereiopods, which scrape the still fluidized sand from underneath and behind the body. In this action, the carapace plays an important role. It is not only perfectly streamlined, but also bears ratcheted burrowing ribs that counteract backslippage. Thus the carapace provides the necessary purchase when pressed against the sand, while easily slipping into the substrate as the pressure is released. In the natural environment of *Emerita*, undertraces are unlikely to be preserved because there are no sand/mud interfacies. Yet the ones produced in an experimental tank reflect the digging directions of the legs.

The fish *Trachinus* (Pl. 22) gets quickly dug-in by undulation of a long ventral fin supported by flattened rays. It would be interesting to test whether its scales are ratcheted or the weight of the body provides sufficient purchase.

To compare an **ophiurid** with a fish and a crustacean appears to be even more silly: stiff fin rays versus hydraulically operated tube feet? Yet, both fulfill the same function of shoveling sediment from underneath the body, and leave transverse grooves in the undertrace. We shall learn more about starfish burrowing in Pl. 24.

Clam-type Burrowing. The locomotory apparatus of most **bivalves** is best characterized by the old name "Pelecypoda", referring to the hatchet-shaped foot. This structure is operated by ring muscles, like the tube feet of echinoderms, but it can be used as a push-and-pull instrument by being narrowed for *penetration* and then hydraulically expanded to allow *protraction anchor*. This, however, is only one part of the burrowing process; the other actor is the two-valved shell. As the foot penetrates into the sediment, the adductor muscles relax, so that the elastically compressed ligament opens the valves into a *penetration anchor*. As in *Emerita*, the wedging function of the shell may be enhanced by *burrowing sculpture*, which is ratcheted to reduce backslippage. At the moment the foot expands into a *protraction anchor*, the adductor muscles close the valves. This reduces the friction of the shell – not only by reducing the cross-sectional area, but also by the water ejected from the mantle cavity: as the jet passes upward along the shell (in the drawing symbolized by bubbles), the surrounding sand becomes fluidized and thus lubricates the shell during protraction. The ichnological result is an almond-shaped undertrace (*Lockeia*), which reflects the shape of the foot in the protractional phase.

While this model applies to most burrowing clams, the foot of **protobranch bivalves** (e.g., *Acila*, Pl. 32) is modified in such a way that it can act like a toggle-bolt: its distal end is split into two flaps that close during penetration and open during protraction. In the corresponding undertrace (*Imbrichnus*) the flapping of this split foot is expressed by chevron ridges (*Protovirgularia*) leading to a resting trace (*Lockeia*) at the end of the burrow.

A split foot is also found in **scaphopods**, but in these molluscs the chevrons in the undertrack are arranged in strings (*Protovirgularia*), rather than being combined with a stationary resting burrow.

Worm-type Burrowing. This kind of burrowing resembles that of clams, but in the absence of a shell, the penetration anchor as well as the protraction anchor must be hydraulically operated.

In burrowing **actinians**, the bottom end of the gastrovascular sac becomes slimmed for penetration by contraction of the circular muscles, whose relaxation produces a penetration anchor in the proximal part. For protraction, the tip swells and the body contracts. During the whole process, the mouth must be closed for maintaining the water pressure inside. The fossil undertraces referred to actinians (*Bergaueria*) typically have a central depression at the base. This is because the swelling end (physis) acts not only as a wedge; it also scrapes sediment radially away from under the body. As we shall see in Pl. 25, pennatulid sea pens burrow similarly using the base of the colony, which makes it difficult to distinguish burrows of the two coelenterate groups.

Some worms, such as *Arenicola*, act in a similar way by everting the proboscis during penetration. A long hydraulic body with a muscular wall also dispenses with the strict temporal alternation between pushing and pulling phase. In *peristalsis*, the inflated penetration anchor passes along the body from the front to the rear end (i.e. opposite to the metachronal wave of the trilobites), pushing the animal ahead in a more or less continuous action, while setae may take over the anchoring function of the clam's ratcheted ribs (an earthworm creeping between your fingers gives you a feeling for this process). Tunnels made by peristalsis (e.g., *Skolithos*) may bear annular tracings.

Some gastropods (*Natica* or the related *Polinices*, Pl. 27) also fall in this category. Their shell is not used directly as a penetration anchor, because it cannot change shape and the hydrostatic foot folds around it as the animal burrows. Nevertheless it assists by being weighted into a ballast skeleton.

Other, less well understood burrowing techniques are not included in this plate. For instance, there may be scales on the dorsal and ventral sides of the animal that act in conjunction like a pair of ratcheted cross-country skis. Or the body may be covered by a multitude of cilia, or spines, whose waves push the animal through the sediment (Pl. 26).

Plate 21 · Burrowing Techniques   65

Burrowing Techniques

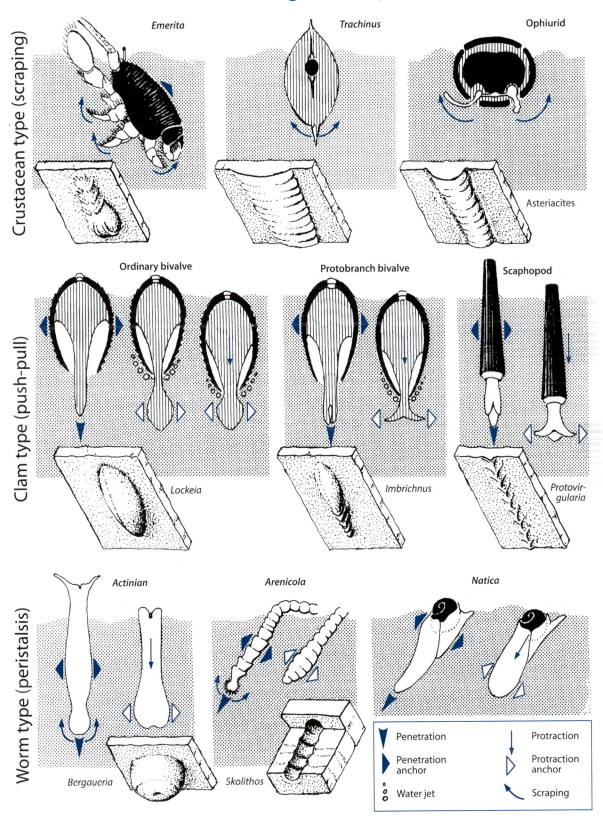

Crustacean type (scraping)

Emerita

Trachinus

Ophiurid

Asteriacites

Clam type (push-pull)

Ordinary bivalve

Protobranch bivalve

Scaphopod

Lockeia

Imbrichnus

Protovir-
gularia

Worm type (peristalsis)

Actinian

Arenicola

Natica

Bergaueria

Skolithos

Penetration

Penetration
anchor

Water jet

Protraction

Protraction
anchor

Scraping

22

Plate 22
Undertrace Experiments

The lifestyle described in the introduction has convergently evolved in various groups of nektobenthic animals that live on, or above, sandy bottoms (epipsammon). They all get quickly buried by removing the sand from under the body with appendages that are otherwise used for swimming or walking. (This is in contrast to the behavior of trilobites (Pl. 11). They needed more time to get dug-in, because instead of pushing the sediment away, they collected it under the body for food processing.) Once the body is hidden, the animal actively smoothens the surface around for perfect concealment. After the animal has left, the impression on top of the sand reflects its leaving rather than the burrowing activity. The latter is better recorded in the exceptional case that the sand was only a veneer on top of a mud layer. Undertraces produced on the interface can be studied in an *undertrace experiment*, in which the bottom of a shallow tank is covered first by a layer of sticky mud, then by sand of appropriate thickness and eventually flooded. After the animal has completed its burrow and left it again, the sand can be gently washed off with a turkey baster. The uncovered undertrace can then be cast with plaster of Paris to obtain a positive hyporelief, as it would appear after the sand had been diagenetically lithified.

Such undertraces do not only show much more detail than the impressions left on the sand surface; they also are more likely to be preserved. The modern examples shown on this plate have – so far – no fossil counterparts, probably because they are too shallow to be commonly preserved. Yet (1) they reflect, to some extent, the shape of the maker and (2) they provide a detailed record of the burrowing process.

Crangon vulgaris is a shrimp living on sandy bottoms. In contrast to relatives from coral reefs it lacks vivid colors. Instead it is speckled in gray tones to blend with the background sand. For hiding within seconds, the animal significantly uses not the walking-legs, but the feathery swimmer appendages below the tail, to wipe the sand from under the body. Surprisingly they do this not by their usual backward beat. Rather the transverse furrows in the undertrace show that, for digging, the rear legs swing at a right angle to the body axis. Nor do they produce a bilobed furrow, as one would expect. After the body is buried, the long antennae smoothen the dug-out sand, so that only the small stalked eyes (looking like sand grains) can be seen at the surface by an attentive observer.

The fish *Trachinus* (Pl. 21) is an ambush predator, as shown by the dorsal displacement of the eyes and the mouth.

Digging is done by an unpaired ventral fin that extends along most of the body. As this fin undulates, the broadened tips of the fin rays shovel the sediment to both sides (Pl. 21). Their transverse scratches are seen in the undertrace.

Although undertraces have not yet been experimentally produced, two other animals sharing this life style should be mentioned.

Flatfish are designed for lying on one side of the body. Accordingly, the eyes and the mouth have moved to the other side. This side also has a camouflage color pattern that can mimic different substrates by contractible chromatophores. In addition, the undulation of the marginal fins stirs up the sand to conceal the outline of the fish. Due to their shallowness, flatfish undertraces would have a very low fossilization potential.

The cuttlefish (*Sepia*) follows very much the same strategy, but in cephalopod style. Again, it can actively adapt the color pattern to the substrate and hide the outline of the body by undulation of the lateral fins to become virtually invisible.

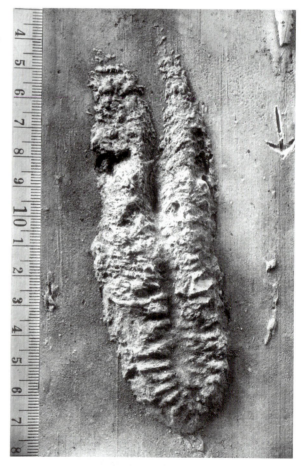

■ Experimental undertrace of *Aphrodite* (motion downward)

Plate 22 · Undertrace Experiments   67

Undertrace Experiments

Trachinus

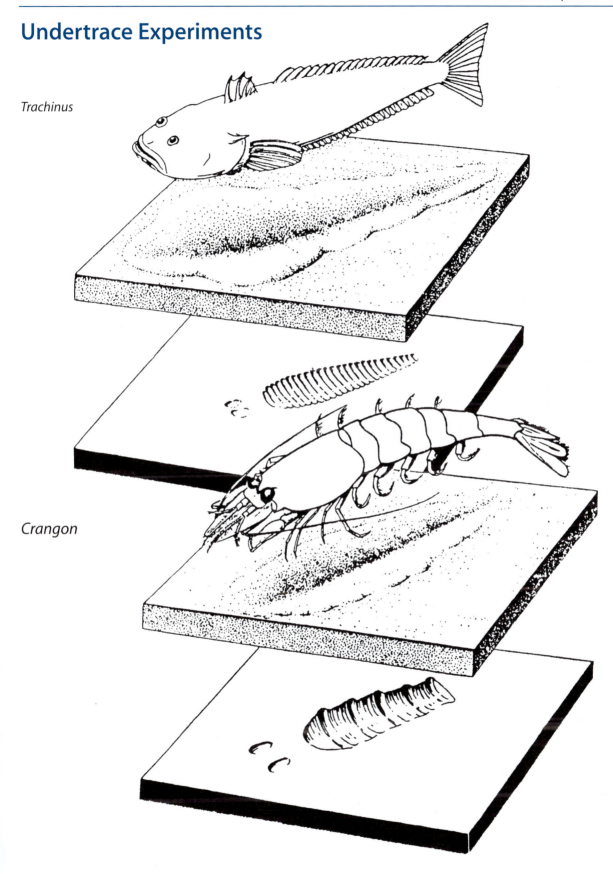

Crangon

23

Plate 23
Bilateral Resting Traces

In this plate, we shall discuss mainly fossil examples. Their makers often remain uncertain, while general symmetries suggest bilateral tracemakers.

As we have seen (Pl. 11), trilobite resting traces hid the animal to some extent, but they served mainly for processing food from the sediment. Accordingly they were dug deeper than necessary for mere hiding, which increased their preservation potential compared to other resting traces. One form from the Upper Cambrian Alum Shales of Sweden (**Cruziana sp.**) differs from this general model in two respects. First, it shows only a few shallow scratches made by legs of the front region plus two furca impressions. Second, it occurs in a low-oxygen environment that would have made most trilobites unhappy. It was this facies context, together with morphological criteria, that led Richard Fortey to propose a chemosymbiotic life style for **olenid trilobites** from the same rocks. Because the sizes of trace and trilobite fit, one may assume that these animals merely scratched the bacterial mat to tap H_2S porewater from below for its symbionts.

The convergent burrowing and feeding behavior of **phyllopod** crustaceans has caused considerable confusion among ichnotaxonomists. Except for their smaller size (hence the name "coffeebean tracks"), phyllopod burrows are virtually indistinguishable from those made by trilobites. They also come in short (rusophyciform) as well as long (cruzianaeform) versions. Still the name *Isopodichnus* is here retained, because the habitat (ephemeral lakes in red-bed facies) and the much longer time range (Cambrian to Recent) makes it easy to resolve this homeomorphy.

Not figured here are resting traces of notostracan phyllopods (*Triops*). Because these animals have a broad head shield, the two lobes of the trace tend to merge in front into a horseshoe. This makes them again difficult to distinguish from the traces of small limulids (the latter with an impression of the single telson), unless one allows for inferences from size, facies and geologic age.

More problematic is the interpretation of somewhat larger coffeebean tracks from the Oldred Sandstone (Devonian) of **Spitsbergen**. As shallow undertracks they may look like *Isopodichnus*, but in deeper ones the serial impressions of pleural spines (compare Pl. 11) speak for a trilobite rather than a phyllopod tracemaker. So it remains an open question, whether one deals with *(a)* a marine horizon within a red-bed sequence, *(b)* a non-marine trilobite, or *(c)* another crustacean (e.g., an isopod) that adopted a trilobite life style.

Non-marine Devonian sandstones have also yielded a number of other crustacean resting traces. **Svalbardichnus** (known from Spitsbergen and claimed Ordovician of Antarctica) and **Chagrinichnites osgoodi** have been referred to phyllocarids, **Chagrinichnites brooksi** to an eocarid. In continental sandstones of later times, insect larvae also produced distinctive resting traces (e.g., **Tonganoxichnus**).

Another Cambrian resting burrow, **Raaschichnus**, has been referred to aglaspids. These were marine trilobite-like creatures, but having a telson they are more closely related to eurypterids and horseshoe crabs.

Resting traces (*Pterichnus isopodicus*, not figured) of a marine isopod (*Archaeoniscus*) from the Upper Jurassic of southwest France provide a rare example from the later times. Significantly they come from a Solnhofen-type *lagerstaette* (Pl. 75), where the presence of biomats allowed the exceptional preservation of body fossils as well as trackways, including the ones of the same isopod.

Cretaceous **Pseudobilobites**, named after its similarity to trilobite-made "bilobites" (Pl. 11), is the deep resting trace of an unknown crustacean. Note that the median furrow is less pronounced than in trilobite burrows and that the scratches are probably made by a latero-posterior beat of the pereiopods.

Experimental undertraces of modern **Aphrodite** show that "bilobites" need not necessarily be made by arthropods. This marine annelid is also called "sea mouse" because of its size, outline and hairy appearance. The latter is due to iridescent parapodial setae that dig by latero-posterior motions.

Another bilobed resting burrow from the marine Miocene Molasse of Switzerland ("**Isopodichnus**" *tugiensis*) cannot be referred to arthropods either. Having a single median faecal string, it could be compared to burrows of spatangoids (*Bichordites*; Pl. 26), but because these echinoids are sediment feeders, stationary burrowing would make little sense. Rather, the trace could result from the escape of such an animal. Using the same technique as in horizontal bulldozing (Pl. 26), it produced a sandy backfill phantom of its body, but without an accompanying mud phantom. Shown in the figures is the alternative interpretation as a gastropod burrow.

Lockeia (formerly *Pelecypodichnus*) has already been discussed on Pl. 21. The undertrace shown here was experimentally made by a large *Tivela* from the Indian Ocean. It is hardly different, however, from the ones left by freshwater clams in the Upper Cretaceous of Alberta dinosaur park. There, escape structures in the sediment above record the reaction to a flood event and the shells of the tracemakers are preserved on top of the bed.

Plate 23 · Bilateral Resting Traces   69

Bilateral Resting Traces

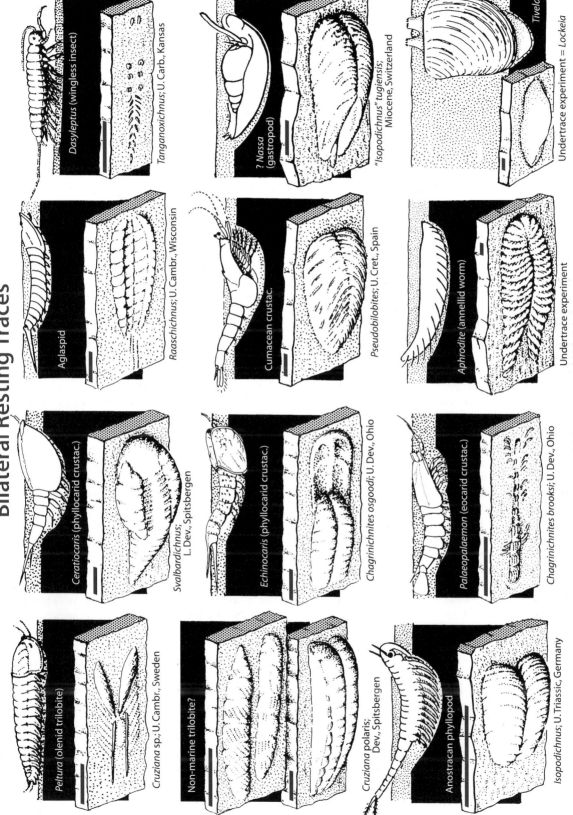

Dasyleptus (wingless insect)
Tanganoxichnus; U. Carb., Kansas

?Nassa (gastropod)
"Isopodichnus" tugiensis; Miocene, Switzerland

Tivela
Undertrace experiment = *Lockeia*

Aglaspid
Raaschichnus; U. Cambr., Wisconsin

Cumacean crustac.
Pseudobilobites; U. Cret., Spain

Aphrodite (annelid worm)
Undertrace experiment

Ceratiocaris (phyllocarid crustac.)
Svalbardichnus; L. Dev., Spitsbergen

Echinocaris (phyllocarid crustac.)
Chagrinichnites osgoodi; U. Dev., Ohio

Palaeopalaemon (eocarid crustac.)
Chagrinichnites brooksi; U. Dev., Ohio

Peltura (olenid trilobite)
Cruziana sp.; U. Cambr., Sweden

Non-marine trilobite?

Cruziana polaris; Dev., Spitsbergen

Anostracan phyllopod
Isopodichnus; U. Triassic, Germany

24

Plate 24
Asterozoan Resting Traces

Undertrace Experiments. Just like shrimp and flounders, many starfishes and brittlestars hide in the sand – except that they get buried more slowly. In fact, the animal looks motionless while burrowing; only a ridge of sand piling up around the contour tells us that the tube feet are active underneath. When the body is buried deeply enough, the rays begin to move in such a way that the sand ridges spill smoothly over the buried animal, with only the arm tips and their sensory organs reaching to the surface. After the animal has left, an indistinct star-shaped depression can be seen at the sand surface. In contrast, the corresponding *undertrace* is much more structured. Because this is the form in which fossil starfish traces are most commonly found, it should not surprise us that they were for a long time (as in their first description in 1755) considered to be body fossils, i.e. direct impressions of carcasses.

In the ichnological perspective, the observed morphological characters of *Asteriacites* reflect mainly the activity of the tube feet, while the taxonomic affiliation of the maker remains less certain. Yet a general distinction can be made between asteroid and ophiuroid burrows.

In asteroids (e.g., *Astropecten*, whose tube feet lack suckers), the arms are usually too broad for a single foot to shovel sediment to both sides. Consequently the scratches of *Asteriacites quinquefolius* are parted.

Ophiuroids crawl by flexing their arms rather than tiptoeing on the tube feet. Therefore the arms are narrow enough for the tube feet to swing to both sides. In burrowing, however, the arms are held stiff. Accordingly ophiuroid undertraces (*Asteriacites lumbricalis*) have straight arms and annulations that could be mistaken for skeletal vertebrae. Only under the central disk is the shoveling one-sided and done by disproportionately large tube feet.

In all asterozoan burrows, a narrow groove accentuates the outline of hypichnial casts, thus testifying that shoveling was indeed towards the margin – in contrast to trilobite burrows.

Modifications in Fossil Forms. The box in the middle shows features of **Jurassic** ophiuroid burrows that have misled earlier workers who considered them as body impressions.

1. Star-shaped impressions on the rippled tops of storm sands reflect ophiurans that had gotten buried during a storm. They escaped as sand deposition came to an end and the suspended mud started to settle. So they are epichnial undertraces, whose broad outlines and straight arms resemble asteroids rather than brittlestars.
2. In contrast, broadened arms of hyporeliefs result from the wiggling action by which the animal smoothened the surface.
3. Split impressions tell us that the buried animal repositioned its arm tips – rather than representing a suspension-feeding basket star (*Euryale*), whose arms branch to enlarge the filter fan.
4. The radial elements in the center of this undertrace are the scratches of larger proximal tube feet rather than impressions of radially arranged skeletal plates on top of the central disk.

Vertical Repetition. Ophiuroids hiding "between meals" will often make another resting burrow close to the previous one. In addition to this *horizontal repetition* there is also **vertical repetition**: in fissile sandstones deposited in the upper flow regime (i.e. without ripple formation) one often observes that the same star-shaped impressions are repeated on several bedding planes. One may first think that they are pressed-through like the Connecticut dinosaur tracks (Pl. 3), but this is mechanically infeasible. In the figured specimen from the Werfen Beds (Lower Triassic) of northern Italy, corresponding impressions also fail to coincide in shapes and positions (lower diagram). So these represent an escape reaction in response to the deposition of new sand layers so that burial and escape must have happened in a matter of hours.

Deep Burrowing. In the previous examples we dealt with species that get buried only for concealment and leave their hiding places to search for food – particularly at night. Today, some ophiuran species (but ones that live in mud rather than sand) are suspension feeders. Accordingly, their arms must be long enough to extend well above the sediment surface. In this way they act as passive filtration fans, in which particles are caught by the sticky tube feet. The burrows of such species must be deeper than mere hiding places in order to allow withdrawal of the arms upon disturbance. The maker of *Asteriacites gugelhupf* (the name referring to a similarly shaped German birthday cake) from the Carboniferous of Egypt was also a suspension feeder, while the bilobed scratch pattern of the upturned arms suggests an asteroid rather than a brittlestar.

Plate 24 · Asterozoan Resting Traces   71

Asterozoan Resting Traces

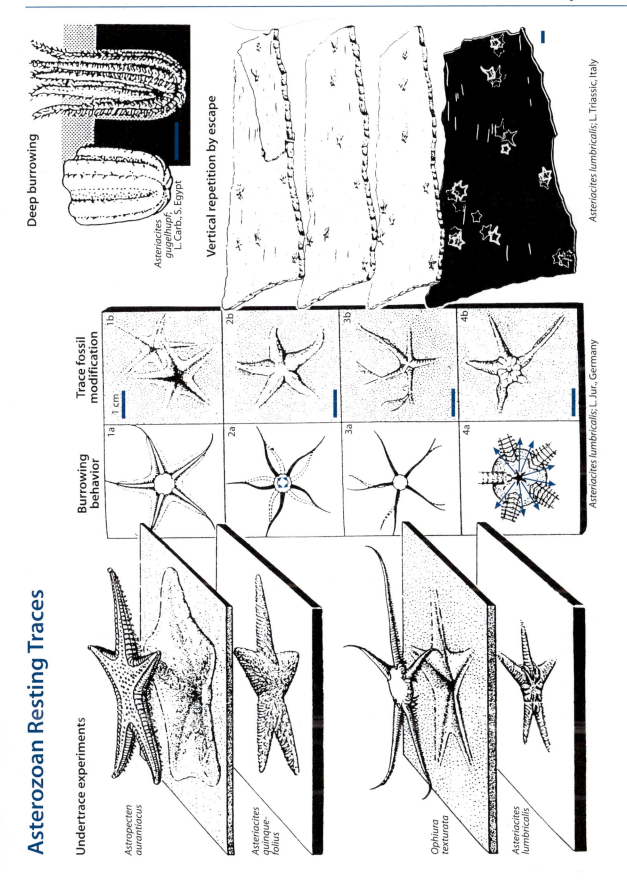

Deep burrowing

Asteriacites gugelhupf, L. Carb., S. Egypt

Vertical repetition by escape

Asteriacites lumbricalis; L. Triassic, Italy

Trace fossil modification

1b 1 cm
2b
3b
4b

Burrowing behavior

1a
2a
3a
4a

Asteriacites lumbricalis; L. Jur., Germany

Undertrace experiments

Astropecten aurantiacus

Asteriacites quinque-folius

Ophiura texturata

Asteriacites lumbricalis

25

Plate 25
Coelenterate Resting Burrows

On Pl. 22 we have already discussed how soft bottom acti-nians burrow. The resulting undertraces (*Bergaueria*) are rather distinctive, but one must be careful not to confuse them with pot casts. These are non-biological and self-enhancing sedimentary structures formed by vortices, in which suspended sand is whirled as in a centrifuge. Consequently *pot casts* have a similar cylindrical shape with a central dimple, but get larger and tend to widen towards the base. As shown by undertrace experiments, smooth *Bergaueria* can also be made by sea pens (e.g., ***Ptilosarcus***). However, such an origin is unlikely in forms that show some kind of radial structures (***Bergaueria radiata, Astropolichnus, Conostichus, Solicyclus***); they probably correspond to the fleshy septa within the acti-nian gastrocoel. The stacked vertical repetition of such structures in the latter three ichnogenera probably re-flects the adjustment to sediment accumulation during a longer timespan, during which the tracemaker grew larger.

Bergaueria sucta deviates from related ichnospecies by being shallow and having a broad concave base. In this case, undertrace formation was probably not simply a preservational accident. Rather the animal appears to have been attached to the buried mud surface like a sucker disk. Multiple impressions due to active disloca-tion (horizontal repetition) also defy a non-biological origin. This detail is important for the specimen from the late Precambrian of **Canada**, which may represent the earliest record of burrowing actinians.

While all other examples appear as casts on the soles of storm sands, *Kulindrichnus* is found three-dimension-ally in dark Jurassic shales. Loose specimens are com-monly mistaken for coprolites, because they are similarly **phosphatized**; *in situ*, however, they always stand verti-cal. Phosphatization of the cortex around a coarser infill was probably induced by bacterial decay of the polyp wall. In an other case, the burrow is preserved within a **cal-careous concretion** that formed before the mud became compacted.

Solicyclus and *Kulindrichnus* are rare exceptions to the rule that actinian (or pennatulid) burrows are largely restricted to Paleozoic tempestites. This is surprising, because the burrowing habit persists in many species to the present day. In view of increasing predator pressure, one would also expect an opposite trend. Thus a tapho-nomic bias appears to be involved, even though biomats are not the answer.

5 cm

■ *Bergaueria phallica* (M. Cambr., Grand Canyon)

Plate 25 · Coelenterate Resting Burrows   73

Coelenterate Resting Burrows

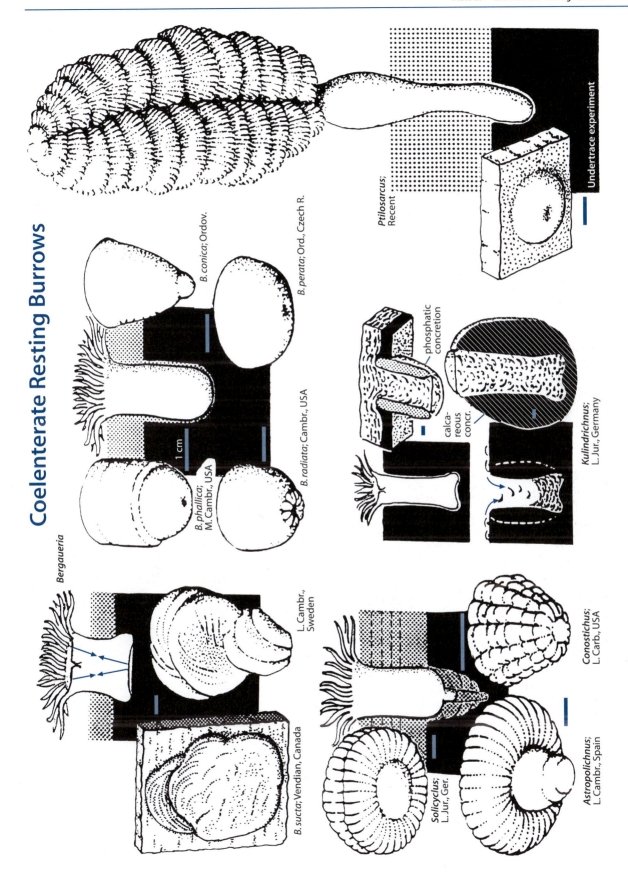

Ptilosarcus;
Recent

Undertrace experiment

Bergaueria

B. conica; Ordov.

B. perata; Ord., Czech R.

B. radiata; Cambr., USA

B. phallica;
M. Cambr., USA

1 cm

B. sucta; Vendian, Canada

Solicyclus;
L. Jur., Ger.

phosphatic
concretion

calca-
reous
concr.

Kulindrichnus;
L. Jur., Germany

L. Cambr.,
Sweden

Conostichus;
L. Carb., USA

Astropolichnus;
L. Cambr., Spain

Asteriacites quinquefolius undertraces (M. Jurassic, Germany)

epirelief

Lockeia

hyporelief

Burrows of Short Bulldozers

In common language, "bulldozing" stands for a kind of activity, in which a vehicle not just rolls over the ground, but forcefully removes quantities of soil in front of it. For the "infaunal bulldozers" discussed in this chapter there is no technological counterpart. Giant drilling machines used in the construction of modern tunnels come closest; yet the comparison fails, because no soft-sediment burrower drills. In tunnel construction, the excavated material is also removed, while animal bulldozers backfill their own tunnel with the material dug out in front of the body.

One might also argue about the distinction of "short" versus "long" bulldozers among the putative tracemakers, because both produce long, worm-shaped backfill burrows. It is only through "phantoms" that the shorter ones (to about four times the width) can be distinguished. In addition, their burrows tend to be wider than high (even in non-compacted sand), in contrast to the long cylindrical "worm" burrows discussed in the next chapter.

Another problem is the *function* of such burrows. For mere protection, the animal would not have to move horizontally and for getting somewhere, locomotion at the surface or simple wedging through the sediment would require much less energy.

In the modern world, three examples come to mind. One are the larvae of moths (Pl. 51). They mine the soft areas (spreite) of plant leaves and fill the tunnel behind with their faeces. The grubs of bark beetles do the same in the sap layer (cambium) between the bark and the wood of trees (Pls. 49 and 51). Both show the tendency to either meander or effectively cover a given surface by close guidance (without intersecting) between the tunnels of siblings.

The third example are modern heart urchins (Spatangoidea). While bulldozing at a safe depth, they can be observed grazing the sediment surface with some overly long specialised tube feet. This would account for the meandering in fossil echinoid burrows (Pl. 26), if there were not an energy problem. Some tellinid bivalves (e.g., *Macoma*) also graze the surface with a long inhaling siphon, while the body moves along at depth. But their shells are well streamlined. In most heart urchins, however, the test is not at all streamlined, in contrast to the related sand dollars. To the contrary, heart urchins are fairly globular and the blunter end is in front rather than the rear! This can only mean that spatangoids essentially bulldoze for food particles. They can be selected by thousands of minute tube feet and spatulate spines, as the sediment loosened in front passes the body, with a mucus sheet acting as a conveyor belt. The loss of the jaw apparatus also attests to a microphagous diet on particles gained mainly by bulldozing, but with a more protein-rich dessert collected at the sediment surface.

Other examples for this kind of infaunal locomotion are, so far, known only from the fossil record. They are tentatively ascribed to a class of soft-bodied molluscs that disappeared in the end-Permian mass extinction (Pls. 27–30), or to flatworms that may still exist (Pl. 30).

In connection with the Tambach vertebrate tracks, we have already come across a short bulldozer when interpreting *Tambia* as the pupa chamber of a large beetle (Pl. 3). In this case the grub dug not only for food. It possibly added a repellent to the backfill in order to secure its future dormitory. This emphasizes that various kinds of function may be combined in this kind of burrowing.

Literature

Chapter VI

Kanazawa K (1992) Adaptation of test shape for burrowing and locomotion in spatangoid echinoids. Palaeontology 35:733–750 (Detailed analysis of spantangoid echinoid functional morphology)

Plate 26: Echinoid Burrows

Bromley RG, Asgaard U (1975) Sediment structures produced by a spatangoid echinoid: A problem of preservation. Bull Geol Soc Denmark 24:261–281 (Classic paper on the burrowing behavior of spatangoid echinoids and their biogenic structures)

Bromley RG, Jensen M, Asgaard U (1995) Spatangoid echinoids: Deep-tier trace fossils and chemosymbiosis. Neues Jahrb Geol P-A 195:25–35 (Documentation of some species of spatangoid echinoids as chemosymbionts)

Chesher RH (1969) Contributions to the biology of *Meoma ventricosa* (Echinoidea: Spatangoida). Bull Mar Sci 19:72–110

Goldring R, Stephenson DG (1970) Did *Micraster* burrow? In: Crimes TR, Harper JC (eds) Trace fossils. Geol J, Special Issue 3:179–184 (*Scolicia* not preserved in chalk)

Götzinger G, Becker H (1934) Neue Fährtenstudien im ostalpinen Flysch. Senckenberg 16(2/3):77–94 (Various preservations of *Scolicia*)

Książkiewicz M (1977) Trace fossils in the flysch of the Polish Carpathians. Paleontologia Polonica 36, 208 p

Mayoral E, Muñiz F (2001) New ichnospecies of *Cardioichnus* from the Miocene of the Guadalquivir Basin, Huelva, Spain. Ichnos 8(1):69–76

Nichols D (1959) Changes in the Chalk heart-urchin *Micraster* interpreted in relation to living forms. Philos Trans R Soc London, Ser B 242:347–437 (Investigates burrowing in *Echinocardium* as analog of Cretaceous *Micraster*)

Noda H (1985) Miocene and Pliocene spatangoid echinoid burrows from Okinawa and Chiba Prefectures, Japan. Ann Rep Inst Geosci Univ Tsukuba (11): 41–44 (*Bichordites*)

Plaziat J-C, Mahmoudi M (1988) Trace fossils attributed to burrowing echinoids: A revision including new ichnogenus and ichnospecies. Geobios 21:209–233 (Taxonomic review of echinoid trace fossils, including definition of the ichnogenus *Bichordites*)

Reineck H-E (1968) Lebensspuren von Herzigeln. Senck Leth 49(4): 311–319 (Modern spatangoid burrows)

Schlirf M (2002) Taxonomic reassessment of *Bolonia* Meunier, 1886 (trace fossil) based on new material from the type area in Boulonnais, northern France. Paläont Z 76:331–338 (Redescription of *Bolonia* = *Palaeobullia* preservation of *Bichordites*)

Seilacher A (1986) Evolution of behavior as expressed in marine trace fossils. In: Nitecki MHG, Kitchell JA (eds) Evolution of animal behavior: Palaeontological and field approaches. Oxford University Press, New York, pp 62–87 (Summary of the evolution of Mesozoic and Cenozoic *Scolicia* in shallow and deep marine environments. See Figure 3-6 for summary)

Smith AB, Crimes PT (1983) Trace fossils formed by heart urchins – A study of *Scolicia* and related traces. Lethaia 16:79–92 (Taxonomic review of spatangoid trace fossils)

Uchman A (1995) Taxonomy and paleoecology of flysch trace fossils: The Marnoso-arenacea formation and associated facies (Miocene, Northern Apennines, Italy). Beringeria 15:1–115 (Updated review of the *Scolicia* group)

Vogeltanz R (1971) Scolicien-Massenvorkommen im Salzburger Oberkreide-Flysch. Verh Geol B-A (1):1–9 (Preservational variants of *Scolicia* in Austrian flysch)

Zapfe H (1935) Lebensspuren grabender Echiniden aus dem Eozän Siebenbürgens. Verh Zool-Bot Ges Wien 85:42–52 (Eocene burrows with swellings: made by stationary echinoid or by crustacean with turning points and flat-bottom broad chambers)

Plate 27: Molluscan Bulldozers

Cloud P, Bever JE (1973) Trace fossils from the Flathead Sandstone, Fremont Country, Wyoming, compared with early Cambrian forms from California and Australia. J Paleontol 47(5):883–885, 1 pl. (*Plagiogmus*)

Elders CA (1975) Experimental approaches in neoichnology. In: Frey RW (ed) The study of trace fossils. Springer-Verlag, New York, pp 513–536

Hofmann H, Patel IM (1989) Trace fossils from the type "Etcheminian Series" (Lower Cambrian Ratcliffe Brook Formation), Saint John area, New Brunswick. Geol Mag 126:139–157 (Description of Tommotian *Psammichnites*)

Jaeger H, Martinson A (1980) The Early Cambrian trace fossil *Plagiogmus* in its type area. Geol Foren Stock For (GFF) 102(2): 117–126

McIlroy D, Heys GR (1997) Palaeobiological significance of *Plagiogmus arcuatus* from the Lower Cambrian of central Australia. Alcheringa 21:161–178 (Detailed analysis of *Plagiogmus* from Australia and its relationship to *Psammichnites*)

Peterson DO, Clark DL (1974) Trace fossils *Plagiogmus* and *Skolithos* in the Tintic quarzite (Middle Cambrian) of Utah. J Paleontol 48(4):766–768 (Transverse ridges broader than in L. Cambrian *Plagiogmus*)

Roedel H (1926) Ein kambrisches Geschiebe mit problematischen Spuren. Ergänzung zu meiner Mitteilung über ein kambrisches Geschiebe mit problematischen Spuren. Z Geschiebeforschung 2,5(1, 1/2):25–26, 48–52

Seilacher A (1997) Fossil art. An exhibition of the Geologisches Institut Tübingen University. The Royal Tyrell Museum of Palaeontology, Drumheller, Alberta, Canada, 64 p (*Psammichnites gigas* from Spain illustrated on p 38–39)

Seilacher A, Gámez-Vintaned JA (1995) *Psammichnites gigas*: Ichnological expression of the Cambrian explosion. Proceedings Sixth Paleobenthos International Symposium, Alghero, pp 151–152 (Illustration of a spectacular example of *Psammichnites gigas* from the Cambrian of Spain)

Seilacher-Drexler E, Seilacher A (1999) Undertraces of sea pens and moon snails and possible fossil counterparts. Neues Jahrb Geol P-A 214:195–210 (Discussion of *Psammichnites* and *Dictyodora*)

Seilacher A, Buatois LA, Mángano MG (2005) Trace fossils in the Ediacaran-Cambrian transition: Behavioral diversification, ecological turnover and environmental shift. Palaeogeog Palaeoclim Palaeoecol 227: 323–356 (Reassignment of *Nereites saltensis* as *Psammichnites saltensis*)

Stanley SM (1970) Relation of shell form to life habits in the Bivalvia. Geological Society of America Memoir 125, 296 p (Functional shell morphologies)

Trueman ER, Brown AC (1992) The burrowing habit of marine gastropods. Adv Mar Biol 28:389–431 (Analysis of gastropod locomotion and burrowing behavior)

Young FG (1972) Early Cambrian and older trace fossils from the Southern Cordillera of Canada. Can J Earth Sci 9(1):1–17 (The new ichnogenus *Didymaulichnus* resembles *Pasmmichnites*, but has a median groove on the lower side. Very large for the suggested Precambrian age)

Plate 28: Paleozoic Psammichnitids

Benton MJ (1982) *Dictyodora* and associated trace fossils from the Palaeozoic of Thuringia. Lethaia 15:115–132 (Redescription and interpretation of *Dictyodora*)

De Quatrefages MA (1849) Note sur la *Scolicia prisca* (A. de Q.). Ann Sci Nat Zool 12(3):365–266

Hakes WG (1976) Trace fossils and depositional environment of four clastic units, Upper Pennsylvanian megacyclothems, Northeast Kansas. Univ Kansas Paleont Contr 5(63):5–46, 13 pls. (Description of a rich ichnofauna. *Chevronichnus* introduced for epireliefs is probably a preservational variant of *Psammichnites*, comparable to *Palaeobullia*)

Seilacher-Drexler E, Seilacher A (1999) Undertraces of sea pens and moon snails and possible fossil counterparts. Neues Jahrb Geol P-A 214:195–210 (Modern naticid undertraces)

Tessensohn F (1968) Unterkarbon-Flysch und Auernig-Oberkarbon in Trögern, Karawanken, Österreich. Neues Jahrb Geol P M (2):100–121 (Fig. 6: *Dictyodora* from L. Carboniferous flysch of southern Austria here drawn)

Zimmermann E (1889) Über die Gattung *Dictyodora*. Z Dtsch Geol Ges 41:165–167 (Synonymizes different genera as parts of one burrow system)

Plate 29: Psammichnitid Behavioral Evolution

Benton MJ, Trewin NH (1980) *Dictyodora* from the Silurian of Peebleshire, Scotland. Palaeontology 23:501–513 (Evolutionary trends)

Doughty PS (1980) Some trace fossils from the Silurian rocks of Co Down. Ir Nat J 20(3):98–104 (*Dictyodora* in greywackes)

Fenton CL, Fenton MA (1937a) Burrows and trails from Pennsylvanian rocks of Texas. Am Midl Nat 18(6):1079–1084 (*Aulichnites parkerensis*)

Fenton CL, Fenton MA (1937b) *Olivellites*, a Pennsylvanian snail burrow. Am Midl Nat 18:452–453 (Diagnosis of *Olivellites* = *Psammichnites*)

Geinitz HB (1867) Über *Dictyophyton ? Liebeanum* Gein. aus dem Culmschiefer vom Heersberge zwischen Gera und Weyda. Neues Jahrb Geol P-A 1867:286–288 (*Dictyodora* as "*Dictyophyton*")

Mángano MG, Buatois LA, Rindsberg AK (2002a) Carboniferous *Psammichnites*: Systematic re-evaluation, taphonomy and autecology. Ichnos 9:1–22 (Review of Carboniferous *Psammichnites* (formerly *Olivellites*))

Mángano MG, Buatois LA, West RR, Maples CG (2002b) Ichnology of an equatorial tidal flat: The Stull Shale Member at Waverly, eastern Kansas. Bulletin of the Kansas Geological Survey 245, 130 p (Analysis of Carboniferous *Psammichnites* from Kansas)

Müller AM (1955) Das erste Benthos (*Planolites*? *vermiculare* n. sp.) aus dem Stinkschiefer Mitteldeutschlands (Zechstein, Staßfurtserie). Geologie 4(7/8):655–659 (Probalbly *Psammichnites*, L. Permian, Germany)

Müller AH (1971) *Dictyodora liebeana* (*Ichnia invertebratorum*, ein Beitrag zur Taxiologie und Ökologie sedimentfressender Endobionten). Monatsbericht der Deutschen Akademie der Wissenschaften 13, pp 136–151 (Functional interpretation)

Pfeiffer F (1959) Über *Dictyodora liebeana* (Weiss). Geologie 8:425–433 (Carboniferous, Germany)

Seilacher A (1967) Fossil behavior. Sci Am 217(2):72–80 (Evolutionary trends in *Dictyodora* behavior)

Zimmermann E (1892) *Dictyodora liebeana* (Weiss) und ihre Beziehungen zu *Vexillum* (Rouault), *Palaeochorda marina* (Gein.) und *Crossopodia henrici* (Gein.). Jahresbericht der Gesellschaft der Freunde Naturwissenschaften, Gera, pp 32–35 (Synonymizes preservational variants)

Plate 30: Bi- and Tripartite Backfill Bodies

Buatois LA, Mángano MG, Mikuláš R, Maples CG (1998) The ichnogenus *Curvolithus* revisited. J Paleontol 72:758–769 (Taxonomic review of *Curvolithus*)

Fritsch A (1908) Problematica Silurica. In: Barrande J (ed) Système Silurien du centre de la Bohême. Barrande Fonds, Prague, 28 p (Diagnosis of *Curvolithus*)

Heinberg C (1973) The internal structure of the trace fossils *Gyrochorte* and *Curvolithus*. Lethaia 6:227–238 (Detailed analysis of the internal structure and modes of backfill)

Kulkarni KG, Ghare MA (1989) Stratigraphic distribution of ichnotaxa in Wagad Region, Kutch, India. J Geol Soc India 33:259–267 (Jurassic ichnocoenosis containing *Curvolithus* backfill phantoms)

Lockley MG, Rindsberg AK, Zeiler RM (1987) The paleoenvironmental significance of the nearshore *Curvolithus* ichnofacies. Palaios 2:255–262 (Paleozoic *Curvolithus* in micaceous sandstone)

Seilacher A (1990) Paleozoic trace fossils. In: Said R (ed) The geology of Egypt. AA Balkema, Rotterdam, pp 649–670 (Description of *Curvolithus* and attribution to flatworms)

Skwarko SK, Seilacher A (1993) Trace fossils and problematica. Bull Geol Soc Western Australia 136:87 and 390–403 (Variations in Permian shallow marine *Psammichnites*)

26

Plate 26
Echinoid Burrows

In general, sea urchins are epibenthic animals. They move over rocks by means of suckered tube feet, scraping off the algal film with the five teeth mounted in the "Aristoteles lantern". On calcareous shells they may also carve deeper to consume endolithic algae. The star-shaped traces of this activity are commonly preserved on belemnite rostra. On rock surfaces, constant grazing may also result in a pit, into which the urchin nestles for protection. One regular echinoid goes one step further. Instead of a circular pit it produces long trenches, in which it can move along with continuous protection. Oviously these trenches are used to farm algae. The elliptic outline of the animal's test facilitates this style, in which motion is possible only in the direction of the longer axis, as in heart urchins.

Some groups left the rocky habitats for soft bottoms, on which suckered tube feet are useless; so irregular echinoids turned spines into stilts and digging organs, while the tube feet serve for respiration and manipulation of food particles The only deep burrowers among echinoids are the heart urchins (Spatangoidea). Their rich trace-fossil record illustrates the burrowing mechanism, but also the evolution of complex search programs.

Backfill Structure. The lamellae of the *Scolicia* backfill contour the rear side of the echinoid. In addition, there is a basal string of faecal material that has been injected by a tuft of long spines around the anus. In deepsea forms, two such funnels produced a *pair* of faecal strings. Vertical sections also reveal a **geopetal** grading within the backfill lamellae: coarser grains are at the base and the finer and darker fractions on top. As grains are mucus-bound when they reach the backfill, this phenomenon can hardly be referred to settling from suspension, but probably reflects active sorting of the sediment while it is passed along the body. As a result of the geopetal asymmetry *Scolicia* looks very different depending on the toponomic context. On top of sand layers (**epichnial**) the coarse bottom parts of the lamellae form a gill-like pattern (*Palaeobullia* preservation). In contrast, the *hypichnial* version of the same burrow (*Subphyllochorda* preservation) has a broad elliptical profile and shows the two faecal strings as a prominent feature. Its sculpture depends on the degree of weathering. If the cortical sand layer is still present, delicate spine scratches form a V that opens backwards in top view, while the chevrons of the underlying backfill lamellae open in the direction of motion. The **endichnial** version shows only the lamellar structure and along the crest a discontinuous ridge tracing the displacement of the inhalant canal.

As the maker of *Scolicia* is already identified, we no longer need indirect clues. Nevertheless it is reassuring that the figured **backfill phantom** (see Pls. 28 and 30) fits an irregular echinoid.

Behavioral Evolution. Now being familiar with the burrowing techniques of spatangoids and the resulting sedimentary structures, we can proceed to their evolutionary history as reflected in the trace fossil record. *Scolicia* first appears in Upper Jurassic shelf sandstones of northern France – at about the same time as spatangoid tests occur in calcareous facies. In *Palaeobullia* preservation (here called **Bolonia**; *Bichordites* is a junior synonym) these burrows show a single faecal string, but no particular behavioral program. In deepsea turbidite series (**flysch facies**), echinoid burrows did not appear before the Late Cretaceous. From then on, however, they have been a dominant element in the **post-turbidite** association (Pl. 72). They also meander, but as yet not very systematically (**c**).

Another strategy to systematically exploit a given area, spiraling, is represented by *Scolicia zumayensis*. Its spirals reach the size of a dinner plate, but they violate the paradigm of optimal foraging by being too tightly coiled. As seen in cross section, turns in fact systematically intersect, so that only one third of the processed sediment has not already been reworked by the same individual during previous turns. This makes only sense if the backfill became microbially fermented and could be harvested when the animal came back on its next turn. So the behavior may represent a kind of bacterial farming.

Scoliciid *meandering* became more efficient only when echinoids entered the exclusive club of the **preturbidite** community (Pl. 72). Having been made in the homogeneous hemipelagic mud that settled during the long intervals between turbidity events, such burrows (*Taphrhelminthopsis* preservation) can only be preserved by the exhuming and casting effect of the next turbidity current (Pl. 52). In spite of having lost their lamellar backfill, they can still be identified by their size and their two-lobed profile.

Eocene representatives preserved in this preservational mode follow a rigid program. They start with a spiral and then continue in tightly guided (but not intersecting!) meanders of increasingly long turns. What made the animal finish after a few turns, rather than plowing-on in the same style, remains a riddle. A *Taphrhelminthopsis* maker from an **Oligocene** flysch near Gibraltar had "learned" to meander *without* a starter spiral. After having completed some tightly guided turns, it could freely proceed without such guidance. This stretch then served as a base for a new meander system.

This evolutionary history, which could not have been derived from body fossils, complements our picture of spatangoid evolution through geologic time. At the same time it exemplifies a pathway that can also be observed in other groups of trace fossils: onshore to offshore shift not only as a retreat from predator pressure, but as the conquest of deepsea bottoms, whose time-stability allowed evolution to proceed to unprecedented levels of specialization.

Plate 26 · Echinoid Burrows   79

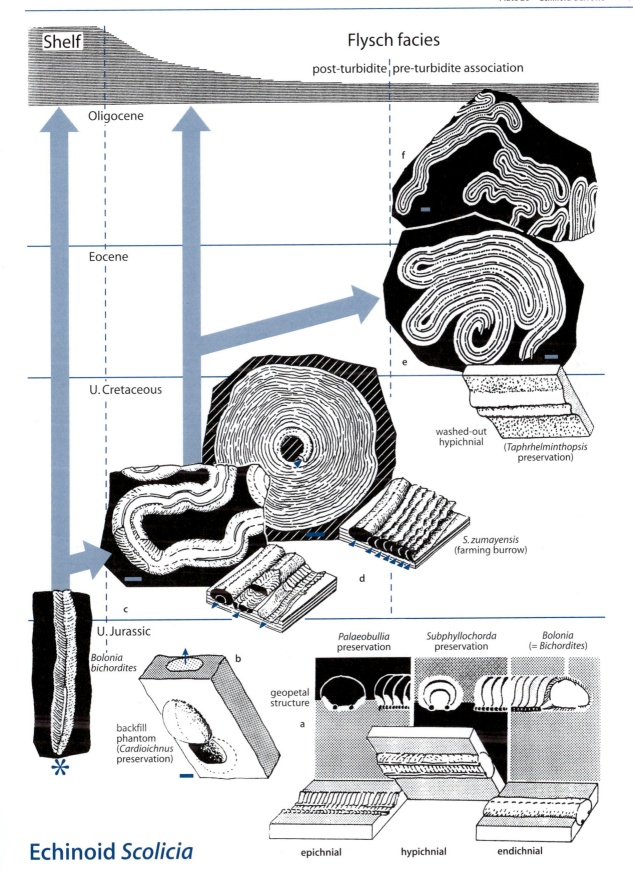

Shelf

Flysch facies

post-turbidite | pre-turbidite association

Oligocene

Eocene

U. Cretaceous

f

e

washed-out
hypichnial

(*Taphrhelminthopsis*
preservation)

S. zumayensis
(farming burrow)

d

c

U. Jurassic

*Bolonia
bichordites*

b

backfill
phantom
(*Cardioichnus*
preservation)

✳

geopetal
structure

a

Palaeobullia
preservation

Subphyllochorda
preservation

Bolonia
(= *Bichordites*)

epichnial   hypichnial   endichnial

Echinoid *Scolicia*

27

Plate 27
Molluscan Bulldozers

In Pl. 21 we have already discussed the push-and-pull technique of burrowing in bivalves and scaphopods. Another kind of molluscan burrowing is derived from the kind of locomotion that can be observed in a chiton or gastropod crawling up the glass wall of an aquarium: as metachronal waves of muscular contraction pass over the sole, the body moves slowly along by local detachment and advance in every wave. But how could this method work on loose sediment?

In fact, many gastropods living on softgrounds have turned to using the front part of the muscular sole (propodium) as an antagonist to the rest of the body. *Strombus* even uses its spiked operculum as a pogo stick for rapid movement. A hydraulically operated push-and-pull action has also been observed in the burrowing of the moon snail *Natica* (Pl. 21), which burrows in the sand in search of its molluscan prey.

Undertrace experiments with the closely related, but larger, genus **Polinices** suggest that this is only half of the animal's locomotory abilities. On the regular underlayer of mud, undertraces were not distinctive. Only when an old and firmer mud was covered by a veneer of softer mud for a new experiment, did a most surprising undertrace result on the interface between the two mud layers: its surface resembles corrugated paper with very high and perfectly parallel grooves and ridges. As the scale of these corrugations does not fit ciliary motion, it probably reflects muscular waves with a much higher amplitude than observed on a glass plate. This makes sense on a substrate, where adhesion must be replaced by increased friction. While the *Polinices* undertrace has, as yet, no direct fossil counterpart, it may be used as a clue to understand a number of problematic trace fossils.

The earliest fossil representative is **Psammichnites gigas** from Early Cambrian (mostly Tommotian) sandstones all over the world. One striking feature is its giant size (by the standards of pre-trilobite times); the other is its scribbling **search behavior**. A two-meter slab in the Department of Geology, University of Zaragoza (Spain) looks as if a cowboy had dropped his lasso on the ancient bedding plane.

Closer inspection of this slab revealed additional details. (1) The ridges are *positive* epireliefs rather than hypichnial casts of gravitational impressions. (2) They must be *shadow traces* pressed through from a lower level, because their lateral boundaries are not sharp, but confluent with the adjacent bed surface. (3) This interpretation is supported by the *overcrossings*. They look like a rope passing over another – the reason being that the layer had been lifted twice at the crossings. (4) A broad *median depression* along the ridge crest suggests partial collapse of a cavity underneath. (5) While all previous structures have the soft outlines indicating a pressed-through overtrace, a conspicuous *sinusoidal line* (**snorkel trail**) along the top of the ridge looks as if it had been cut by a knife.

From these details it was possible, first, to determine the *direction* in which this trace was made. Because in pressed-through crossings, the later trace appears like the rope lying on top, the loops can be oriented. Building on this information, a consistent asymmetry of the sinusoidal groove can also be gauged: its steeper slopes always face forward. Through these directional criteria it can be shown that the animal entered the sediment near the upper corner of the slab to first make two clockwise loops, then two counterclockwise ones and so on – in contrast to the consistent handedness of the trilobite pirouettes (Pl. 14).

With regard to burrowing technique, the pressed-through overtrace suggests that the animal moved by *wedging* through the sediment (an echinoid-type bulldozer would not have lifted the layer on top). As the overtrace did not totally collapse after the animal had passed, the burrow must nevertheless have become partly backfilled with sediment imported from the surface – probably by the snorkel that produced the sine-wave line.

This model could be elaborated in a contemporaneous siltstone of central Australia, where fine lamination allows the same overtrace to be split at several levels. Here the amplitude of the sinusoidal line increases upward, i.e. away from the body. This conforms to a *snorkel* swinging sideways like a pendulum. This snorkel, however, must have been able to cut its own way through the sediment, probably by cilia covering its outer surface. At the same time it is clear that it served not only for ventilation (for which a cheaper straight course would have been sufficient), but at the same time acted as a vacuum cleaner collecting detritus and sand grains coated by bacteria from the surface.

In Australia it was also possible to prepare the *actual* burrow fills. They are rather flat with sharp lateral edges and readily intercut without vertical deflection, as expected. Their **bottom sides** also show two incongruent sculpture patterns, (1) a coarser one consisting of broadly undulating chevrons – probably diverging in the direction of movement; (2) a much more delicate pattern of *transverse corrugations* reminiscent of the *Polinices* undertrace. In one variant (**Plagiogmus** preservation), the bottom shows regularly spaced cross ridges. They are possibly related to backfill lamellae, a counterpart to the *Palaeobullia* preservation in *Scolicia* (Pl. 26).

There are two options with regard to the taxonomic affiliation of the *Psammichnites* maker. The cross section of the burrow might suggest a flatworm, in which case the fine corrugation would be related to ciliary waves. On the other hand, no modern flatworm has a long tubular siphon, while such siphons have independently evolved in various molluscan lineages. My best guess at the moment is a mollusk related to the extinct halkieriids that could do without a dorsal armor by becoming infaunal. In either case the animal would have been relatively short. Unfortunately the wedging mode of penetration did not create backfill phantoms from which body shapes could be inferred, as was possible in the following examples from later deposits.

Plate 27 · Molluscan Bulldozers   81

Polinices lewisii (Recent)

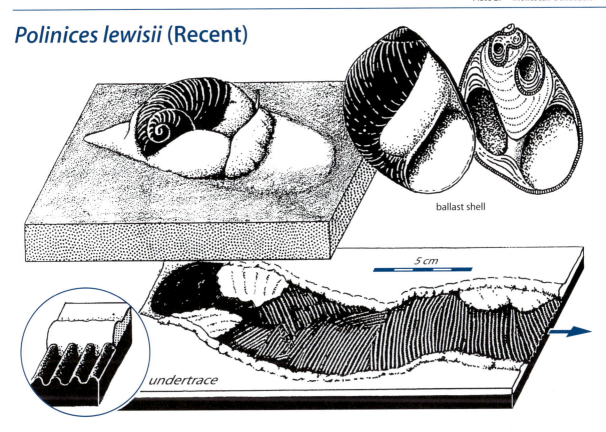

ballast shell

5 cm

undertrace

Psammichnites gigas (Tomotian)

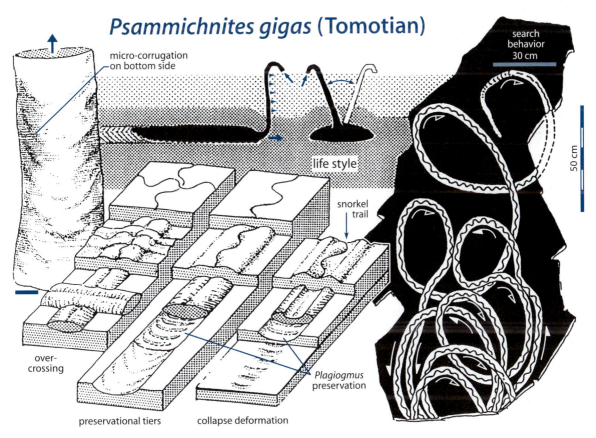

micro-corrugation
on bottom side

life style

snorkel
trail

search
behavior
30 cm

50 cm

over-
crossing

Plagiogmus
preservation

preservational tiers

collapse deformation

28

Plate 28
Paleozoic Psammichnitids

Psammichnites gigas marks the beginning of an ichnological tradition that lasted throughout the Paleozoic. Described as different ichnogenera, its successors were generally smaller. Nevertheless they show the typical bipartition into the broad trail of the bulldozing body and the foliate spreite made by the siphon on top of it. They also provide cues ("phantoms") to the shapes of the otherwise unknown soft-bodied tracemakers.

Psammichnites **Phantoms.** The burrows presented on this plate resemble the Lower Cambrian *Psammichnites gigas* by their backfill structure, an oval cross section, and a snorkel trace along the top; but neither do they produce an elevated overtrace, nor do the snorkel traces undulate. This suggests that the tracemaker did not wedge through the sediment, but bulldozed in echinoid fashion by removing sediment in front, and passing it over the body for terminal backfilling.

This mode of burrowing allows, as in *Scolicia* (Pl. 26), the formation of **backfill phantoms**: as the animal crosses the interface between mud and sand layers, the nature of the backfill changes accordingly. The result is a tongue-like sand body (**sand phantom**), whose length corresponds to the length of the tracemaker. In *Psammichnites* (formerly *Olivellites) plummeri*, for example, it indicates a slug-like animal about three times as long as wide. When crossing again from sand into mud on the other side of the ripple, the animal produced an equivalent **mud phantom**. It is less obvious, but expressed by a depression in the bed surface due to compaction of the backfilled mud underneath.

The surface of the lower slab, from a **Silurian** storm sand in Argentina, looks as if an animal had been crawling along the ripple troughs on an exposed sand flat. The trails, however, preserve details (including overhanging slopes) too fine to have been exposed to tidal flooding and sedimentation. More probably they are epichnial burrows made by an infaunal bulldozer that preferred the buried troughs for their higher food content, not for their relief. When it moved along a ripple slope (**D**), the lateral ridge of sand became higher on the upslope side. The animal also meandered at times to better utilize the broad valley floors (**F**). But even on a straight course (**E**) there are conspicuous **wiggles** that could only be made by a short animal. The meniscate sections fit this interpretation, because they reflect neither backfill lamellae nor muscular waves, but the contour of the rounded rear edge of the sole sinking down after a wave had passed. Therefore each wiggle starts with a discordance.

Modification in the Deepsea: *Dictyodora.* During the Paleozoic, the snorkel became shorter and gave up its pendulum swing in *Psammichnites* makers that continued in the shallowmarine realm. Deepsea forms (*Dictyodora*) used the same burrowing technique, but their snorkel became increasingly longer. By Carboniferous times it had gotten so long that the spreite, produced by a siphon less than 1 mm wide, became the most prominent feature. The long snorkel allowed the animal to explore deeper tiers, while no attempt was made to enlarge the area covered at the sediment surface. To the contrary, the tip of the siphon tended to remain fixed as the animal cruised at depth, so that the spreite surfaces are conical. Obviously, the *Dictyodora* animal extracted its food at depth, rather than from the surface.

Another problem is the snorkel's cutting mechanism. The delicate sculpture of the spreite surface looks like the trace of a tight string tied at both ends, because the lines are smoothly curved and never show any kink. Yet, cross sections show that the spreite had the crescentic structure of an active transversal backfill, which is incompatible with a mucus-lined cleft that would have closed behind a *wedging* string. This siphon must have been covered all along by a ciliated epithelium that actively moved sediment from the cutting to the trailing edge. We shall come back to a similar mechanism in the discussion of gyrochortids (Pl. 35) and *Daedalus* (Pls. 43–45), except that in these cases the string was the *body* of a wormlike tracemaker.

The meandering epirelief of a **body backfill** from a **heterolithic** flysch can hardly be told apart from scoliids in the same mode of preservation. As in the *Palaeobullia* preservation of *Scolicia* (Pl. 26c), the chevron sculpture of the outer surface (marginal zones) opens backwards, while the laminae of the geopetal backfill open in the direction of motion. The lower part of the backfill lamellae is also divided by what appears to be a median faecal string. The identity of the second specimen with *Dictyodora* is only revealed by the associated **snorkel spreite** that cuts across sand and mud layers.

The connection between the two structures is more obvious in **mud facies** (Austria, Germany, Spain), even though specimens tend to be tectonically deformed. Part of a body backfill (top and bottom views; oriented by conical snorkel spreiten) also shows a median snorkel seam, expressed by an elevated strip with a central furrow. In the lower figure, snorkel spreiten progressing from left to right are shaded, reverse ones white. They reflect concentric meanders, as shown in the following plate.

Plate 28 · Paleozoic Psammichnitids 83

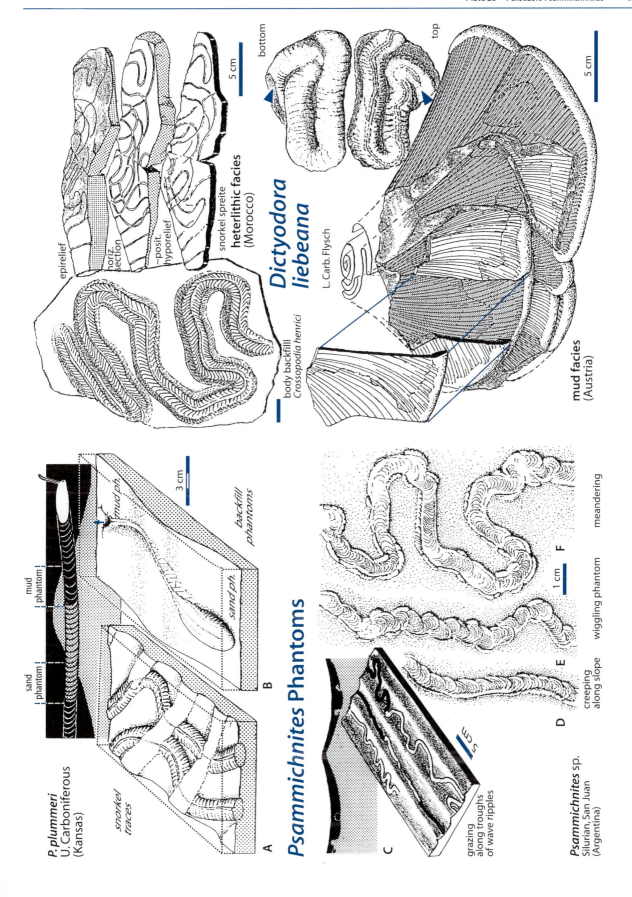

Dictyodora liebeana

L. Carb. Flysch

epirelief
horiz. section
~posit. hyporelief
snorkel spreite **heterlithic facies** (Morocco)

body backfill *Crossopodia henrici*

mud facies (Austria)

bottom
top

Psammichnites **Phantoms**

P. plummeri U. Carboniferous (Kansas)

snorkel traces

mud phantom
sand phantom
mud ph.
sand ph.
backfill phantoms

Psammichnites sp. Silurian, San Juan (Argentina)

grazing along troughs of wave ripples

creeping along slope
wiggling phantom
meandering

29

Plate 29
Psammichnitid Behavioral Evolution

Having tentatively identified psammichnitid trace fossils as the burrows of slug-like bulldozers with a snorkel (possibly representing an infaunal and naked branch of halkieriids), we can now trace their behavioral evolution, as we did for the analogous echinoid burrows (Pl. 26).

The story possibly begins with the minute *Aulichnites* from the Vendian (= Ediacaran) of Russia, even though the trace of the snorkel is replaced by a median furrow. Anyway the sharp bend in the figured specimen suggests a relatively short rather than a wormlike tracemaker.

In *Psammichnites gigas* (here from Lower Cambrian of **Sweden**), we see that in **Russia** the scribbling soon changed to a more efficient *meandering* program. Such meanders continue into the Carboniferous (Auernig Beds) of **Austria** and the Permian of Western Australia (Pl. 30). In this **shallow-marine** line, the snorkel became shorter and gave up its own wiggling – perhaps because food was derived more at the level of the body than from the surface.

A similar trend to tighter meanders is observed in **deep-sea** sediments, except that the animals became even smaller. At the same time the snorkel became lengthened, allowing the animal to bulldoze at deeper levels. Still the snorkel maintained its ability to cut its own way through the sediment, only lagging somewhat behind the course of the body at depth. As a consequence, the thin and conical spreite of the *snorkel* becomes the most eye-catching part of the fossil burrow (Pl. 28). It becomes visible either on vertical fracture surfaces, which expose the delicate and slightly oblique lines of progression. Alternatively, it can be studied in bedding plane sections (*Myrianites* or *Palaeochorda* preservation), in which the same meander pattern is repeated at subsequent levels, but becomes successively smaller higher up. This pattern is the result of the snorkel lagging behind the body. So, whereas the tip of the snorkel scanned a larger area than the body in Cambrian *Psammichnites* (Pl. 27), the roles became reversed in the *Dictyodora* lineage. The trace of the *body* contouring the lower edge of the *Dictyodora* spreite resembles *Psammichnites*. It was originally described under a separate name (*Crossopodia henrici*).

Forms from **Ordovician** to **Devonian** flysches still follow a simple meander course of unlimited length. Carboniferous members added a starter spiral – not only for establishing a guideline for the meanders, but as a winding stairway to the deeper level of bulldozing.

The three **Carboniferous** examples of *Dictyodora* come from different areas in Germany and Austria, so that their stratigraphic relationship is not established.

Nevertheless they could represent a lineage in which the two stages of the burrowing program became successively modified. First, the meanders turned free from the starter spiral after a few hugging lobes, so that later parts resemble pre-Carboniferous forms. At a second stage, all meanders kept contouring. Eventually, the meandering part of the program was dropped alltogether and the whole burrow consisted only of the spiral, but with many more and tighter coils. Such corkscrew variants resemble "*Spirophyton*" *eifeliense* (Pl. 39) except that they have a marginal backfill structure.

To test this evolutionary model, one would have to study more specimens collected in stratigraphic order. The dark shales in which *Dictyodora* is found also lend themselves to serial sectioning and reconstruction in the form of glass or computer models. This would be an interesting task, because heterochronic models of evolutionary change could be applied to behavioral programs instead of anatomical bauplans. Also similar is the evolution of simpler and less constrained programs (continous meandering) into more complex ones. It resembles the phenomena of terminal growth and countdown morphologies in marginally growing shells. After all, morphogenetic and behavioral programs may be considered as different expressions of a similar genomic control.

■ Hypichnial trilobite track deformed by shadow trace of *Psammichnites saltensis* (Puncoviscana Fm., Salta)

Plate 29 · Psammichnitid Behavioral Evolution   85

Psammichnitid Behavioral Evolution

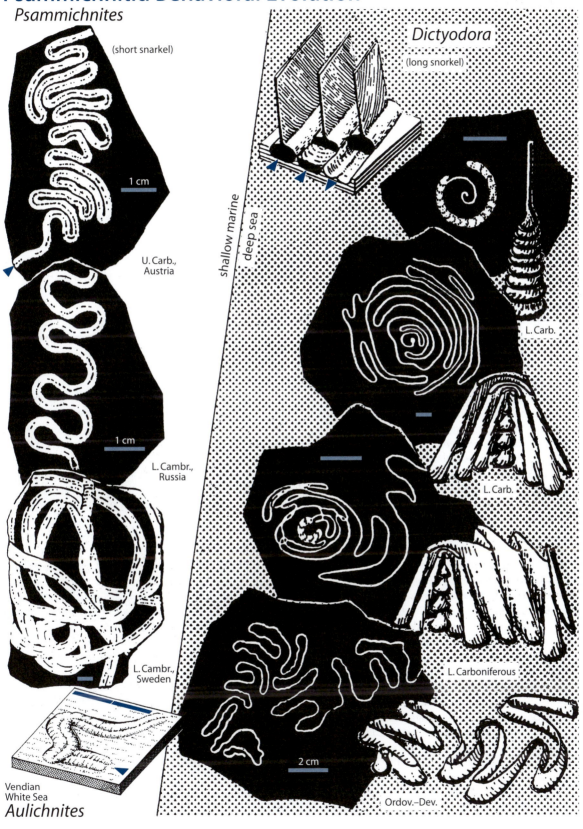

Psammichnites

(short snarkel)

1 cm

U. Carb.,
Austria

1 cm

L. Cambr.,
Russia

L. Cambr.,
Sweden

Dictyodora

(long snorkel)

shallow marine

deep sea

1 cm

L. Carb.

L. Carb.

L. Carboniferous

2 cm

Ordov.–Dev.

Vendian
White Sea
Aulichnites

30

Plate 30
Bi- and Tripartite Backfill Bodies

Due to the scarcity of deepsea turbidite series (which are normally lost by subduction and are preserved only when pushed up on continents) and to their heavy tectonization, the stratigraphic sequence and range of *Dictyodora* is poorly known. In any case, it has never been found in a post-Paleozoic flysch. In the shallowmarine realm *Psammichnites* extends into the Permian, but has neither been found in younger sediments. So we may assume that the makers (possibly slug-like infaunal halkieriid molluscs) vanished in the end-Permian extinction. In the present plate we deal with the last representatives of *Psammichnites* and with another distinctive kind of backfill burrow (**Curvolithus**) that extends from the Paleozoic into the Tertiary and possibly to the present day. In both, the backfill body is divided into longitudinal strings, two in the Permian *Psammichnites*, three in *Curvolithus*.

Permian *Psammichnites*. In the previous plates we modeled *Psammichnites* after echinoid burrows, in which the backfill consists of lamellae in the shape of watchglasses or contact lenses. In this perspective, psammichnitid bilobality was introduced only by the trace of the snorkel running along the crest. In Permian examples from the Canning Basin (**Western Australia**), however, the whole backfill body appears to have been bipartite. This becomes evident not by sectioning, but by the comparison of epichnial specimens, in which only the lower part of the backfill lamellae has been sandy (geopetal backfill). As the muddy upper part of the backfill got lost with the surrounding shale, weathered top surfaces provide us with natural CAT-scan profiles of the burrow fill.

The difference starts in the deeper (**A**), where the midline is traced by a furrow instead of a crest corresponding to the snorkel, similar to *Aulichnites* (Pl. 29). At higher levels (**B**), bipartition becomes more pronounced. Even when the sandy backfill peters out (mud phantom situation; **C**), the lower surface does not show transverse hatching or a ladder-like *Plagiogmus* pattern (Pl. 27), but a delicate median ridge zigzagged by alternating backfill packages on either side. Probably the waves transporting sediment along the right and left half of the body were no longer synchronized, but acted in alternation. This does not necessarily imply a completely different maker, because a similar switch to bipartite activity of the sole muscles also happened in some gastropods. Perhaps this arrangement facilitated steering, which was essential in making the smooth meanders observed in the same beds.

***Curvolithus*.** A different kind of maker, but a similar burrowing technique, is expressed in this characteristic trace fossil. The name was coined for an occurrence in the Ordovician of Bohemia and refers to a typical behavior: on rippled top surfaces of storm sands, *Curvolithus* tends to emerge and dive down again like a dolphin at play. In Swabia, collectors of the last century called Jurassic forms *Ordensband*, referring to another characteristic: the distinct parallel bands resemble the colored ribbons that decorated heroes' chests. In the trace, the three bands differ not in color, but in sculpture. The two marginal lobes bulge and commonly show irregular swellings. In contrast, the median zone is completely smooth and evenly vaulted. Well preserved epichnial specimens also show a delicate median line, but instead of forming a ridge or a furrow, it consists of a **double groove** that looks as if the burrow fill had been decorated from outside with a minute two-pronged fork. Possibly the feature represents a median faecal string that was mucus-lined and placed at the crest of the backfill rather than on the base. Hypichnial expressions of *Curvolithus* are also tripartite, but with a somewhat flatter profile. They lack the dolphin behavior as well as the median grooves.

With regard to the potential tracemaker, the jumps across the sand/mud interface provide us not only with a directional clue, but also with perfect backfill phantoms. In longer jumps, the **sand phantom** emerges from the top surface of the bed like a minute cat's tongue, while the corresponding **mud phantom** can be seen at some distance (see Pl. 28). Both indicate a flat body that was three to five times longer than wide. In shorter jumps, the tip of the sand tongue telescopes into the mud phantom, remaining sharply separated from the sand around. A further clue to the shape of the animal is a crenulated **tail print** on the top of a sand phantom. There may have also been some median organ (but not a snorkel!) on the dorsal side that is responsible for the paired groove. So far, our description would well fit a flatworm bulldozing through the sediment by cilia too small to produce a visible transverse sculpture. But why the thick marginal lobes?

Normally, the backfill lamellae of *Curvolithus* cannot be seen, because the sand is too uniform and because there is never a geopetal mud part of the lamellae that would have made them visible, as in the Permian *Psammichnites* (above). Here, an occurrence of *Curvolithus* in the Jurassic of Greenland comes in handy, because the host rock is so micaceous that backfill structures are expressed by the orientations of mica flakes. In this exceptional material, Heinberg showed that the marginal bulges represent separate backfill bodies. They were probably produced by undulation of the flatworm's margin, while the rest of the backfill was fitted into them upon passage of the rear end. This model fits the shape of the sand phantoms, where the marginal lobes begin only at a distance in front of the rear tip: as their backfill was deposited along the flanks of the body rather than at the rear end of the animal, their phantoms cannot extend as far back. ▶

Plate 30 · Bi- and Tripartite Backfill Bodies   87

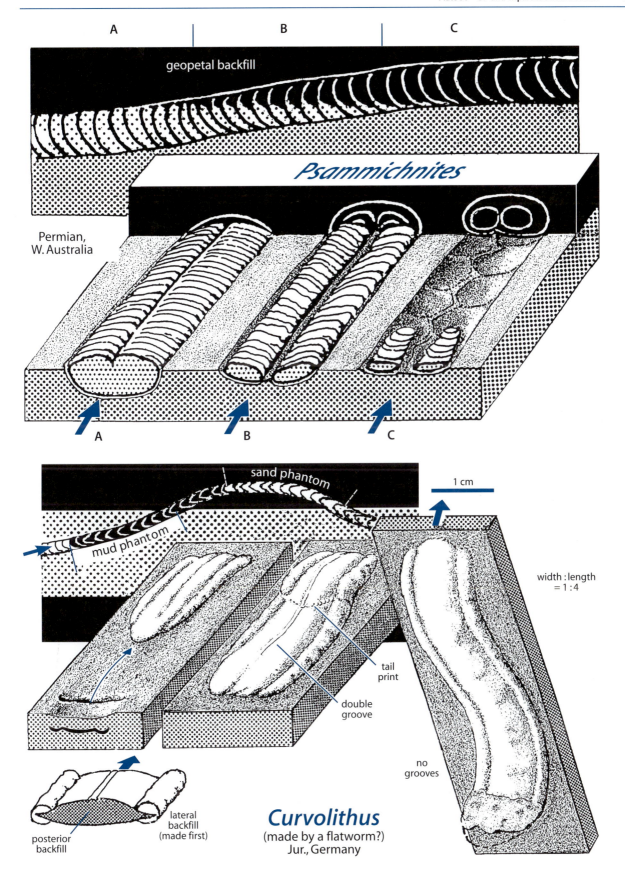

A   B   C

geopetal backfill

Psammichnites

Permian,
W. Australia

A   B   C

sand phantom

1 cm

mud phantom

width : length
= 1 : 4

tail
print

double
groove

no
grooves

lateral
backfill
(made first)

posterior
backfill

Curvolithus
(made by a flatworm?)
Jur., Germany

The other claim derived from the Greenland *Curvo-lithus*, that the central backfill body was bipartite as in the Permian *Psammichnites* (see above), is less convincing. If this had been the case, the tip of the phantom tongue should also be asymmetric. More likely the backfill lamellae are chevron-shaped, conforming to the rear end of the sand phantom. In any case, the Greenland example underlines the usefulness of highly micaceous sediments for revealing otherwise invisible structures. This should be remembered in the field, as well as in laboratory experiments

Backfill phantoms, however, may be deceiving. A film shown by Steve Hasiotis at the Goldring symposium in Reading (July 2006) shows the nymph of a terrestrial insect (hemipteran) at work. It removes the substrate in front of its burrow, but unlike subaquatic bulldozers it does not redeposit it right behind the body. Instead it turns with a somersault and carries the load for some distance through the still open tunnel before it is backstuffed in the typical meniscoid fashion. Evidently the open chamber behind the body is needed by the insect for breathing air with its tracheal lungs. Combined with a change in substrate, the resulting backfill phantom would thus suggest an overly long, wormlike trace maker. This must be kept in mind when studying trace fossils in continental deposits (Pl. 32).

■ *Curvolithus* (L. Jurassic, Germany)

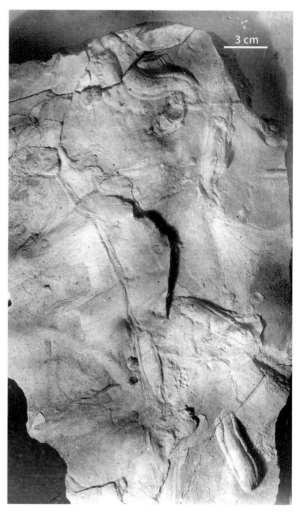

■ *Curvolithus* (M. Jurassic, Germany)

Burrows of Wormlike Bulldozers

When we say worms, we mean long, soft and usually cylindrical animals be-longing to many different phyla. They all have evolved this shape and a peri-staltic mode of movement for burrowing in soft sediments. Therefore, instead of trying to shoehorn fossil worm burrows into zoological systematics, we shall dis-tinguish informal groups that show particular modes of penetration, backfilling, and pattern formation.

This is a less exciting chapter, because except for spirals and meanders (Pls. 33 and 34), there are no complex behavior patterns involved. Still this is what one most commonly finds; so it requires the distinction of minor details to say more than just "worm burrows". Plate 35 also introduces a kind of infaunal locomotion that is so far unknown from modern environments. By moving at an angle to the body axis, these animals produced a kind of spreite that is neither connected with a stationary tunnel system, nor with an obvious grazing strategy.

At this point, we also have to deal with the difficult problem of ichnotaxonomy. So far, the general affiliation of the maker in terms of phyla was no real problem. This changes as we approach the world of "worm trails". First, we are less familiar with these "lower" creatures. Second, they are not represented by body fossils and, third, anatomical fingerprints (bioglyphs) tend to become wiped out during fabrication of the trace. Yet we may rely on the lesson of previous chapters: what we find on bedding planes are most likely undertraces or burrows, rather than surface trails.

Literature

Chapter VII

Fauchald K, Jumars PA (1979) The diet of worms: A study of polychaete feeding guilds. Oceanogr Mar Biol Annu Rev 17:193–284 (Morphology of polychaete families and genera)

Howell BF (1944) A new *Skolithos* from the Cambrian Hardyston Formation of Pennsylvania. Bull Wagner Free Institute of Science 19(4):41–46

Howell BF (1945) *Skolithos, Diplocraterion*, and *Sabellidites* in the Cambrian Antietam Sand-stone of Maryland. Bull Wagner Free Institute of Science 20:4, 2 pls

Howell BF (1954) Burrows of *Foralites* from the Cambrian of New York. Bull Wagner Free Insti-tute of Science 29(1):1–4, 1 pl.

Howell BF (1958) *Skolithos woodi* Whitefield in the Upper Cambrian of Minnesota and Wiscon-sin. Bull Wagner Free Institute of Science 33(2):17–20, 2 pls.

Hyman LH (1959) The invertebrates, 5. Smaller coelomate groups. McGraw-Hill, New York, 783 p (Standard work on morphology of wormlike phyla)

Schäfer W (1972) Ecology and palaeoecology of marine environments. University of Chicago Press, Chicago, 568 p (Very useful review of burrowing mechanisms among worm-like infaunal animals)

Plate 31: Trace Fossil Classification

Bromley RG (1996) Trace fossils: Biology, taphonomy and applications, 2nd edn. Chapman & Hall, London, 361 p (Excellent review of taxonomical, preservational and ethological principles of classification)

Desio A (1949) Sulla nomenela tura delle vestigia problematiche fossili. Serie P(59):1–5 (Proposes to give monomial scientific names to physical pseudofossils)

Frey RW (1973) Concepts in the study of biogenic sedimentary structures. J Sediment Petrol 43:6–19 (Consensus on English, French and German terminology of trace fossils)

Frey RW, Seilacher A (1980) Uniformity in marine invertebrate ichnology. Lethaia 13:183–207

Häntzschel W (1975) Trace fossils and problematica. In: Teichert C (ed) Treatise on invertebrate paleontology, Part W, Suppl. 1. Geological Society of America and Univerity of Kansas, W1–W269 (The treatise summarizes and describes available trace fossil names alphabetically)

Haubold H (1996) Ichnotaxonomie und Klassifikation von Tetrapodenfährten aus dem Perm. Hallesches Jahrbuch für Geowissenschaften (B) 18:23–88 (Permian tetrapod tracks)

International Commission of Zoological Nomenclature (1999) International code of zoological nomenclature, 4th edn. International Trust for Zoological Nomenclature, London (The Code states nomenclature rules)

Krejci-Graf K (1932) Definition der Begriffe Marken, Spuren, Bauten, Hieroglyphen und Fucoiden. Senckenberg 14(1/2):19–39 (Classification of physical and biogenic sedimentary structures)

Magwood JPA (1992) Ichnotaxonomy: A burrow by any other name? In: Maples CG, West RR (eds) Trace fossils. Short Courses in Paleontology 5:15–33 (Review of ichnologic nomenclature and taxonomic issues)

Martinsson A (1970) Toponomy of trace fossils. In: Crimes TP, Harper JC (eds) Trace fossils. Geol J, Special Issue 3, pp 323–330 (Proposes a preservational classification of trace fossils)

Miller W III (2002) Complex trace fossils as extended organisms: A proposal. Neues Jahrb Geol P M 2002, pp 147–158 (Distinction between incidental and intentional behavior)

Miller MF (2003) Styles of behavioral complexity record by selected trace fossils. Palaeogeog Palaeoclim Palaeoecol 192:33–44 (Discusses the possibility that vertical Skolithos, horizontal Palaeophycus, chevron burrows and Isopodichnus in an Antarctican Triassic sandstone stem from the same animal; i.e., they would be parts of a compound trace)

Pickerill RK (1994) Nomenclature and taxonomy of invertebrate trace fossils. In: Donovan SK (ed) The palaeobiology of trace fossils. Johns Hopkins University Press, Baltimore, 308 p (Review of ichnologic nomenclature and taxonomic issues)

Sarjeant WAS, Kennedy WJ (1973) Proposal of a code for the nomenclature of trace-fossils. Can J Earth Sci 10(4):460–475

Seilacher A (1953) Studien zur Palichnologie. I. Über die Methoden der Palichnologie. Neues Jahrb Geol P-A 96:421–452 (Proposal of an ethologic classification of trace fossils. See Plate 21 for a summary)

Seilacher A (1964) Sedimentological classification and nomenclature of trace fossils. Sedimentology 3:253–256 (Proposes of a preservational classification)

Vjalov OS (1963) Zur Frage der Klassifikation von Spuren der Lebenstätigkeit von Organismen und von Texturspuren in den Molasse- und Flyschschichten. Akad Wissensch Ukrainischen SR Geol J 23(1):16–30 (in Russian)

Plate 32: Unbranched Burrows

Bayer FM (1955) Zoology: Remarkably preserved fossil sea-pens and their recent counterparts. J. of the Washington Academy of Sciences 45(9):294–300 (Pennatulites interpreted as body fossil)

Buckman JO (2001) Parataenidium, a new Taenidium-like ichnogenus from the Carboniferous of Ireland. Ichnos 8:83–97 (Diagnosis of Parataenidium)

Chamberlain CK (1971) Morphology and ethology of trace fossils from the Ouachita Mountains, southeast Oklahoma. J Paleontol 45:212–246

Claus H (1965) Eine merkwürdige Lebensspur (Protovirgularia? sp.) aus dem Oberen Muschelkalk NW-Thüringens. Senck Leth 46(2/3):187–191

Clifton HE, Thompson JK (1978) Macaronichnus segregatis: A feeding structure of shallow marine polychaetes. J Sediment Petrol 48: 1293–1301 (Diagnosis of Macaronichnus)

D'Alessandro A, Bromley RG (1987) Meniscate trace fossils and the Muensteria-Taenidium problem. Palaeontology 30(4):743–763 (Review of meniscate backfill burrows)

Fillion D (1989) Les critères discriminant à l'íntérieur du triptyque Palaeophycus-Planolites-Macaronichnus: essai de synthèse d'un usage critique. Comptes Rendus de l'Académie des Sciences, Paris 309, pp 169–172 (Discussion on preservational differences)

Frey RW, Howard JD (1981) Conichnus and Schaubcylindrichnus: Redefined trace fossils from the Upper Cretaceous of the Western Interior. J Paleontol 55(4):800–804 (Schaubcylindrichnus = communal arched mantle burrows)

Frey RW, Pemberton SG, Fagerstrom JA (1984) Morphological, ethological, and environmental significance of the ichnogenera Scoyenia and Ancorichnus. J Paleontol 58(2):511–528

Gevers TW, Frakes LA, Edwards LN, Marzolf JE (1971) Trace fossils in the Lower Beacon sediments (Devonian), Darwin Montains, southern Victoria Land, Antarctica. J Paleontol 45:81–94 (Beaconites antarcticum, probably a non-bipartite Psammichnites)

Häntzschel W (1938) Quer-Gliederung bei rezenten und fossilen Wurmröhren. Senckenberg 20(1/2):145–154 (Transversal corrugations in worm burrows)

Häntzschel W (1958) Oktokoralle oder Lebensspur? Mitt Geol Staatsinst Hamburg (27):77–87 (Trace fossil interpretation of Perinatalites)

Ireland RJ, Pollard JE, Steel RJ, Thompson DB (1978) Intertidal sediments and trace fossils from the Waterstones (Scythian-Anisian?) at Daresbury, Cheshire. Proc Yorkshire Geol Soc 41,4(31):399–436, 3 pls. (Scoyenia in Triassic redbeds)

Linck O (1949) Lebens-Spuren aus dem Schilfsandstein (mittl. Keuper km 2) NW-Württembergs und ihre Bedeutung für die Bildungsgeschichte der Stufe. Verein für Vaterländische Naturkunde in Württemberg, Jahreshefte 97–101, 1–100 (Ichnocoenosis in Upper Triassic fluvial sandstones, Germany)

Macsotay O (1967) Huellas problematicas y su valor paleoecologico en Venezuela. Universidad Central de Venezuela, Escuela de Geologia, Minas y Metalurgia, Geos 16:7–79

Pemberton SG, Frey RW (1982) Trace fossil nomenclature and the Planolites-Palaeophycus dilemma. J Paleontol 56:846–881 (Taxonomic review of Planolites and Palaeophycus)

Pemberton SG, Spila M, Pulham AJ, Saunders T, MacEachern JA, Robbins D, Sinclair IK (2001) Ichnology and sedimentology of shallow to marginal marine systems. Ben Nevis and Avalon Reservoirs, Jeanne d'Arc Basin. Geological Association of Canada, Short Course 15, 343 p (Discussion of different Macaronichnus ichnospecies)

Peneau J (1941) Die Anwesenheit von Tomaculum problematicum im Ordovicium West-Frankreichs. Senckenberg 23(1/3):127–132

Pollard JE, Lovell JPB (1976) Trace fossils from the Permo-Triassic of Arran. Scott J Geol 12(3):209–225, 2 pls. (Siphonites = Ancorichnus with micaceous marginal backfill in fluvial sandstone)

Radig F (1964) Die Lebensspur Tomaculum problematicum Groom 1902 im Landeilo der Iberischen Halbinsel. Neues Jahrb Geol P-A 119(1):12–18 (Occurrences in Spain)

Reineck H-E (1955) Marken, Spuren und Fährten in den Waderner Schichten (ro) bei Martinsstein/Nahe. Neues Jahrb Geol P-A 101(1):75–90 (Scoyenia in Permian redbeds)

Richter R, Richter E (1939a) Die Kot-Schnur Tomaculum Groom (= Syncoprulus Rud. and E. Richter), ähnliche Scheitel-Platten und beider stratigraphische Bedeutung. Senckenberg 21(3/4):278–291

Richter R, Richter E (1939b) Eine Lebensspur (Syncoprulus pharmaceus), gemeinsam dem rheinischen und böhmischen Ordovicium. Senckenberg 21(1/2):152–168 (Tomaculum)

Richter R, Richter E (1941) Das stratigraphische Verhalten von *Tomaculum* als Beispiel für die Bedeutung von Lebensspuren. Senckenberg 23(1/3):133–135

Schäfer W (1972) Ecology and palaeoecology of marine environments. University of Chicago Press, Chicago, 568 p (Very useful review of burrowing mechanisms by worm-like infaunal animals)

Schwab K (1966) Ein neuer Fund von *Scoyenia gracilis* White 1929. Neues Jahrb Geol P M (6):326–332

Seilacher A, Buatois LA, Mángano MG (2005) Trace fossils in the Ediacaran-Cambrian transition: Behavioral diversification, ecological turnover and environmental shift. Palaeogeog Palaeoclim Palaeoecol 227:323–356 (Redescription of *Nenoxites*)

Stanley DSA, Pickerill RK (1994) *Planolites constriannulatus* isp. no. from the Late Ordovician Geogian Bay Formation of southern Ontario, eastern Canada. Ichnos 3:119–123 (Meniscoidally backfilled burrow with bioglyphs like *Scoyenia*, but marine)

Tavani G (1941) Nuovi ritrovamenti di *Uintacrinus* nelle "Argille scagliose" dell'Appenino reggiano e bolognese. Mem Soc Tosc Sc Nat 49:3–7, 1 pl. (Meniscoid burrows considered as crinoid arms)

Volk M (1941) Die Lebensspur *Tomaculum problematicum* Groom auch im Griffelschiefer des Thüringer Ordoviciums. Senckenberg 23(1/3):123–126

Volk M (1961) *Protovirgularia nereitarum* (Reinhard Richter), eine Lebensspur aus dem Devon Thüringens. Senck Leth 42(1/2):69–75, 2 pls. (Spicate burrow made by protobranch bivalve or scaphopod)

Plate 33: Spiral Burrows

Hatai K, Noda H (1971) Peculiar markings on a sandstone layer of the Hagino Formation, Nagano Prefecture. Trans Proc Palaeont Soc Japan N S (83):162–165 (Miocene *Cochlichnus*)

Hitchcock E (1858) Ichnology of New England: A report on the sandstone of the Connecticut Valley, especially its fossil footmarks. William White, Boston, 220 p (Diagnosis of *Cochlichnus anguineus*)

Jensen S (1997) Trace fossils from the Lower Cambrian Mickwitzia sandstone, south-central Sweden. Fossils and Strata 42:1–111 (Description of *Gyrolithes polonicus = Spiroxcolex*)

Kim JY (1996) Behavioral patterns expressed in scribbling trace fossils from Ordovician strata of Yeongweol, Korea. Ichnos 4:219–224 (Study of Ordovician scribbling trace fossils)

Mason TR, Stanistreet IG, Tavener-Smith R (1983) Spiral trace fossils from the Permian Ecca Group of Zululand. Lethaia 16:241–247 (Description of very regular *Spirodesmos archimedes*)

Plate 34: Nereitids

Chamberlain CK (1971) Morphology and ethology of trace fossils from the Ouachita Mountains, southeast Oklahoma. J Paleontol 45:212–246 (Description of various ichnospecies of *Nereites*)

Chiplonkar GW (1972) A new trace fossil from the Upper Cretaceous of South India. Curr Sci 41(20):747 (*Scalaritubu*-like meniscate burrows, but in bundles without marginal backfill on opposed stuffing)

Conkin JE, Conkin BM (1968) *Scalarituba missouriensis* and its stratigraphic distribution. University of Kansas, Paleontological Contributions, Paper 31, 7 p (*Nereites missouriensis*)

Crimes TP, Crossley JD (1991) A diverse ichnofauna from Silurian flysch of the Aberystwyth Grits Formation, Wales. Geol J 26:27–64 (Description of various ichnospecies of *Nereites*)

D'Alessandro A, Bromley RG, Stemmerik L (1987) *Rutichnus*: A new ichnogenus for branched, walled meniscate trace fossils. J Paleontol 61(6):1112–1119 (Resembling *Neonereites* by knobby marginal backfill, but branching and with smooth terminal backfill)

Häntzschel W (1964) Spurenfossilien und Problematica im Campan von Beckum (Westf.). Fortschr Geol Rheinld Westf 7:295–308, 4 pls. (*Dreginozoum nereitiforma* tentatively interpreted as egg cases of marine gastropods)

Mángano MG, Buatois LA, West RR, Maples CG (2002) Ichnology of an equatorial tidal flat: The Stull Shale Member at Waverly, eastern Kansas. Bulletin of the Kansas Geological Survey 245, 130 p (Analysis of shallow-marine Carboniferous *Nereites* from Kansas)

Netto RG (1987) Sobre a ocorréncia de *Neonereites* Seilacher 1960 no permiano do Rio Grande do Sul. Anais do X Congr. Bras. Paleontologia, Rio de Janeiro, pp 285–296

Orr P, Pickerill RK (1995) Trace fossils from Early Silurian flysch of the Waterville Formation, Maine, USA. Northeastern Geology and Environmental Sciences 17:394–414 (Description of several ichnospecies of *Nereites*)

Perdigao JC (1961) *Nereites* do Baixo Alentejo. Comun Serv Geol Portugal 45:339–363

Seilacher A (1960) Lebensspuren als Leitfossilien. Geol Rundsch 49:41–50 (Description of *Neonereites*)

Seilacher A (1986) Evolution of behavior as expressed in marine trace fossils. In: Nitecki MHG, Kitchell JA (eds) Evolution of animal behavior: Palaeontological and field approaches. Oxford University Press, New York, pp 62–87 (Evolution of Paleozoic, Mesozoic and Cenozoic *Nereites* in shallow and deep marine environments, still including *N. saltensis*)

Seilacher A, Meischner D (1965) Fazies-Analyse im Paläozoikum des Oslogebietes. Geol Rundsch 54:596–619 (Describe nereitids)

Uchman A (1995) Taxonomy and paleoecology of flysch trace fossils: The Marnoso-arenacea formation and associated facies (Miocene, Northern Apennines, Italy). Beringeria 15:1–115 (Review of *Nereites*)

Wetzel A (2002) Modern *Nereites* in the South China Sea – Ecological association with redox conditions in the sediment. Palaios 17:507–515 (Modern *Nereites* and their relationship with the redox surface)

Plate 35: Gyrochortids

Bradshaw MA (1981) Palaeoenvironmental interpretations and systematics of Devonian trace fossils from the Taylor Group (Lower Beacon Supergroup), Antarctica. N Z J Geol Geophys 24:615–652 (*Heimdallia* Diagnos.)

Gibert JM de, Benner JS (2002) The trace fossil *Gyrochorte*: Ethology and paleoecology. Rev Esp Paleontol 17:1–12 (Updated review of *Gyrochorte*)

Hallam A (1970) *Gyrochorte* and other trace fossils in the Forest Marble (Bathonian) of Dorset, England. In: Crimes TP, Harper JC (eds) Trace fossils. Geol J, Special Issue 3, pp 189–200 (Interpretation of *Gyrochorte* as arthropod trace)

Heinberg C (1970) Some Jurassic trace fossils from Jameson Land (East Greenland.) In: Crimes TP, Harper JC (eds) Trace fossils. Geol J, Special Issue 3, pp 227–234 (Three-dimensional reconstruction of *Gyrochorte* backfill structure)

Heinberg C (1973) The internal structure of the trace fossils *Gyrochorte* and *Curvolithus*. Lethaia 6:227–238 (Detailed analysis of the internal structure and mode of backfill of *Gyrochorte*)

Karaszewski W (1974) *Rhizocorallium*, *Gyrochorte* and other trace fossils from the Middle Jurassic of the Inowlódz Region, Middle Poland. Bull Acad Pol Sci 21(3–4):199–204, 4 pls.

Seilacher A (1955) Spuren und Fazies im Unterkambrium. In: Schindewolf O, Seilacher A (eds) Beiträge zur Kenntnis des Kambriums in der Salt Range (Pakistan). Akademie der Wissenschaften und der Literatur, Mainz, Abhandlungen der mathematischnaturwissenschaftlichen Klasse, 10, pp 261–446 (Block diagram of *Gyrochorte*)

Seilacher A, Alidou S (1988) Ordovician and Silurian trace fossils from northern Benin (W. Africa). Neues Jahrb Geol P-A 7:432–439 (Diagnosis of *Gyrochorte zigzag*)

Seilacher A, Cingolani C, Varela C (2003) Ichnostratigraphic correlation of early Paleozoic quartzites in central Argentina. In: Salem MJ, Oun KM, Seddig HM (eds) The geology of Northwest Libya. Earth Science Society of Libya 1, Tripoli, pp 275–292 (*Gyrochorte zigzag*, here referred to *Heimdallia*)

Weiss W (1940) Beobachtungen an Zopfplatten. Z Dtsch Geol Ges 92:333–349 (Early analysis of Jurassic *Gyrochorte comosa*)

Weiss W (1941) Die Entstehung der "Zöpfe" im Schwarzen und Braunen Jura. Nat Volk 71:179–184 (Origin of *Gyrochorte*)

31

Plate 31
Trace Fossil Classification

In previous chapters, it was still possible to affiliate given trace fossils with particular kinds, or groups, of animals. From now on, such affiliation will be rarely possible, even though the trace fossils concerned may be very distinctive. New taxonomic links will undoubtedly be discovered as neoichnologists proceed from studying only surface features to explore the burrows and undertraces that are more likely to be preserved. Also, established ichnogenera may be recognized as preservational variants of the same kind of burrow. Nevertheless it would be useful to agree on some general system by which trace fossils can be ordered – be it in textbooks or in reference collections.

The safest way out of the dilemma appears to be an *alphabetic* order of ichnogenera, as was practiced in the Treatise on Invertebrate Fossils – provided that these names have some stability. Unfortunately this is not the case. In parataxa that potentially span the whole geologic record, senior synonyms often remain unnoticed. Also, specialists have their own unique field experience and their own interpretations, so that there will be much disagreement about grouping. As a result, there is still a vast number of synonyms to be weeded out, while new ones are introduced at an even higher rate. Unfortunately names will and must change, even though the Rules of Zoological Nomenclature were established to keep them constant.

Many ichnologists support a purely morphological classification of trace fossils, disregarding their interpretation: short bilobites are attributed to *Rusophycus* and long ones to *Cruziana* – even if we know that both were made by the same trilobite in one case and by a branchiopod crustacean in another. The ultimate goal would be to use the computer as an impartial arbiter.

The core of the problem at hand is the incompatibility of two systems. *Nomenclature* attempts to unmistakably label natural objects for easier communication; therefore it is primarily concerned with their identity and with the constancy of the name by which they happened to be christened. Therefore one continues to use names such as *Arthrophycus*, although the fossil is certainly not the impression of a seaweed, as the Greek root suggests. But Latin names, like personal names, must be binominal to become valid. So the genus name (like the family name in our passport) already implies a hypothesis about relationship that goes beyond pure identification. In practice, however, ichnologists often use only ichnogeneric names and forget the ichnospecies name after the baptism ceremony. As most ichnogenera are monospecific, this neglect is pardonable and has also been followed in parts of the present text.

In contrast to nomenclature, *classification* goes beyond description by implying hypothetical relationships between nonidentical objects. These relationships may be of different kinds and it is only a matter of agreement which kind is chosen as the base of an encompassing classification. In systematic biology this is the relation by descent (phylogeny). As we have seen, the uncertain authorship makes it difficult to apply this principle to all trace fossils. Nor can any other kind of relation do justice to the whole spectrum of trace fossil morphologies, from dinosaur tracks to worm burrows.

It is probably realistic to abandon the dream of a unified and logically consistent ichnotaxonomy. Rather, we should decide from case to case whether taxonomic, behavioral, preservational or stratigraphic relationships are most adequate to order ichnogenera and ichnospecies (or subspecies) in each particular group. Accordingly, the "classifications" used in this book should be understood as coordinates in a multidimensional morphospace, rather than universal ordering systems.

After having defined trace fossils as contrasted to body fossils (even our **definition** is arguable; where do coprolites, egg shells, or rock borings belong?), we can order them in an **ethological** scheme according to function. Instead of listing all thinkable activities (sitting, flying, mating etc.), the diagram on the lower left contains only six activities that are most commonly expressed in trace fossil morphologies. Given forms can be placed in one of the sectors, closer to the core if the tracemaker can be identified, or nearer the periphery if it is unknown. Multiple functions are common, but usually one dominates the morphology more than others. The callianassid tunnel system in Pl. 18, for instance, is primarily a dwelling structure (domichnion), but it may also contain fodinichnial elements in the spreite section and an agrichnial structure in the corkscrew.

Equally important is the **preservational** scheme describing the position and relief of the trace fossil relative to a layer of coarser sediment (commonly sand deposited during a high-energy event). Photographs of trace fossils are often difficult to judge. Unconventional lighting (other than from the upper left) makes grooves appear as ridges and vice versa. It is also important to know whether one is viewing a top or a sole surface (epi- versus hyporeliefs). In particular cases, the topological scheme needs to be elaborated. In *Psammichnites gigas*, for instance (Pl. 27), the soft outline of the positive epirelief indicates that it was pressed-through from below (shadow trace), while the sharp sinus line on its top suggests the cutting effect of the snorkel at this very level. On the other hand, soft contours in burrows preserved as positive hyporeliefs may indicate that they were exhumed in the erosive phase of a high-energy event and cast in its depositional phase (*Taphrhelminthopsis*, Pl. 26; graphoglyptids, Pl. 52).

Intercutting relationships between associated trace fossils form another important aspect of topology. In mixed-tier communities, burrows that intersect mostly belong to a deeper tier relative to the ones penetrated. Thus, the sharp reentrants around worm tunnels into the positive hyporeliefs of trilobite burrows tell us that they belong to a deeper tier and were made after the trilobites had moved to a higher level (Pl. 74).

Another possible classification is by behavioral **complexity**. But is a trilobite's trackway incidental, its burrow intentional?

Plate 31 · Trace Fossil Classification   93

Preservation

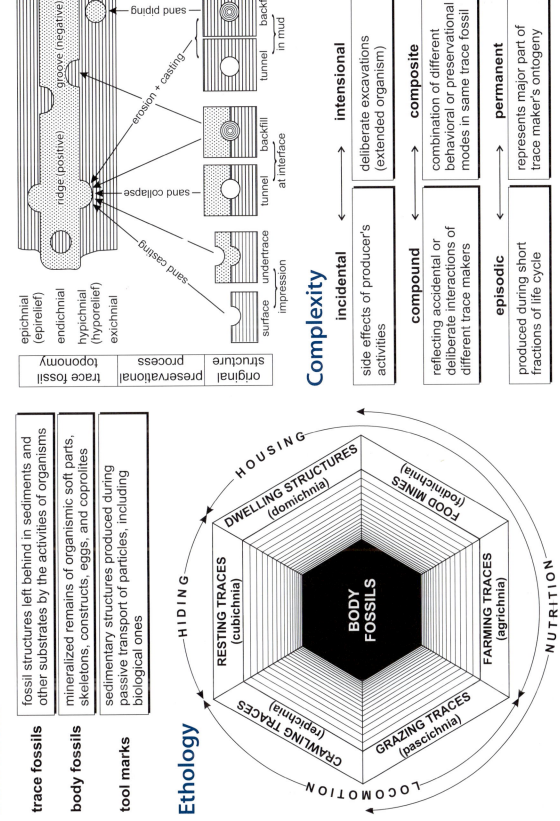

	epichnial (epirelief)	trace fossil toponomy
	endichnial	preservational process
	hypichnial (hyporelief)	
	exichnial	original structure

ridge (positive)
groove (negative)
sand piping
tunnel | backfill | in mud
erosion + casting
backfill | at interface
sand collapse
tunnel
sand casting
undertrace
surface | impression

Complexity

incidental	intensional
side effects of producer's activities	deliberate excavations (extended organism)
compound	composite
reflecting accidental or deliberate interactions of different trace makers	combination of different behavioral or preservational modes in same trace fossil
episodic	permanent
produced during short fractions of life cycle	represents major part of trace maker's ontogeny

Definitions

trace fossils — fossil structures left behind in sediments and other substrates by the activities of organisms

body fossils — mineralized remains of organismic soft parts, skeletons, constructs, eggs, and coprolites

tool marks — sedimentary structures produced during passive transport of particles, including biological ones

Ethology

HOUSING

HIDING

NUTRITION

LOCOMOTION

DWELLING STRUCTURES (domichnia)

FOOD MINES (rodinichnia)

RESTING TRACES (cubichnia)

BODY FOSSILS

FARMING TRACES (agrichnia)

CRAWLING TRACES (repichnia)

GRAZING TRACES (pascichnia)

32

Plate 32
Unbranched Burrows

There are three ways to wedge a worm-shaped body through the sediment: undulation, peristalsis and ciliary or parapodial waves. *Undulatory* movements are used for swimming in fluids by bacteria, flagellates and sperm, as well as by fish. In larger animals, undulation is assisted by compressed body shapes (sea snakes, flatfish, salpid colonies).

Burrowing *nematodes* are able to "swim" through soft sediment (or intestinal contents) in this manner in spite of their circular cross section. Accordingly their surface is perfectly smooth, but they do not exceed a certain size compatible with the viscosity of the medium in which they burrow (Pl. 33). *Peristaltic* motion has been discussed in Pl. 21. In earthworms, it is assisted by setae that act like a ratchet: they rise up in the expanded and fold down in the contracting parts. Rhythmic constrictions and longitudinal scratches in the intervening parts of *Scoyenia* could be referred to such a kind of locomotion; but the question remains, whether pure peristalsis is compatible with active backfill.

The third case, burrowing by appendages such as *parapodia* activated in metachronal waves, resembles bulldozing by a pelt of spines in spatangoids (Pl. 26).

The present plate unites wormlike trace fossils that do *not* branch, either because the animal could not move backwards, or because the tunnel was actively backfilled behind it. Therefore this is a very heterogeneous group that cannot claim the status of an ichnofamily.

A trace from the Silurian of Russia (cf. *Petalichnus*) resembles *Scolicia* (Pl. 26) in that the meniscate backfill and the chevron sculpture open in opposite directions. But the chevrons are pustulate like those made by the crenulated margin of a molluscan foot. Perhaps *Psammichnites* (Pl. 30) would look similar if preserved as a positive hyporelief?

Granularia has already been mentioned (Pl. 18) as a miniaturized deepsea ophiomorphid. Our diagram only emphasizes the oblique scratches in a *Spongeliomorpha*-like preservation. This is in contrast to the more longitudinal scratches of the backfilled *Scoyenia*, which could have been made by *insect larvae* or earth worms.

As shown in Pl. 21, spicate trails are made by the split foot of *molluscs* (protobranch bivalves and scaphopods), as shown by experiments in which a small bivalve (*Acila*) produced undertraces identical to *Protovirgularia*. A similar form (*Uchirites*) from Carboniferous and Paleocene turbidites is more deeply impressed and sometimes shows the trace of the shell dragged behind as a smooth core. The angularity of the specimen from Venezuela is not true branching. It could have been produced by withdrawing the foot before protracting the shell and penetrating again in a different direction.

The cylindrical burrows in the third row lack characteristic surface sculptures, but differ in the structure of the backfill. In *Muensteria* the meniscoid backfill packages consist of host sediment, but are smaller than the lamellae of scoliciid bulldozers. The larger form from Italy incorporates *faecal pellets* that are smaller relative to the tunnel diameter than in worms. *Tomaculum* may originally have been similar, but due to compaction it appears as a string of elongate pellets in Ordovician shales. *Scolecocoprus*, in contrast, is stuffed with faecal pellets almost the diameter of the tunnel. Because it occurs in nonmarine deposits, it may well be produced by earthworms.

Mantle burrows (*Biformites, Angorichnus*) have a marginal backfill (Pl. 37). As it consists of displaced host sediment, it may represent a backfill phantom of the evertible proboscis that made it. In the right specimen from the German Keuper, two individuals passed obliquely from clay into an overlying sand and the terminal backfill inside the tunnel produced a much longer phantom of the body itself (see diagram). *Ancorichnus* ranges widely in time and environments. In the Ediacaran *Nenoxites* it is not clear whether the transversal segmentation is caused by marginal backfill or elongated pellets. In any case, meandering is not as "Greek" as commonly claimed.

■ *Muensteria* (U. Cretac. flysch, Italy) Epichnial groove with geopetal backfill, going deeper to lower right (fecal pellets!)

Plate 32 · Unbranched Burrows   95

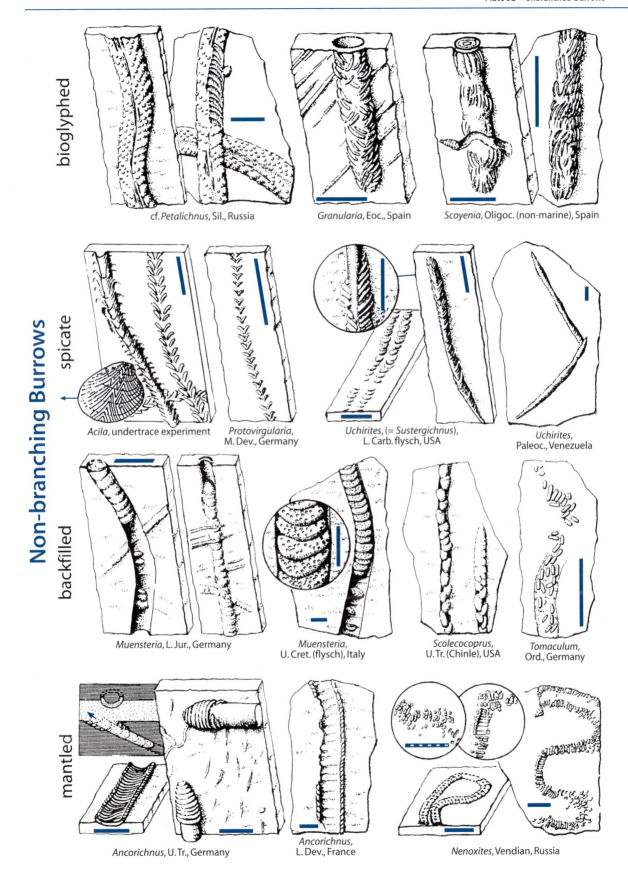

Non-branching Burrows

bioglyphed

cf. *Petalichnus*, Sil., Russia

Granularia, Eoc., Spain

Scoyenia, Oligoc. (non-marine), Spain

spicate

Acila, undertrace experiment

Protovirgularia, M. Dev., Germany

Uchirites, (= *Sustergichnus*), L. Carb. flysch, USA

Uchirites, Paleoc., Venezuela

backfilled

Muensteria, L. Jur., Germany

Muensteria, U. Cret. (flysch), Italy

Scolecocoprus, U. Tr. (Chinle), USA

Tomaculum, Ord., Germany

mantled

Ancorichnus, U. Tr., Germany

Ancorichnus, L. Dev., France

Nenoxites, Vendian, Russia

33

Plate 33
Spiral Burrows

From early Polynesian petroglyphs on, humans perceived spirals as ornaments, sometimes with magical underpinnings. In nature, they appear as fabricational noise, functional design, or both. Molluscs secrete logarithmically coiled spiral shells, which conform to the growth of the contained soft bodies. Spiral traces, in contrast, keep the same distance between whorls (Archimedean spirals), because they have the basic function to uniformly fill a given space without overcrossing. At the same time, spirals can be produced with a simpler behavioral program than guided meanders: "Just turn right (or left) and keep the same distance from your previous trail or burrow."

In the first category, **corkscrew burrows**, not even such a program is required for the simplest forms, because they derive from the locomotory principle of nematode worms. In a three-dimensional substrate, the undulation of *Cochlichnus* can easily switch to corkscrew motion. In the two specimens from the **Lower Jurassic** of Helmstedt, northern Germany, the epichnial one follows the bedding plane in a sinuous line, while only the lower whorls are seen in the hypichnial corkscrew variant. As we shall see in Pl. 53, such a mode could have developed into complex graphoglyptid burrowing programs. In contrast, the vertical burrows of the Early Cambrian *Spiroscolex* are too tightly coiled for a locomotory (repichnial) interpretation. As open tunnels they may have been used to farm bacteria (see *Gyrolithes*, Pl. 17) and could later become passively filled with sand from above. On the other hand, spiral programs may serve quite different purposes. Some beetle grubs use it for constructing safe pupa chambers (*Tambia*, Pl. 3), while small intertidal crabs in India vault their burrows during low tide with pellets of wet sand in a spiral fashion. Their techniques may be compared to pre-wheel pottery or to the way in which beehives were made of straw bundles in earlier times. Note, however, that whorl diameters must systematically change during execution of the program. Also, *Tambia* as well as the crab cupolas start with wide whorls and end in the center.

The second category, horizontal **scribbles**, may be viewed as a sloppy version of planar spirals: by allowing for some double coverage in overcrossings, their makers bypassed the difficulty to maintain constant whorl distances. We already discussed this strategy in trilobite (Pl. 14) and molluscan bulldozers (Pl. 27) of Cambrian times. So it is probably no coincidence that our examples of scribbling worm burrows are also very old (*Gordia* from the Vendian, Upper Cambrian of USA and Lower Ordovician flysch deposits of Barrancos, Portugal), while no counterparts are known from later times. In all three cases, the burrows became actively backfilled.

The **guided spirals** of our third category also follow the bedding plane. An anthropogenic example are the spiral plowings in North Africa, by which German troops made their airstrip unusable during *World War II*. The two fossil examples, from turbidite soles in a lower Carboniferous (Kulm) flysch of northern Germany, keep a wide and very regular distance between whorls. Whereas the trace is continuous in **Spirodesmos archimedes**, it is segmented in **Spirodesmos interruptus**. The regular obliquity of the segments throughout the spiral excludes a tectonic artifact. Rather it suggests that the segments represent rod-like and relatively stiff faecal pellets, whose distal ends became pushed outward during backfilling. Thus, *Spirodesmos* cannot be compared with the open tunnel systems of graphoglyptids in the same facies (Chap. XII). More properly they may be compared to spirals on *modern deepsea floors*. Some of the available photographs show that they are faecal strings left by much wider browsing holothurians. As such surface structures would have been erased by a turbidity current, the fossil forms may reflect a similar behavior of an *undermat miner* (Pl. 49).

The **starter spirals** in the lower bracket do not properly belong here, because they result from a general dilemma in the performance of guided meandering (Pl. 51), rather than requiring a distinct program: in the absence of a previous turn, the animal uses the point at which it entered the bedding plane as initial reference. All four fossil examples come from flysch deposits. *Nereites*, *Dictyodora* and *Taphrhelminthopsis* are discussed on Pls. 34, 29 and 26, respectively. Not represented in other plates is **Spirophycus**, whose preservation (as washed-out casts on the soles of Early Tertiary turbidites) resembles that of the associated washed-out version of spatangoid burrows (note irritation of *Taphrhelminthopsis* by *Spirophycus*). Therefore it probably represents the backfilled cylindrical burrow of an unknown and fairly large animal. The species name (**Spirophycus bicornis**) refers to a specimens in which two starter spirals made by different individuals happened to be symmetrically arranged.

In spite of the similar combination of a starter spiral with regularly spaced meanders, the open tunnel systems of the annelid *Paraonis* (Pl. 51) in modern beach sands serve yet another function. As H. Roeder found out, they are regularly revisited and made anew after every tide. They possibly function as a kind of intrasedimentary spider webs for catching vertically migrating interstitial organisms. The absence of starter spirals in graphoglyptid meanders (Pls. 53, 54) suggests that they are derived from undulating or corkscrew locomotion. Note also that a starter spiral still occurs in the Devonian *Nereites*, while the younger *Helminthoida* (Pl. 34) could start without it. A similar behavioral improvement can be observed in Oligocene representatives of *Taphrhelminthopsis* (Pl. 26).

Plate 33 · Spiral Burrows   97

Spiral Burrows

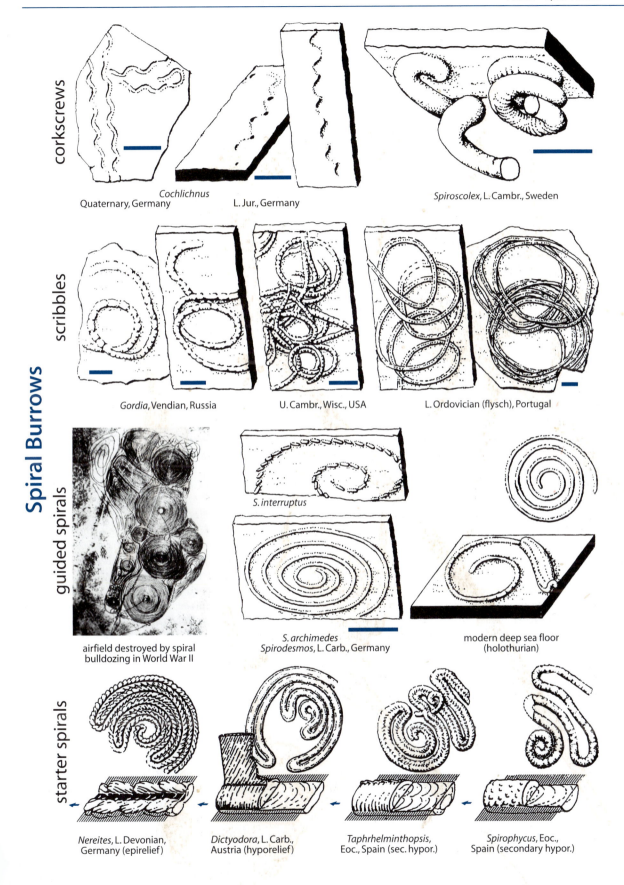

corkscrews

Cochlichnus
Quaternary, Germany

L. Jur., Germany

Spiroscolex, L. Cambr., Sweden

scribbles

Gordia, Vendian, Russia

U. Cambr., Wisc., USA

L. Ordovician (flysch), Portugal

guided spirals

airfield destroyed by spiral
bulldozing in World War II

S. interruptus

S. archimedes
Spirodesmos, L. Carb., Germany

modern deep sea floor
(holothurian)

starter spirals

Nereites, L. Devonian,
Germany (epirelief)

Dictyodora, L. Carb.,
Austria (hyporelief)

Taphrhelminthopsis,
Eoc., Spain (sec. hypor.)

Spirophycus, Eoc.,
Spain (secondary hypor.)

34

Plate 34
Nereitids

The name *Nereites* refers to the annelid worm *Nereis*, for whose body impression this trace was originally held. On the top face of thin sand layers (**epichnial** = *Phyllodocites* preservation), this trace appears as a central groove flanked on either side by an elevated lobate zone. It is analogous to the *Palaeobullia* preservation of *Scolicia* (Pl. 26), where the geopetal transition from a sandy into a muddy "facies" within meniscate backfill lamellae causes a similar gill-like structure.

Without such preservational effect, the lateral backfill zones easily pass unnoticed. On split surfaces (**endichnial** = *Scalarituba* preservation), the median string becomes the dominant feature. It is also segmented in a meniscoid fashion. But as the segments are much thicker than usual backfill lamellae, and because they consist of a finer fraction than the host sand, they probably correspond to large faecal pellets. Nevertheless the lateral backfill zones were always present. This can be seen at intersections, where the penetrated faecal string ends short of the penetrating one.

The full three-dimensional structure of *Nereites* is revealed by rather inconspicuous variants on sole surfaces (**hypichnial** = *Dreginozoum* preservation). Because sand sinks to the bottom of the backfill lamellae, no "gill" structure can develop at the interface with the underlying mud. Instead one sees pustules on the outer surface of the trace. One may assume that the animal backfilled processed sediment radially in all directions.

In summary, *Nereites* and related ichnogenera are produced by animals that removed the sediment in front, but not by radial eversion of a voluminous proboscis. Rather they probed and backfilled the rejected sediment laterally around the burrow by bending the head (probably with a small proboscis) in various directions at regular intervals. The finer and more nutritious fraction of the excavated sediment was ingested in each bite, passed through the gut and backfilled at the rear end. The fact that the faecal pellets are almost as wide as the body, points to a worm rather than an arthropod. It is also clear that food was extracted from the sediment, because the reworked halo is relatively more voluminous in sand than in silt or foraminiferal ooze. With this very distinctive mode of sediment feeding as a guideline, we can now follow the behavioral evolution of the unknown trace-makers through Earth history.

The record possibly starts with *Nenoxites* (Pl. 32) in shallowmarine sediments of the late Proterozoic of Russia. It already shows a central string of backfilled faecal pellets and a reworked marginal zone. While the environment and the behavioral interpretation would fit nereitids, the interaction of *"Nereites" saltaensis* with tri-

lobite tracks (Pl. 65) and the deformation of adjacent bedding planes suggest that it should better be affiliated with *Psammichnites*.

Our first two examples come from Silurian turbidites. They fall out of line by their relatively small size, but also because they are behaviorally more advanced than Devonian *Nereites*. In fact, the loops contour each other so tightly that little or no unused space is left between them, while the reworked halo is wider than the faecal string, as corresponds to the siliclastic nature of the sediment. With distance/ length ratios of 1/16 and 1/22 of they were already very efficient in covering a given area. Yet there is a significant difference between occurrences. The meanders from the Silurian flysch of Aberystwyth (**Wales**) maintain the arcuate course that probably initiated with a starter spiral. They may also bend back around the turns of previous loops, but much less so than *Helminthoida labyrinthica* of later flysches. In contrast, the form from an equivalent facies of the Ouachita Mountains (**Oklahoma**) tends to give up contact with previous loops before turning (white arrows). Thereby the loops become straight, except for the turns that tend to bulge due to a limited turning radius. It would be worthwhile to excavate more complete systems of this unusual form. Middle Devonian **Nereites loomisi** from the Variscan fold belt in Germany still burrowed in silts and had more loosely coiled meanders.

Helminthoida labyrinthica from Upper Cretaceous to Eocene turbidites in the Alps differs in four respects. (1) It is much smaller than the Paleozoic forms; (2) it is more tightly coiled with narrower lateral backfill zones; (3) it occurs in foraminiferal ooze rather than siliciclastic sediments; (4) individual loops bend back over previous ones. Although such overlapping disrespects the limitation of loop length, it improved the ability to fill unused corners of the nutritious bedding plane. While the reworked halo is narrower than the faecal string, elevated lobe zones can be seen between the lighter-colored faecal strings in meanders that were made on top of a somewhat coarser layer (block diagram). So there is no doubt about the affiliation with *Nereites*.

As a whole, nereitids in flysch facies show a trend towards miniaturization and more complex behavior, as observed in other deepsea lineages (Pls. 26, 29, 55). Still they seem to have never made it into the exclusive pre-turbidite club (Pl. 72).

Shallowmarine (*shelf*) representatives, in contrast, retained open meanders. The Paleozoic *Scalarituba* occurs in impure sandstones of the *Zoophycos* facies, while the Jurassic **Neonereites** is found on the tops of thin storm sands. In this ichnogenus the faecal pellets are well separated. They appear as a string of beads in the Lower Jurassic (**Neonereites uniserialis**), but alternating in two rows (**Neonereites biserialis**) in the Middle Jurassic.

Plate 34 · Nereitids   99

Nereitids

Shelf

Flysch facies

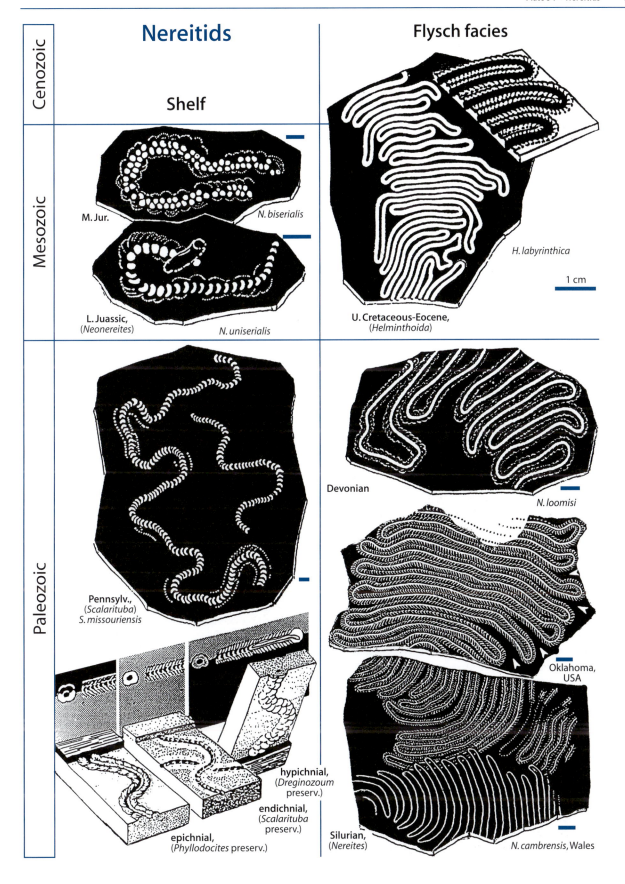

M. Jur.

N. biserialis

L. Juassic,
(*Neonereites*)

N. uniserialis

U. Cretaceous-Eocene,
(*Helminthoida*)

H. labyrinthica

1 cm

Pennsylv.,
(*Scalarituba*)
S. missouriensis

Devonian

N. loomisi

Oklahoma,
USA

hypichnial,
(*Dreginozoum*
preserv.)

endichnial,
(*Scalarituba*
preserv.)

epichnial,
(*Phyllodocites* preserv.)

Silurian,
(*Nereites*)

N. cambrensis, Wales

Cenozoic

Mesozoic

Paleozoic

35

Plate 35
Gyrochortids

So far we have tacitly assumed that wormlike bulldozers burrow head-on in the direction of their body axes and that sediment is backfilled either in a proboscideal mantle layer and/or behind the rear end of the animal. If there had been no other information available, this model might also have been applied to *Gyrochorte*: a worm trace with terminal backfill burrowing along the bedding plane. Yet, this interpretation is incompatible with a number of strange phenomena.

Vertical Repetition. If the sand layer bearing the positve epirelief of *Gyrochorte* is not too thick, the sole face regularly shows a corresponding hypichnial groove. This might remind us of the vertical repetition in *Psammichnites* (Pl. 27), which is due to the wedging effect of the animal below; or to the vertical repetition in escape burrows of *Asteriacites* (Pl. 24). Yet the case of *Gyrochorte* is different. Here, the corresponding epi- and hyporeliefs are equally sharp and follow a similar, but not identical course: if both are superimposed in a drawing, the hypichnial expression (shaded) always wiggles less than the corresponding epirelief. Why should that be?

Imagine the tracks of a bicycle on a beach: the rear wheel will follow the excursions of the front wheel, but in a smoother curve and offset by the wheel distance. By the same token, the epichnial ridge of *Gyrochorte* must have been made by the front part, and the hypichnial groove by the rear part, of the worm. In contrast to the bicycle, however, these parts did not move at the same level; i.e. the animal bulldozed through the sediment like an inclined jigsaw, oblique to its body axis!

Backfill Structure. More information comes from the structure of the backfill, which can normally be seen only where the trace intersects a bedding plane. In *Gyrochorte comosa*, the positive epirelief looks like a maiden's braid with a median groove, while the orientation of the chevrons is not immediately clear. In the former interpretation as crustacean trackway, they would have opened backwards, but with the *offset* of the hypichnial curve as a directional criterion, a forward divergence can be established.

In conclusion, *Gyrochorte* was made by an animal moving sediment from the ventral to the dorsal side of the body. As in an oblique posture this motion implied a vertical component, the backfill became elevated relative to the surrounding bedding plane. In *G. comosa*, each segment probably corresponds to the action of one pair of parapodial shovels.

Crossing-over. As in *Psammichnites* (Pl. 27), double elevation makes epichnial intersections look as if one string had been laid on top of another. In *Gyrochorte*, however, elevation was not caused by vertical pressure, but by an oblique action of parapodia. So we should also expect a slight *backward* displacement in the hump of the later trace. This can actually be seen in plan views of the crossings.

Inclination. Unfortunately, few sandstones are as micaceous as in the Jurassic of Greenland, where three-dimensional backfill structures can be directly observed not only in *Curvolithus* (Pl. 30), but also in *Gyrochorte*. But one can estimate body inclination also from the offset of top versus sole-face wiggles, or kinks, and the thickness of the slab. It turns out that the body was inclined by about 30° in *Gyrochorte comosa* – less steeply than in *Heimdallia*, as dicussed below.

Number of Body Segments. In epichnial *Gyrochorte comosa* one commonly observes brushlike **fannings**. As they (1) are less common in the hypichnial version, (2) coincide with kinks of the trail and (3) always feather out in forward direction, it may be assumed that they correspond to a stepwise lateral swing of the animal's head end. In one of these brushings one can distinguish not only the radial impressions, but also concentric rings (**circling**). If interpreted as traces of the parapodia swung around, they tell us that at least 20 body segments were involved. As the whole body must have been at least double as long as the swinging front part, we can safely assume that the maker was a polychaete annelid.

Shape of Tail End. As may be expected in a tapering worm, the hypichnial grooves are narrower and less continuous than their epichnial counterparts. But where the hypichnial grooves terminate, in either direction, there is a strange *inversion* of relief. More exactly, a sandy tail in positive relief appears to extend into the broader groove. Rather than being the direct impression of the animals' tail, it probably traces its passage through the sand/mud interface. As actively burrowing parapodia are unlikely to have been present to the very last segment, the tail part of the body must have passively dragged behind. It thus produced a momentarily open tube of reduced diameter. At the interface, this tube collapsed asymmetrically, because the sand was more mobile than the mud below. Thus the tail trace became a positive hyporelief.

Reconstruction. The block diagram combines all these clues into a picture of the animal cutting its way through the layered sediment. Only the position of the animal's head is questionable: did it emerge at the surface and did it have tentacles for collecting food? In any case the tops of rippled storm sands were already covered by mud when the traces were inscribed, otherwise their steep or overhanging contours could not have survived.   ▶

Plate 35 · Gyrochortids   101

Gyrochortids

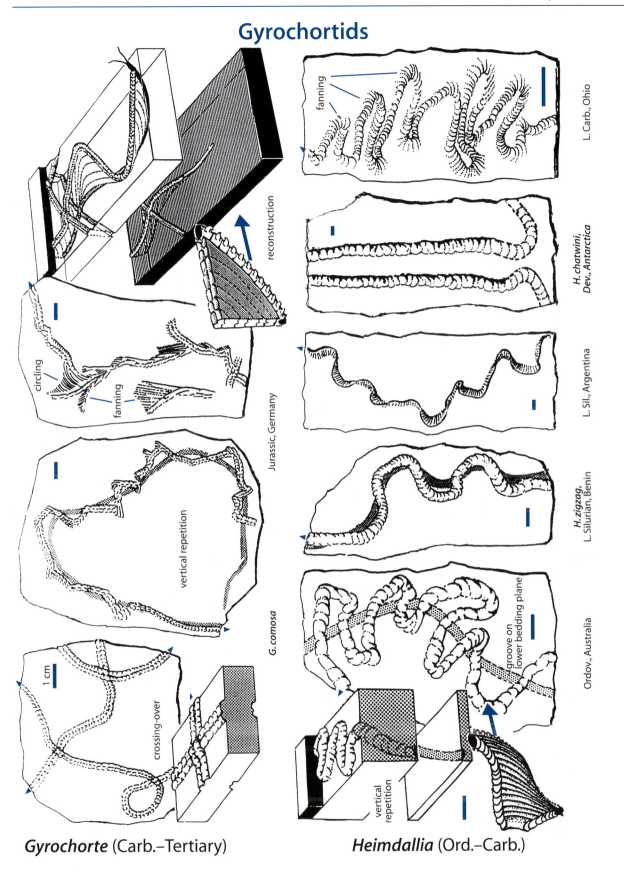

fanning

L. Carb., Ohio

reconstruction

H. chatwini,
Dev., Antarctica

circling

fanning

Jurassic, Germany

L. Sil., Argentina

vertical repetition

G. comosa

H. zigzag,
L. Silurian, Benin

1 cm

crossing-over

groove on
lower bedding plane

Ordov., Australia

vertical
repetition

Gyrochorte (Carb.–Tertiary)   *Heimdallia* (Ord.–Carb.)

There remains another open question: Why did this animal employ such an expensive mode of locomotion? Probably it selected food particles from the burrowed sediment, as other infaunal bulldozers do. Yet, if this were the case, why wasn't the animal affected by the difference between sand and mud layers, unlike the associated *Curvolithus*? And why did the maker of *Gyrochorte comosa* never evolve more complex search programs (such as guided meanders), not even when immigrating into Eocene deepsea environments?

These questions may be answered in the future. Having survived several major extinctions, the maker of *Gyrochorte* is probably still with us and most likely has already a scientific name. Will it ever be caught in the act?

Early Paleozoic Representatives (*Heimdallia*). While *Gyrochorte comosa* is not known before the Upper Carboniferous (Hartshorne Sandstone, Arkansas), somewhat similar traces are common in some shallowmarine Ordovician and Silurian sandstones. The vertical repetition of their epichnial ridges and hypichnial grooves (shaded in specimens from **Australia** and **Benin**), as well as increasing amplitudes at higher levels, reflect an identical burrowing technique. The main difference from *Gyrochorte comosa* is the lack of a braid-like segmentation. Instead, the positive epireliefs show a crescentic backfill structure that, together with the vergence of the meanders, leaves no doubt about the direction in which the animal moved. Specimens from the Georgina Basin (**Australia**) also show that the animal was more steeply inclined than in the Mesozoic forms. Accordingly there are no fannings.

Meandering is most highly developed in a single specimen from the Lower Carboniferous of **Ohio** (courtesy of Steve Stanley, Baltimore). Like the less meandering form from the Silurian Balcarce Quartzite of **Argentina**, it shows the backfill lamellae fanning out in the curves. This indicates that the animal was *laterally* inclined when turning.

A similar trace fossil from the Lower Silurian of **Benin**, West Africa, was originally described as *Gyrochorte zigzag*. Here it is referred to *Heimdallia*, coined for a closely related form from the Devonian of Antarctica.

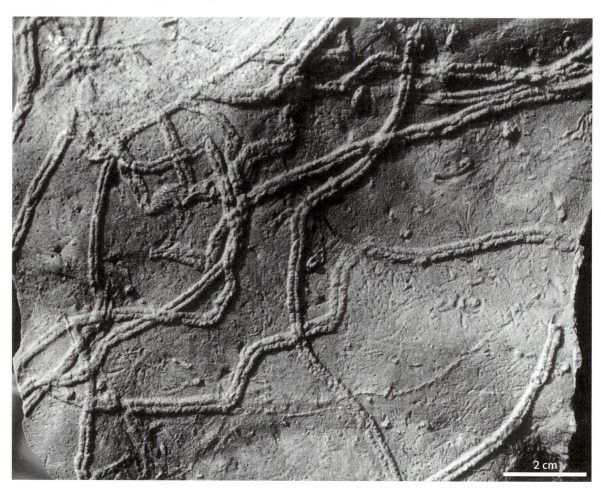

2 cm

■ *Gyrochorte comosa*. Epirelief (L. Jur., Germany)

Burrows of Stripminers

The principle of the U-tube is used by a variety of infaunal organisms, because it allows continuous ventilation. The ventilation current may also be used for filter feeding – either from the suspension imported from outside, or from dug-out sediment suspended within the burrow. We have already come across U-shaped feeding burrows made by arthropods in Pl. 18. The present chapter deals with similar spreite burrows made by wormlike organisms, but it also introduces stripmining systems, in which the generating tubes were either J-shaped or straight with dead ends.

In this connection, a general problem must be discussed. In a strictly *fabricational* sense, *Zoophycos* (Pl. 38) may well be compared with rhizocoralliid spreite-burrows (Pls. 19, 20): a U-tube is enlarged by expansive displacement and thereby produces a backfill structure. The analogy is less clear in a *functional* sense.

1. The tube of *Zoophycos* is much narrower and longer than in *Rhizocorallium* relative to the width of the spreite. This means that its ventilation costs more energy and may have been virtually impossible in very large and lobate burrow systems. More probably the owner did not flush this tube, but had to return to the sediment surface for breathing.
2. *Zoophycos* is never vertical and its spreite is never retrusive. So it never was a pure domicile and spreite construction was intentional rather than the incidental outcome of growth.
3. It is assumed to be a typical feeding burrow (fodinichnia) with the tacit assumption that food was derived from the burrowed sediment.
4. As seen in fine calcareous muds, the spreite is indeed stuffed with faecal pellets, whose elliptical shape and relatively large size fits a worm rather than a crustacean.
5. As Kotake has shown in a burrow that looks rather daedaloid, the spreite also contains material imported from above, including volcanic ash that happened to cover the sediment. His conclusion was that the animal was actually feeding at the surface and stuffed its excrements into the sediment in order to keep the dinner table clean; i.e. the trace represents, in his opinion, a kind of *sanitary burrow*.

This idea is rather compelling, particularly in quiet-water environments, where the garbage is not automatically swept away. It can also be applied to *Chondrites* (Pl. 50) and other "fucoid" burrows, which could hardly be seen without an alien component. My only objection: Why did these animals evolve such elaborate programs just to stow garbage away and why did they participate in the competitive expansion into deeper tiers (Pl. 72)?

The solution to this dilemma can be sought in two directions. (1) The faecal material was used as a substrate for bacteria that could break down otherwise indigestible organic materials. The difficulty is that neither the *Zoophycos* nor the *Chondrites* animal ever returned to the dump for a second harvest. (2) These animals had a mixed diet, with the bulk food coming from the burrowed sediment, plus a more protein-rich dessert from the sediment surface.

Literature

Plate 36: Modern U-Tubes

Bromley RG (1996) Trace fossils: Biology, taphonomy and applications, 2nd edn. Chapman & Hall, London, 361 p (General discussion of U-tubes and their ecological significance)

Frey RW, Cowles J (1969) New observations on *Tisoa*, a trace fossil from the Lincoln Creek Formation (Mid Tertiary) of Washington. The Compass 47(1):10–22

Gottis C (1953–1954) Sur un *Tisoa* très abondant dans le Numidien de Tunisie. Bull Soc Sci Nat Tunisie 7:183–192 (Very long hairpin tubes without spreite)

Häntzschel W (1939) Die Lebens-Spuren von *Corophium volutator* (Pallas) und ihre paläontologische Bedeutung. Senckenberg 21(3/4):215–227 (U-burrows and star-shaped feeding traces of modern *Corophium*)

Hertweck G, Reineck H-E (1966) Untersuchungsmethoden von Gangbauten und anderen Wühlgefügen mariner Bodentiere. Nat Mus 96(11):429–438 (Preparation of modern burrows, including those of *Notomastus*)

MacGinitie GE (1949) Natural history of marine animals. McGraw-Hill, New York (A Californian neoichnologic classic that discusses life styles of *Chaetopterus* and *Urechis*)

Mayer G (1951) *Balanoglossites eurystomus* Mägdef. und andere Lebensspuren aus dem Unteren Hauptmuschelkalk (Trochitenkalk) von Bruchsal. Jber Mitt Oberrh Geol Ver 33(1):126–132 (Triassic U-burrows)

Noda H (1987) First discovery of trace fossil *Tisoa* from the Pliocene Shinzato Formation in Okinawa-jima, Okinawa Prefecture, southwest Japan. Ann Rep Inst Geosci Univ Tsukuba (13):100–104

Rieth A (1931) Neue Beobachtungen an U-förmigen Bohrröhren aus rhätischen und oberjurassischen Schichten Schwabens. Centralbl Mineral Geol Paläontologie B (8):423–428 (*Solemyatuba*)

Schäfer W (1972) Ecology and palaeoecology of marine environments. University of Chicago Press, Chicago, 568 p (Discussion of *Arenicola, Lanice, Notomastus* and *Corophium*)

Seilacher A (1951) Der Röhrenbau von *Lanice conchilega* (Polychaeta), ein Beitrag zur Deutung fossiler Lebensspuren. Senckenberg 32:267–280 (Morphogenesis of modern terebellid U-tubes)

Seilacher A (1953) Studien zur Palichnologie. I. Über die Methoden der Palichnologie. Neues Jahrb Geol P-A 96:421–452

Seilacher A (1990) Aberrations in bivalve evolution related to photo- and chemosymbiosis. Hist Biol 3:289–311 (Discussion of *Solemya* as a producer of *Solemyatuba*)

Trusheim F (1930) Sternförmige Fährten von *Corophium*. Senck 12(4/5):254–260 (Star-shaped surface patterns)

Wells GP (1945) The mode of life of *Arenicola marina* L. J Mar Biol Assoc UK 26:170–207

Wells GP (1949a) Respiratory movements of *Arenicola marina* L.: Intermittent irrigation of the tube, and intermittent aerial respiration. J Mar Biol Assoc UK 28:447–464

Wells GP (1949b) The behaviour of *Arenicola marina* L. in sand, and the role of spontaneous activity cycles. J Mar Biol Assoc UK 28:465–478

Wells GP (1952) The proboscis apparatus of *Arenicola*. J Mar Biol Assoc UK 31:1–28

Wells GP, Albrecht EB (1951) The integration of activity cycles in the behaviour of *Arenicola marina* L. J Exp Biol 28(1):40–50

Wells GP, Dales RP (1951) Spontaneous activity patterns in animal behaviour: The irrigation of the burrow in the polychaetes *Chaetopterus variopedatus* Renier and *Nereis diversicolor* O.F. Müller. J Mar Biol Assoc UK 29:661–680

Ziegelmeier E (1952) Beobachtungen über den Röhrenbau von *Lanice conchilega* (Pallas) im Experiment und am natürlichen Standort. Helgoland Wiss Meer 4(2):107–129 (Tube construction)

Ziegelmeier E (1969) Neue Untersuchungen über die Wohnröhren-Bauweise von *Lanice conchilega* (Polychaeta, Sedentaria). Helgoland Wiss Meer 19:216–229 (Orientation of fan to currents)

Plate 37: Backfill Structures

Gaillard C, Hennebert M, Olivero D (1999) Lower Carboniferous *Zoophycos* from the Tournai area (Belgium): Environmental and ethologic significance. Geobios 32:513–524 (New backfill model)

Kotake N (1997) Ethological interpretation of the trace fossil *Zoophycos* in the Hikoroichi Formation (Lower Carboniferous), southern Kitakami Mountains, northeast Japan. Paleontol Res 1:15–28 (Description of specimens with backfill derived from surface)

Legrand R (1948) Observations à propos des *Spirophyton* du Tournaisis. Bull Soc Belge Geol Paleont Hydrol 57(2):397–406, 1 pl.

Miller MF (1991) Morphology and paleoenvironmental distribution of Paleozoic *Spirophyton* and *Zoophycos*: Implications for the *Zoophycos* ichnofacies. Palaios 6:410–425 (Evaluation of shallow marine *Spirophyton* versus *Zoophycos*)

Osgood RG (1972) The trace fossil *Zoophycos* as an indicator of water depth. B Am Paleont 62(271):5–22

Reineck H-E (1958) Wühlbau-Gefüge in Abhängigkeit von Sediment-Umlagerungen. Senck Leth 39(1/2):1–23 (Modern backfill structures)

Sarle CJ (1906) Preliminary note on the nature of *Taonurus*. Proc Rochester Acad Sci 4:211–214 (Discovery of backfill mechanism of *Zoophycos* spreite)

Seilacher A (1957) An-aktualistisches Wattenmeer? Paläont Z 31:198–207 (Backfill structures unseen in soft sediment)

Yabe H (1950) *Taonurus* from the Lower Permian of Western Hills of Taiyuan, Shansi, China. Proc Japan Acad 26(8):36–40 (*Zoophycos*)

Plate 38: Post-Paleozoic *Zoophycos*

Badve RM, Ghare MA, Kulkarni KG (1985) Ethological interpretation of ichnogenus *Zoophycos* Massalongo. Curr. Sci. 54(15):723–727

Bellotti SP, Valeri P (1978) L'influenza dell'ambiente sedimentario sull'assetto elicoidale delle strutture a *Zoophycos*. Boll Soc Geol It 97:675–685

Bischoff B (1968) *Zoophycos*, a polychaete annelid, Eocene of Greece. J Paleontol 42(6):1439–1443

Bromley RG (1991) *Zoophycos*: Strip mine, refuse dump, cache or sewage farm? Lethaia 24:460–462 (Concise evaluation of the different ethologic models for *Zoophycos*)

Bromley RG, Ekdale AA (1984) Trace fossil preservation in flint in the European chalk. J Paleontol 58(2):298–311 (Cork screw *Zoophycos*)

Ekdale AA (1977) Abyssal trace fossils in worldwide Deep Sea Drilling Project cores. In: Crimes TP, Harper JC (eds) Trace fossils 2. Geol J, Special Issue (9):163–181 (Includes *Zoophycos*)

Ekdale AA, Lewis DW (1991) The New Zealand *Zoophycos* revisited: Morphology, ethology, and paleoecology. Ichnos 1:183–194 (*Zoophycos* from the Amuri Limestone of New Zealand interpreted as a series of *Rhizocorallium* burrows)

Fuchs T (1909) Über einige neuere Arbeiten zur Aufklärung der Natur der Alectoruriden. Mitt Geol Ges Wien 2:335–350 (Discusses origin of Zoophycos and introduce term alectorurids)

Gaillard C, Olivero D (1993) Interprétation paléoécologique nouvelle de *Zoophycos* Massalongo, 1855. C R Acad Sci (II)316:823–830 (Describe Jurassic *Zoophycos* from France as formed around a vertical shaft)

Girotti O (1970) *Echinospira pauciradiata* g.n., sp.n., ichnofossil from the Serravallian-Tortonian of Ascoli Piceeno (Central Italy). Geologica Rom 9:59–62 (Lobate *Zoophycos*)

Häntzschel W (1960) Spreitenbauten (*Zoophycos* Massal.) im Septarienton Nordwest-Deutschlands. Mitt Geol Staatsinst Hamburg (29):95–100, 1 pl. (Fragments of Oligocene *Echinospira*)

Kotake N (1989) Paleoecology of the *Zoophycos* producers. Lethaia 22:327–341 (*Zoophycos* as sanitary burrow produced by surface detritus feeders)

Kotake N (1992) Deep-sea echiurans: Possible producers of *Zoophycos*. Lethaia 25:311–316 (Attribution of *Zoophycos* to echiuran worms)

Lewis DW (1970) The New Zealand *Zoophycos*. N Z J Geol Geophys 13:295–315

Löwemark L, Schäfer P (2003) Ethological implications from a detailed X-ray radiograph and ^{14}C study of the modern deep-sea *Zoophycos*. Palaeogeog Palaeoclim Palaeoecol 192:101–121

Lucas G (1938) Les *Cancellophycus* Jurassique sont des Alcyonaires. (Presenter par L. Cayeux). C R Acad Sci 200:1914–1916 (*Zoophycos* interpreted as soft coral)

Maubeuge P-L (1961) Un *Cancellophycus* remarquable du Bajocien de la province du Luxembourg Belge. Bull Soc Belge Geol Paleont Hydrol 69(3):328–333 (*Zoophycos*)

Olivero D (1994) La trace fossile *Zoophycos* dans le Jurassique du sud-est de la France: Signification paléoenvironmentale. Documents des Laboratoires de Géologie Lyon 129, 329 p (Environmental significance)

Olivero D, Gaillard C (1996) Paleoecology of Jurassic *Zoophycos* from south-eastern France. Ichnos 4:249–260 (Proposal of a deep deposit feeder origin for Jurassic *Zoophycos* from France)

Plicka M (1965) Origin of fossil '*Zoophycos*'. Nature 208(5010):579 (Comparison with sabellid worms, but in wrong geopetal orientation)

Plicka M (1968) *Zoophycos*, and a proposed classification of sabellid worms. J Paleontol 42(3):836–849, 2 pls. (Interpretation as body fossil elaborated, under the name *Spirographis*)

Plicka M (1969) Methods for the study of "*Zoophycos*" and similar fossils. N Z J Geol Geophys 12:551–573

Seilacher A (1986) Evolution of behavior as expressed in marine trace fossils. In: Nitecki MHG, Kitchell JA (eds) Evolution of animal behavior: Palaeontological and field approaches. Oxford University Press, New York, pp 62–87 (Evolutionary diversification of *Zoophycos*)

Stevens GR (1968) The Amuri fucoid. N Z J Geol Geophys 11(1):253–261, 236–264 (Detailed description of giant New Zealand *Zoophycos*. Nevertheless referred to sabellid worms)

Taylor BJ (1967) Trace fossils from the Fossil Bluff Series of Alexander Island. Brit Antarct Surv B 13:1–30 (L. Cretaceous *Zoophycos* from Antarctica studied in polished sections; extensive bibliography)

Uchman A, Demíran H (1999) A *Zoophycos* group trace fossil from Miocene flysch in southern Turkey: Evidence for U-shaped causative burrow. Ichnos 6:251–259 (New name *Echinospira* for large lobate forms)

Voigt E, Häntzschel W (1956) Die grauen Bänder in der Schreibkreide Nordwest-Deutschlands und ihre Deutung als Lebensspuren. Mitt Geol Staatsinst Hamburg (25):104–122 (*Zoophycos* in Cretaceous chalk, Germany)

Wetzel A, Werner F (1981) Morphology and ecological significance of *Zoophycos* in deep-sea sediments off NW Africa. Palaeogeog Palaeoclim Palaeoecol 32:185–212 (Modern *Zoophycos* in sediment cores)

Plate 39: Daedaloid Burrows

Antun P (1950) Sur les *Spirophyton* de l'Emsian de l'Oesling (Grand-Duché de Luxembourg). Ann Soc Geol Bel (B)73:241–262 (Analysis of Devonian "*Spirophyton*" *eifeliense*)

Jensen S (1997) Trace fossils from the Lower Cambrian Mickwitzia sandstone, south-central Sweden. Fossils and Strata 42:1–111 (Description of a Cambrian *Spirophyton*-like spreite burrow)

Plate 40: Lophocteniids

Chamberlain CK (1971) Morphology and ethology of trace fossils from the Ouachita Mountains, southeast Oklahoma. J Paleontol 45:212–246 (Description of *Lophoctenium*, Pl. 33)

Ekdale, AA, Bromley, RG (2001) A day and a night in the life of a cleft-foot clam: *Protovirgularia-Lockeia-Lophoctenium*. Lethaia 34:119–124 (Composite trace fossil referred to sediment-feeding bivalve)

Fu S (1991) Funktion, Verhalten und Einteilung fucoider und lophocteniider Lebensspuren. Courier Forschungsinstitut Senckenberg 135, pp 1–79 (Review of Lophocteniids)

Goldring R, Pollard JE, Taylor AM (1991) *Anconichnus horizontalis*: A pervasive ichnofabric-forming trace fossil in post-Paleozoic offshore siliciclastic facies. Palaios 6:250–263 (Preservational variant of *Phycosiphon* forming distinct ichnofabrics)

Miller MF (2003) Styles of behavioral complexity record by selected trace fossils. Palaeogeog Palaeoclim Palaeoecol 192:33–44 (Perfect examples of "*Spirophyton*" *eifeliense* from Devonian Catskill delta in New York state)

Pfeiffer H (1966) O nálezu Ichnofosilie *Phycosiphon* Fischer-Ooster, 1858 v roblínskych vrstách (Givet) ceského devonu (Über den Fund der Ichnofossilie *Phycosiphon* Fischer-Ooster, 1858, in den Roblíner Schichten (Givet) des Böhmischen Devons). Cas Národniho Muzea, Paleozoologie 85:135–136 (Identification questionable)

Stepanek J, Geyer G (1989) Spurenfossilien aus dem Kulm (Unterkarbon) des Frankenwaldes. Beringeria 1:1–55 (Introduction of the lophocteniid *Falcichnites*)

Wetzel A, Bromley RG (1994) *Phycosiphon incertum* revisited: *Anconichnus horizontalis* is its junior subjective synonym. J Paleontol 68:1396–1402

Plate 36
Modern U-Tubes

Sediment Feeders. We have already discussed the U-shaped burrow of **Corophium volutator** and its spreite in Pl. 19. In contrast, the tubes of the closely related **Corophium arenarium** (right block) are J-shaped without spreite. As this species inhabits cleaner sands, it can pump the waste water right into the pore space.

In connection with a particular feeding process, the same principle of ventilation is used by the lugworm **Arenicola**. Although often described as the prototype of a U-tube, the open burrow is actually J-shaped. Within the tube, the head of the animal is positioned at depth to swallow sand from above by means of its large eversible proboscis. The sand thus ingested has been enriched in organic food particles by two effects. On the one hand, ingestion at depth produces a crater at the sediment surface, in which drifting organic particles accumulate. On the other hand, the water pumped down from the tail end for respiration becomes filtered by the sand as it circulates back to the surface. Occasional displacement of the head shaft also opens new areas of unreworked sediment. In the fossil state, one would thus expect a J-tube with one or more cones of disturbed sediment on the other side, plus a horizon of rejected shell fragments at the level of the mouth.

Balanoglossus from the Adriatic Sea appears to employ a similar feeding strategy, even though it belongs to a different phylum (Hemichordata). Like *Arenicola*, it can be easily spotted by the faecal strings of digested sand that heap up on top of the tail shaft like toothpaste. The only difference seems to be that *Balanoglossus* exploits new areas by *branching* the head shaft, rather than displacing it as a whole.

Suspension Feeders. As a sand-living suspension feeder does not need to dislocate its burrow to get new food, it can afford to stabilize the tube not only by mucus, but in a more permanent way. Accordingly, all three forms shown here produce "chitinous" wall linings that look like sandpaper when washed out of the sediment. However – like textile garments – such linings cannot expand as the organism grows bigger. In order to save energy and to avoid leaving the protective sediment, all three tube dwellers shown here use the same trick: they bite a hole near the base of the previous U-tube and add a new limb of adequate size and diameter. In this way, one half of the old U-tube can still be used as a tail shaft. The other half is abandoned and decays with time, but it could still be seen in the fossil state. Thus, fossil W-burrows tell us that the maker was a suspension-feeder.

Filtering techniques differ in detail. **Chaetopterus** (an annelid) and **Urechis** (an echiurid) use mucus nets suspended across the tube and digested when full. Note that tube entrances are in both cases constricted in order to keep predators and oversized particles out (see similar constriction in *Callianassa*, Pl. 18). The terebellid worm **Lanice**, in contrast, extends the agglutinated wall into an elevated straining device. It consists of two broad lappets with radiating sand "tentacles". Being oriented to face tidal currents, this instrument (enlarged view) passively strains potential food, which the owner harvests at intervals with its tentacles. If the strainer has been smothered during a storm, a new one can be added at a higher level.

Bacteria Farmers. Like *Corophium*, *Arenicola* and *Lanice*, the thread-like worm **Notomastus** is common in the Wadden Sea. It can be easily spotted by a small heap of faecal pellets on top of the rear entrance, but in contrast to the faecal piles of *Arenicola*, those of *Notomastus* contrast with the surrounding sand by their black color. This means that food is extracted at a depth, where the sediment is black and anoxic. This view is corroborated by the corkscrew course of the U-tube's base. As in other corkscrew burrows with adequately distanced coils (Pl. 18), this part is probably used as a bacterial garden.

While *Notomastus* cultivates the bacteria probably in the burrow wall, the symbionts of **Solemya** live in the gill tissues of the animal itself. In contrast to other burrowing bivalves, it has not internalized the U-shaped ventilation (as expressed by inhaling and exhaling siphons in other burrowing bivalves). Instead it digs an open U-tube, in which the elongate shell can be moved freely by a piston-like foot. It differs from other U-tubes by having an oval cross section with the long axis in the plane of the U. Fossil tubes, **Solemyatuba**, are known in dysoxic environments from the Ordovician onward. The figured specimen from the Upper Triassic of Germany also has an oblique *appendix* tube. This tube probably reached into deeper and more anoxic zones, from where the bivalve could pump H_2S for its bacterial endosymbionts with its foot and still spend most of the time in the continuously aerated U-tube.

These modern examples will help us to understand the more complex fossil burrow systems with which the remaining part of this chapter is concerned.

Plate 36 · Modern U-Tubes   107

Modern U-tubes and Nutrition

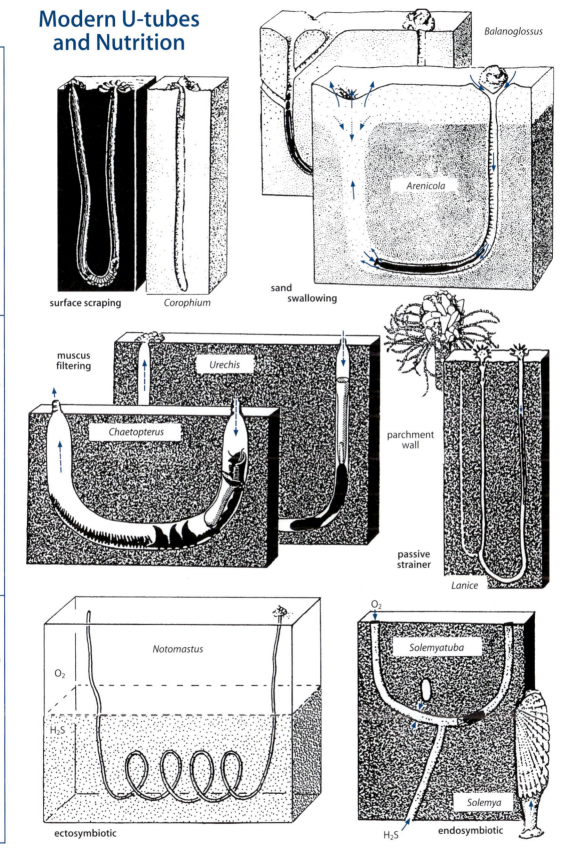

sediment feeding

surface scraping   *Corophium*

Balanoglossus

Arenicola

sand swallowing

suspension feeding

muscus filtering

Urechis

Chaetopterus

parchment wall

passive strainer

Lanice

bacteria farming

Notomastus

O₂

H₂S

ectosymbiotic

O₂

Solemyatuba

Solemya

H₂S   endosymbiotic

37

Plate 37
Backfill Structures

As shown in previous chapters, the active displacement of tunnels within the substrate – be it for growth, infaunal locomotion, or sediment feeding – results in meniscoid *backfill structures*. With regard to their geometry, one can distinguish: **terminal, marginal, radial,** and **transversal backfills.** It should be noted, however, that terminal and transversal ones are difficult to tell apart in vertical sections. Only cuts at right angles, or serial sectioning, reveal their true three-dimensional character.

Another important feature of backfill structures (except the radial ones), is their potential *geopetality.* The very fact that meniscate lamellae can be seen means that they contain varying proportions of coarser and finer material. In sediment feeders this is due to rejection of the coarser and ingestion of the finer fraction rich in edible detritus. Thus, intermittently released faecal material may form finer-grained backfill lamellae. As mud is more cohesive than sand portions, the latter will tend to gravitationally slide towards the bottom part. The result is a **geopetal** asymmetry between sand and mud wedges in cross-sectional view, which also accounts for the gill-like appearance of *Scolicia* in *Palaeobullia* preservation (Pl. 26).

In fresh sediment samples, backfill structures tend to escape attention, whereas they are greatly enhanced by diagenetic processes during fossilization. This is another reason (in addition to the low fossilization potential of surface traces) why ichnological investigations in modern tidal flats have yielded very few direct counterparts to known trace fossils. None of the backfill burrows discussed in the following chapters would be recognized only from the shapes of the generating tubes or their resin casts (black in examples shown). This does not mean that actualistic observations are useless, but instead of providing the direct keys to the past, they can tell us how the locks work.

Paleozoic *Zoophycos*. *Zoophycos* is important not only because of its long time range (Ordovician to Recent) and local abundance, but also because of its environmental significance. In principle, it is a spreite burrow comparable to *Rhizocorallium* and *Diplocraterion* (Pl. 19). In contrast to rhizocoralliid burrows, however, the marginal tube describes a widely rounded rather than a hairpin curve (except in the tonguelike lobes of derived forms). The generating tube is also much narrower relative to the width of the spreite. This diagnostic difference between *rhizocoralliid* and *alectorurid* spreite burrows (a term referring to the similarity with a cock's tail and the junior synonym *Alectorurus*) reflects a difference in behavioral monitoring. The *Zoophycos* animal does not need a body bend to start spreite construction (Pl. 19); new lobes may arise from a straight, or even a concave, stretch of the marginal tunnel. Spreite construction has also a tangen-

tial component. In spiral forms, the resulting second-order cross lamination may run towards the margin (*centrifugal*) or towards the center (*centripetal*). Alternatively, cross lamination may be *ambivalent*, suggesting that the animal turned around in its tube. This technique allows for more versatility in utilizing a given surface. At the same time, alectorurid burrows are ill-suited for mere dwelling, because they are not vertical and hard to ventilate. Therefore, *Zoophycos* has traditionally been considered a feeding burrow. This is in accordance with field evidence. True *Zoophycos* is never found in clean, ripple-laminated event layers. Rather it prefers impure and structureless sands and muds corresponding to quiet-water paleoenvironments, where the sediment became readily mixed by bioturbation. As such conditions prevail on the deep shelf below the reach of storm waves and above the zone of turbidity currents, *Zoophycos* may, to a certain extent, be used as a depth indicator (Pl. 71). But this should be done with care, because protected lagoonal environments may provide similar conditions at moderate depth. Derived forms of *Zoophycos* have also successfully invaded the deepsea floor, where they survive to the present day. Nevertheless, no core has ever captured the tracemaker.

The **Paleozoic** history of the group is somewhat confused by the presence of other spreite burrows. "*Spirophyton*" *eifeliense* (Pl. 39), for instance, superficially resembles *Zoophycos*, but it has an unlobed edge without marginal tube and is therefore assigned to the daedaloid burrows.

Also confusing is the winged *Zoophycos* from the Devonian of Southwest **Libya** (Murzuq Basin). First, because it appears in thinly laminated silts, together with plant debris indicating a near-shore facies. Second, each arcuate strip consists of very fine centrifugal cross laminae. Thus spreite construction proceeded away from the center as a sequence of bites and oblique backfills made by a defined head region. Third, the marginal tunnel is *discontinuous*. Rather than contouring the whole spreite, it rhythmically jets-out to initiate a new "wing". This **alate** ichnospecies can be interpreted as an alectorurid burrow system, in which the maker avoids the marginal tunnel to become overly long and difficult to flush by making new entrances.

In Southeast Libya (Kufra Basin), Devonian *Zoophycos* burrows are never winged, but describe a continuous **spiral.** Yet, they show the same centrifugal cross lamination of the spreite. Associated other burrows (*Lennea schmidti*, Pl. 45; "*Spirophyton*" *eifeliense*, Pl. 39) suggest a Middle Devonian age. Notably, the environment was poorly oxygenated, but still reached by occasional storms.

Ambivalent *Zoophycos* burrows with a continuous marginal tube appear in impure and highly bioturbated ("churned") sandstones from the Ordovician on (Cambrian occurrences are questionable; Pl. 39). Shown here is a specimen from the Upper Carboniferous of Austria. Morphologies of this type are little different from those in the shallowmarine lineage of the next plate.

Plate 37 · Backfill Structures   109

Backfill Structures

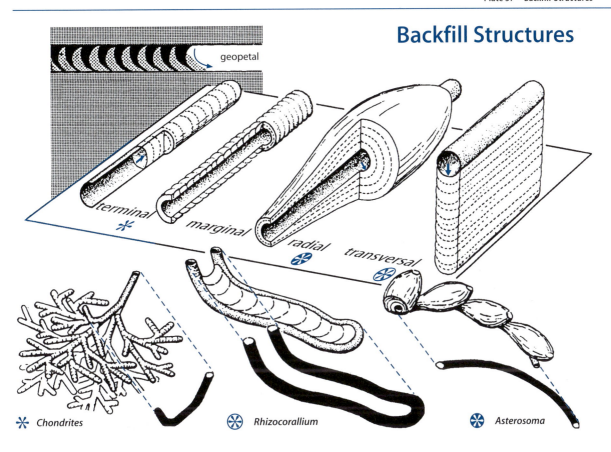

geopetal

terminal *

marginal

radial ✸

transversal ✸

✳ *Chondrites*   ✸ *Rhizocorallium*   ✸ *Asterosoma*

Paleozoic *Zoophycos* (shallow marine)

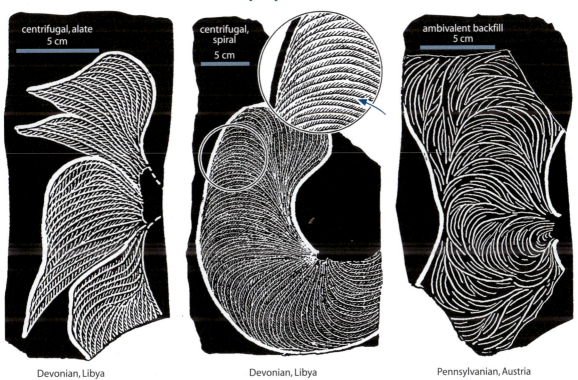

centrifugal, alate
5 cm

centrifugal,
spiral
5 cm

ambivalent backfill
5 cm

Devonian, Libya   Devonian, Libya   Pennsylvanian, Austria

38

Plate 38
Post-Paleozoic *Zoophycos*

After the Paleozoic, the shallowmarine (**shelf**) lineage of *Zoophycos* continues into the Triassic and Jurassic with little change in shapes and habitats (**b**; **c**). Only in the Upper Cretaceous chalks do they develop multistory structures whose preservation is sometimes enhanced by selective chertification. To my knowledge, no *Zoophycos* has ever been found in shallowmarine deposits of Cenozoic age.

More complex burrowing programs developed in connection with an onshore/offshore shift. In the relatively deep "Grès à *Cancellophycus*" (the name is another junior synonym of *Zoophycos*) of the Middle Jurassic in southern France (**d**) there is an abundance of burrows. They differ from coeval forms in shallower facies of southern Germany (**c**) by that the **sigmoidal** (asterisk) secondary lamination resembles the Libyan forms (Pl. 37), but now with a systematically **centripetal** sequence. There is also a strong tendency for the whole spreite to proceed in a flat spiral, whose later turn passes on *top* of the earlier one. As such a spiral might eventually intersect the outer limb of a U-shaped marginal tube, French workers suggested that the spiral spreite originated by midway extension of a dead-end vertical tube. If one gives up the idea of continuous ventilation, this argument is irrelevant.

The real conquest of deepsea floors appears to have happened in the Cretaceous. In the corresponding **flysch** facies, *Zoophycos* appears in two versions. One group penetrates the sediment in multi-storied spirals, with the slightly conical turns often proceeding in *lobate* rhythms at lower levels (**e**). In another form found together with *Scolicia zumayensis* (Pl. 26d), such narrow lobes grew into large arrays resembling staghorns. In the figured specimen (**f**) all lobes turn counterclockwise; but as new lobes originate always on the convex sides (as in lobate *Rhizocorallium*; Pl. 19), the whole system developed in a clockwise direction. Each lobe is also inclined, so that tips pass underneath other lobes. Before compaction, these must have been spectacular three-dimensional underground edifices.

In a calcareous **Eocene** flysch near Florence, Italy (**h**) still larger spirals start in a *lobate* fashion, while mining proceeds *sigmoidally* along previous stretches in later parts (**a**). As a result, the marginal tube is still lobate, but shorter than it would have been with separated lobes.

The famous *Zoophycos* from the **Oligocene** Amuri Limestone of New Zealand (**i**) is impressive because of its size. While so far we dealt with structures the size of saucers or dinner plates, this form is better compared to a large umbrella. This seems to contradict the rule that animals tend to miniaturize in deepsea environments. But this rule applies mainly to body sizes in the preturbidite community; burrow size may have as much to do with *lifespan* as with body size.

In the Amuri *Zoophycos*, burrowing proceeded in the sigmoidal mode from the beginning. Nevertheless there is a two-stage program. The *starter program* is expressed by a steep *Rhizocorallium*-like spreite, in which the secondary cross lamellae are nevertheless centrifugal. After having reached a certain depth, the animal switched instantaneously to the *spiral* program, with the proximal side (relative to cross-lamellar progression) of the starter U-tube serving as a baseline. In the following spiral part, each radial field was built by sigmoidal backfill lamellae prograding from the center to the margin. Another program change is expressed by individual projecting lobes that are made after completion of a certain rotational angle. These lobes not only lack the backward curvature of earlier forms (**f**, **h**); they also had another function: besides securing the real estate against competitors, they reset the *radius* of the fields, which continuously shrank due to the obliquity of the sigmoidal bends. After having completed one turn, the spreite arrived at a higher level than the starter spreite, but seems rarely to continue further.

The Miocene form from Turkey (**k**) extended all sigmoidal fields into long lobes. They made the marginal tube virtually unflushable, so that the animal had to go to the entrance for breathing. Its name, **Echinospira**, is here considered as an ichnosubgenus.

Different fabricational models have been proposed to explain these complex burrow systems. One of them interprets the spreite of the Amuri *Zoophycos* as a series of radial U-burrows, of which only the last was functional at any one time. This model is falsified by the cross-sectional structure of the spreite, which confirms a sigmoidal, rather than lobate, progression of the radial sectors. It has also been suggested that very different kinds of animals were responsible for different ichnospecies. But even if *Zoophycos* is a coherent group, we should not forget the functional problems addressed in the introduction.

At the moment, there is no simple explanation for all the intricacies of *Zoophycos* and its evolutionary modifications. The problem will probably stay with us for a while.

Plate 38 · Post-Paleozoic *Zoophycos*   111

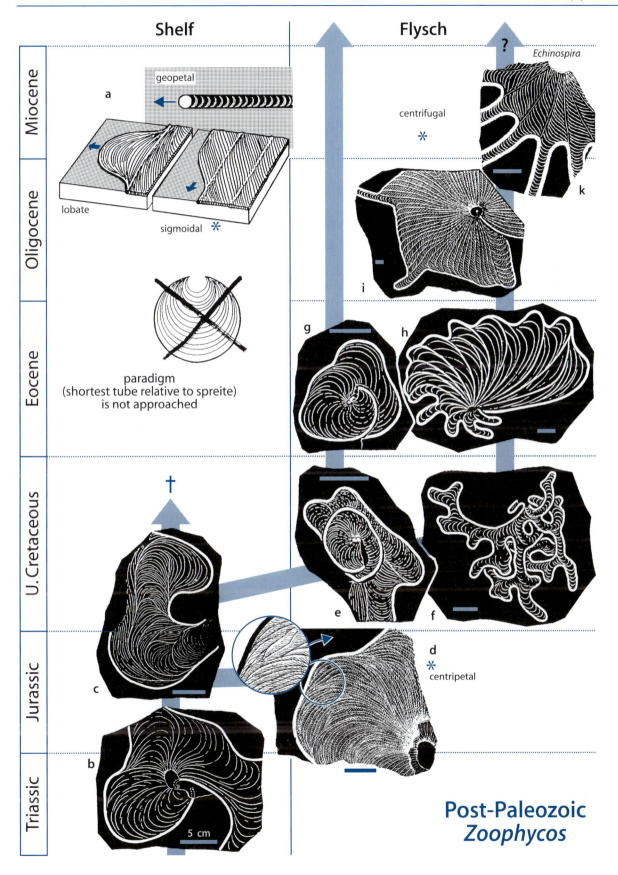

Shelf

Flysch

Miocene

Oligocene

Eocene

U.Cretaceous

Jurassic

Triassic

geopetal

a

lobate

sigmoidal *

paradigm
(shortest tube relative to spreite)
is not approached

†

c

b

5 cm

d
*
centripetal

e

f

g

h

i

*
centrifugal

Echinospira

k

?

Post-Paleozoic
Zoophycos

39

Plate 39
Daedaloid Burrows

Vertical strip mining is not restricted to U-tubes; it can also be executed by the lower end of a J-shaped tunnel. By turning around the vertical shaft, this part may develop into a spiral spreite whose three-dimensional shape depends on the geometry of the tube.

The term "daedaloid" refers to *Daedalus*, in which this burrowing technique is most typically expressed. The fact that this ichnogenus is discussed in the next chapter (Pl. 44) illustrates the real dilemma of trace fossil classification: in a fabricational sense, *Daedalus* should be grouped here, but by its fingerprints it belongs to the arthrophycids, in which a variety of mining techniques has been employed.

"Spirophyton" eifeliense. Formally, the name *Spirophyton* is no longer available, because it was originally created for spiral variants of *Zoophycos* from the Upper Devonian of New York. "*Spirophyton*" *eifeliense* from the Lower and Middle Devonian of Libya, Germany and Luxembourg would much better fit this name, because it looks like a pinwheel firework in plan view and has many spiral turns widening towards the base. However, two features distinguish this form from similarly high-spired *Rhizocorallium* and *Zoophycos* (Pls. 19, 20 and 38).

(1) The spiral spreite is gutter-shaped (a model could be used to let marbles roll down) and its margin is neither lobate nor has it a marginal tunnel. Together with a protrusive structure, this geometry suggests that a dead-end tube in the shape of a skin-diver's snorkel rotated around the vertical shaft, growing in size on the way down. (2) The diameter of the central shaft exceeds the thickness of the spreite, probably because it served for ventilation and housed the main part of the animal's body. With up to about 20 spiral turns growing in diameter, each system probably represents the work of a lifetime. (3) Like the coeval Libyan *Zoophycos* (Pl. 37), these burrows occur in rather shallowmarine sandstones and are much smaller than *Zoophycos*. The same is true for spiral spreite burrows from the *Mickwitzia Sandstone* (Lower Cambrian) of Sweden. Being still smaller than "*Spirophyton*" *eifeliense*, they differ from it by having a marginal tunnel and were therefore compared with *Zoophycos* in Sören Jensen's marvelous monograph. On the other hand, the spreite lamellae are very delicate and there is no sign of a vertical shaft in the center. Therefore these burrows were more probably generated by a J-tunnel, but one in which the curved part was horizon-tally deflected, so that it could strip-mine the interface with the underlying mud. As the coil became at the same time wider, it resembles the outline of an involute ammonite shell.

The spiral spreite burrows from the Miocene of Japan, in which Kotake found faecal pellets derived from an ash layer on top may also belong into this group. In contrast to *Zoophycos* they have no marginal tube, while the diameter of the central shaft by far exceeds the thickness of the spreite. The sideritized burrow from a Miocene flysch of Borneo corresponds in principle to *S. eifeliense*, only that the generating tube was relatively wider. It was found together with a teichichnoid shrimp burrow (Pl. 18) and *Cycloichnus* (Pl. 48).

A very strange spreite burrow was found by H. Hölder (pers. comm.) in Upper Jurassic pelagic limestones of southern France. From outside it looks like a giant high-spired gastropod shell, but sections show the typical meniscate backfill structure. Had this burrow been found in shallowmarine sandstones of the Lower Ordovician, it would have been called *Daedalus desglandei* (Pl. 44), but in view of the difference in age and facies, the makers of the two were certainly unrelated. So the question is again whether they should be grouped in the same ichnogenus and thereby confuse stratigraphers?

The diagram compares the fabricational modes of *spreite formation*. More modifications are represented among Ordovician and Silurian arthrophycids (Chap. IX).

For lack of a better place, a very distinctive trace fossil, *Paradictyodora*, is added here. It has been independently discovered (and named) in the Cretaceous of Antarctica and in the Tertiary of Southern Italy. Had specimens not been found *in situ*, one might have oriented them upside-down in the manner of daedaloids. The surface of the figured concretion also shows that the tube generating this spreite moved in meanders rather than a spiral. In a fabricational sense, this structure might be compared to *Dictyodora* (Pls. 28, 29), only that the body of the animal at depth remained stationary, while the snorkel on top made the meandering excursions. The idea that the tracemaker was a burrowing bivalve is convincing, because smaller modern tellinids (e.g., *Macoma*) in fact swing their inhaling siphons around to scan the sediment surface for small food particles. It would be nice to find a shell of the producer in connection with *Paradictyodora*. A detailed study of modern representatives could be even more interesting, because it might tell us how long snorkels managed to cut through the sediment in other molluscan trace fossils, such as *Psammichnites gigas* (Pl. 27) and *Dictyodora* (Pl. 28).

Plate 39 · Daedaloid Burrows   113

Daedaloids

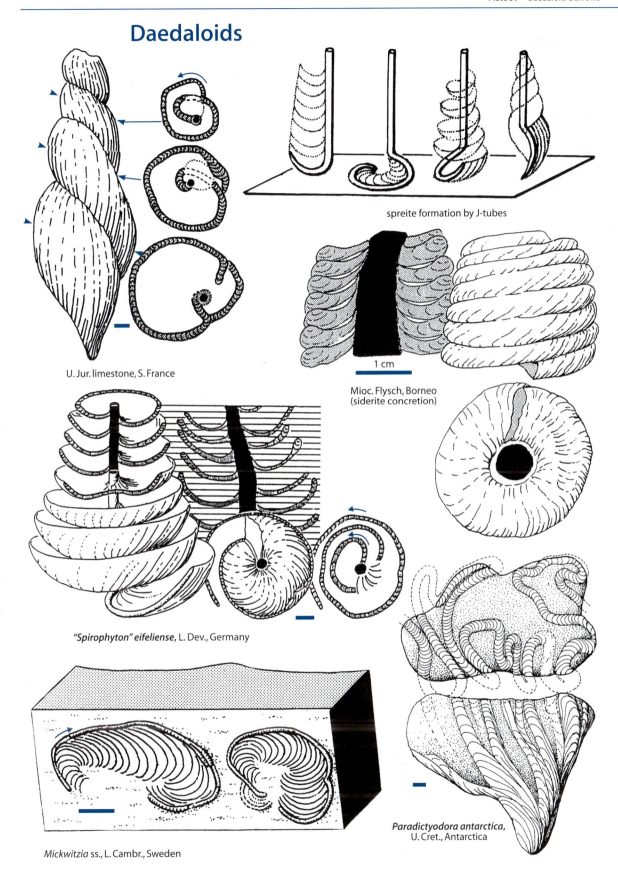

U. Jur. limestone, S. France

spreite formation by J-tubes

1 cm

Mioc. Flysch, Borneo
(siderite concretion)

"Spirophyton" eifeliense, L. Dev., Germany

Mickwitzia ss., L. Cambr., Sweden

Paradictyodora antarctica,
U. Cret., Antarctica

40

Plate 40
Lophocteniids

Whereas the trace fossils shown on the previous plate share a certain burrowing technique rather than kinship, the lophocteniids (as here defined) are a more homogeneous group, characterized by a very distinctive kind of horizontal stripmining. So far we have only considered a generating tube that opened to the sediment surface at one or at both ends. Accordingly the backfill was generally transversal, even though backstuffing oblique to the body axis has been involved in the sigmoidal subprogram of some *Zoophycos* (Pls. 37, 38).

Lophoctenium and *Phycosiphon* have a clearly offset marginal string surrounding the strip-mined area. Still it would be misleading to call it a "generating tube", because its course is an outcome of the burrowing activity rather than its determinant. It is not even likely to have been used for ventilation and may just as well represent a terminal backfill behind the tail. In fact, cases in which earlier parts of the same marginal tunnel are intersected by later ones (**B**) speak against an open connection with the surface. Nevertheless, comparison with structures and behavior patterns in alectorurids is fruitful by way of analogy.

Lophocteniids may also be used as a counterexample for the celebrated onshore-offshore trend in ecological evolution. If we believe the fossil record and the kinship here proposed, the first *Lophoctenium* occurs in an Ordovician flysch of Portugal (where it is associated with *Oldhamia*; Pl. 49) and the first *Phycosiphon* comes from a similar turbidite facies in the Devonian. Yet it is only in the Jurassic that *Phycosiphon* appears also in shallow-marine sands and silts.

Phycosiphon. The most obvious feature of *Phycosiphon*, whether preserved as epirelief or on split bedding surfaces, is its marginal tube. Being hairpin-shaped, individual lobes look like minute *Rhizocorallium*, while larger arrays may resemble meandering burrows (**A**). The difference becomes clear when the spreite is preserved. Its lamellae do not fully contour the lobe, but are asymptotic to the marginal tunnel on the concave side of the lobe and discordant on the other (**B**). Sometimes (**C**) the spreite lamellae even protrude *beyond* the marginal burrow on the convex side of the lobe. Thus the spreite was formed by protrusion of a wormlike animal bent like a "J". By this very process the lobes always come to turn backwards. At the distal end of each lobe, spreite construction was discontinued and the animal returned head-on in the opposite direction. While doing so, it more or less contoured the spreite just made and then turned to the side to start a new lobe. Spreite construction may also proceed in series without turnbacks, as shown by the specimen from Kentucky.

In more complex systems (**A, B**), all spreiten are still made on the same side of the marginal burrow (either right or left) and eventually bend back towards the origin. At the tips, the spreite lobes swing around without changing their backward curvature.

By learning this program ourselves, we can produce virtual *Phycosiphon* systems on paper. This exercise makes us aware of another necessary instruction: after the terminal swing-around, the distance from the previous tunnel must be sensed without direct guidance by a free spreite edge. Probably it was the same sensor that controlled distancing of the return tube in every lobe. The *length* of individual spreite lobes seems to be limited as well. Had this anything to do with body *length*; i.e. can these lobes be considered as *phantoms* of a wormlike animal?

With regard to the function of the *marginal tube*, several details suggest that it became actively backfilled rather than remaining open. (1) It is filled with finer material than the surrounding sediment– similar to the faecal string in associated *Helmithoida* burrows (Pl. 34). (2) The accidental intersection with an earlier part of the tube (asterisk in **B**) would have obstructed connection with the surface. (3) When cutting across fine silt/mud laminae (*Anconichnus* preservation in "paper shales"), the two tubes have opposite reliefs, resembling the backfill phantoms discussed in Pl. 28.

In summary, *Phycosiphon* is perceived as the work of a small wormlike bulldozer that moved more or less horizontally through the sediment without an obvious connection to the sediment surface. Being able to periodically discontinue spreite production with the laterally bent front end, it could strip-mine a given surface with some efficiency, but not without leaving areas unused between the radiating spreite lobes.

Lophoctenium. In a strictly morphological categorization, *Phycosiphon* and *Lophoctenium* would certainly fall into different groups: an open meshwork of spreite lobes in the first and gapless spreite mosaics in the second ichnogenus. In a fabricational sense, however, the basic procedure appears to be the same in both cases: discontinuous stripmining by a head end turned to one side.

The morphological difference between the two ichnogenera is probably due to a minor behavioral change. The *Phycosiphon* animal made its bites and the bite series freely and was controlled by a kind of internal clock. It then returned to the base *head-on* in a sharp turn before beginning a new bite series. The *Lophoctenium* maker, in contrast, was guided by contact with previous elements at every step and returned to the base by *backing up* rather than moving on in a hairpin curve, before starting a new bite series. As in meanders, increasingly external guidance allows for more effective utilization of an available area, but it also results in greater variability. On the other hand, the method used by *Lophoctenium* allowed for higher orders of behavioral complexity that allow us distinguish various ichnospecies. These are as yet unnamed, ▶

Plate 40 · Lophocteniids   115

Lophocteniids

Phycosiphon

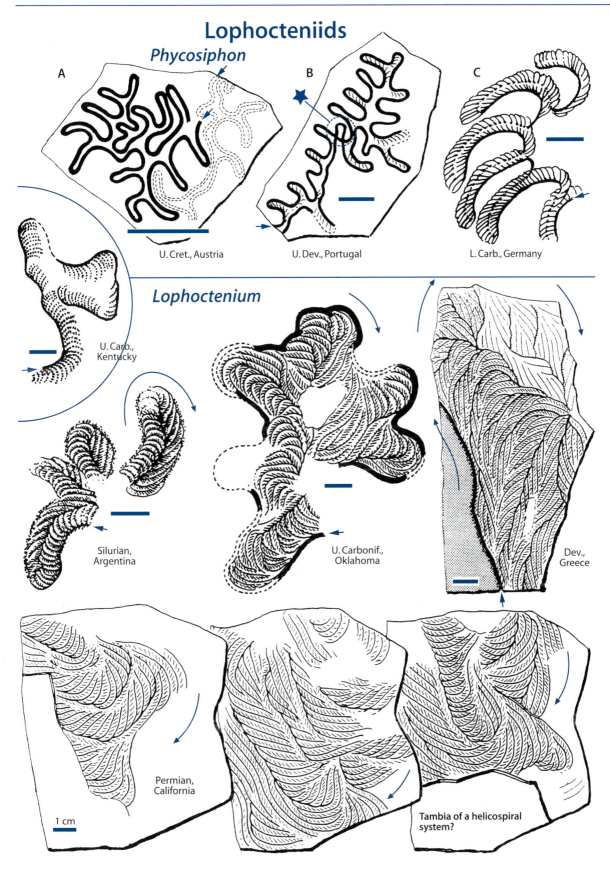

A

U. Cret., Austria

B

U. Dev., Portugal

C

L. Carb., Germany

Lophoctenium

U. Carb., Kentucky

Silurian, Argentina

U. Carbonif., Oklahoma

Dev., Greece

Permian, California

1 cm

Tambia of a helicospiral system?

Lophoctenium comosum being the most representative ichnospecies name. Better distinction may also allow ichnostratigraphic application.

Before discussing the figured examples, we also need to understand the *preservation* of *Lophoctenium*. It normally occurs on turbidite soles in the form of positive hyporeliefs, but it can also break off, leaving a negative hyporelief of the spreite's upper surface. This clearly shows that *Lophoctenium* is a member of the *post-turbidite community* (Pl. 72). It may also be found on split laminae within a turbidite. This allows to compare different levels of the same system, as in the **Permian** example. An unusual kind of preservation is on top of sandy turbidites (Oklahoma example). In this case, analysis is complicated by the geopetal asymmetry of the backfill (Pl. 37) and by the tendency of our eye to overemphasize protruding structures. To overcome the latter problem, it is useful to analyze rubber casts instead of the original specimens.

Let us begin programs with the small form from the Upper Carboniferous of Kentucky. Like *Phycosiphon*, it consists of distinct bite series, but these are serially arranged and there is no marginal tube on the convex sides of the spreite lobes.

In a similarly small **Silurian** *Lophoctenium* from Argentina, spreite construction continued after completion of one series, so that each lobe contains two spreiten protruding in opposite directions. In the return stack of each lobe, probes tend to be longer and less curbed. This, however is an effect of the new reference frame: instead of stopping after having reached a certain length (as in *Phycosiphon* and on the way out), bite series continue until they collide with the outgoing stack. The behavioral switch was induced by the turn at the end of each lobe. As can be seen in the first and second lobe, transition to the next lobe is continuous and automatic: as soon as the contact with the previous outgoing stack is lost, the length of the bite series becomes again controlled by an internal command. Because of the larger size and perfect preservation, one can also recognize oblique bitelets within the arcuate probes, which cannot be seen in *Phycosiphon*. They attest for a centripetal construction (see *Zoophycos*, Pl. 38).

In the next specimen (top surface from an Upper Carboniferous flysch in the Ouachita Mountains, Oklahoma), the program is very similar to the one just discussed, but only as far as the outgoing stacks are concerned. The difference lies in the return stacks, where contact is not maintained by collision, but by *contouring*. This allows for larger spreite areas, greater flexibility and a more effective utilization of the reentrants between lobes. Also the system becomes much larger and probably continued spirally into a lower level.

The fourth example, based on a single fragment found by Rudolf Höll in a Devonian flysch on Chios, Greece is still more complex. It has no radiating lobes. Instead, subsequent bite series form a mosaic of leaflike arrays. Using the same terminology as in the lobate forms, already the outgoing stack (here actually ingoing) keeps contact with preceding leaf. This can alternatively be done by collision or by contouring. But for making the return stack, the animal did not turn around in a sharp curve, as in the lobed forms. Rather it continued to contour the previous probes, but withdrew the body step by step before adding a new bite series in the "regressive" sequence. This means that the program controlled not only the lengths of the leaves, but also the length of the bite series in the return stack. It is a tempting thought that the basic measure stick was the animals's own body. If this is the case, body length was at least equal to maximum leaf length, because the tail region would probably have remained in the part that eventually became the marginal tube (or the toilet) to the whole system.

Having integrated the leaves, the maker of the Greek *Lophoctenium* could proceed by employing the same strategy (that produced lobes at a lower level of the behavioral hierarchy) in the Ouachita form: as shown by the opposing leaves (where bite lamination has been left out to be completed by you!), the animal turned around to add a return stack of leaves from the other side. It is left to our imagination, how this complexity went on in more complete systems. Perhaps the superlobes were themselves elements of spiral arrays, which extended into deeper levels? The answer can only be found in Chios.

Our last example, collected by the late Simon Muller, comes from a Permian flysch in an exotic terrane of California. It proves the existence of multi-floor *Lophoctenium* spirals, because the only available specimen could be split at several levels. On all four bedding planes (of which only three are shown) there are *Lophoctenium* systems spiraling in the same sense and around more or less corresponding centers. So they are probably parts of the same edifice. Even more interesting in the present context is the program. Following an evolutionary prejudice, we might have expected an increase in behavioral complexity compared to the Devonian form, but instead there is an apparent simplification. In most of the leaves, the outgoing stack is reduced to one contouring bite series of indefinite length. It provided a base for the tremendously enlarged regressive stack, in which uniform lengths of the series guaranteed an unusually smooth reference front for the next leaf.

If we trust the relationship between leaf and body lengths, the maker of the Permian form was very long. It may also be considered as more advanced, because the reduced rigor of its behavioral program (there is no clear marginal tube to be seen) allowed it to effectively cover large areas and to expand its strip mining to deeper bedding planes in a continuous spiral.

To our knowledge, *Lophoctenium* did not survive the end-Permian extinction. Its last appearance is on tectonized turbidite soles near the southern tip of New Zealand. What has formerly been quoted as a later representative (*Criophycus*, Pl. 48) is a homeomorph with a different program and different preservation.

Arthrophycid Burrows

Arthrophycids, like lophocteniids, are a kind of taxonomic island in a sea of homeless worm burrows. In their case, however, the principal shared character is not feeding behavior, but a distinctive kind of *fingerprints*. They consist of evenly spaced transverse rings that are discontinuous and thereby produce longitudinal rows of prominent knobs. Two such rows can be seen on the exposed side of each tunnel cast, but two more rows must be assumed on the other side to account for the squarish cross section. In addition to these knobs, there is a much more delicate kind of ornamentation (Pl. 42, top right) that can only be seen in large and exceptionally well preserved burrows. It consists of a very fine transverse lineation reminiscent of wrinkles in a flexible cuticle. So we may imagine the maker as a very large wormlike animal (burrows reach the diameter of a finger, but the transverse backfill provides no phantoms of body length) with a flexible cuticle and parapodial protuberances. No matter how different the burrow systems may look from one ichnospecies to the other, these fingerprints signal a taxonomically coherent group of burrowers, which nevertheless cannot be affiliated to an established animal order, class, or phylum.

The second characteristic of arthrophycid burrows is their tendency to produce transversal *backfill structures* (Pl. 37), but in contrast to the burrows treated in the previous chapter, the backfill consists only of sand and proceeds mostly in a vertical plane.

Thirdly, the tube generating the spreite bodies ended *blindly*, even if shapes are reminiscent of rhizocoralliid U-burrows (Pls. 19, 20).

Such taxonomic coherence is the prerequisite for the biostratigraphic use of trace fossils, because it reduces the noise caused by behavioral convergence in unrelated groups of animals and brings us closer to an evolutionary measuring stick. In fact, arthrophycids have been successfully used to correlate otherwise unfossiliferous sandstones in now separated fragments of the old Gondwana continent. As yet, arthrophycids provide a lower stratigraphic resolution than associated trilobite burrows (Chap. XIV). On the other hand, the larvae of their makers seem to have had less difficulty in crossing oceanic barriers. So there is a good chance that a scheme established in Gondwana can also be used in shallowmarine sandstones on other paleocontinents, such as Laurentia and *vice versa*. In the following plates we shall first discuss unrelated teichichnoid burrows of various ages and then proceed to true Ordovician and Silurian arthrophycids.

Literature

Chapter IX

Mángano MG, Carmona NB, Buatois LA, Muñiz F (2005) A new ichnospecies of *Arthrophycus* from the Upper Cambrian-Tremadocian of northwest Argentina: Implications for the arthrophycid lineage and potential in ichnostratigraphy. Ichnos 12:179–190 (First description of *Arthrophycus minimus,* probably belonging to *Phycodes*)

Neto de Carvalho C, Fernandez ACS, Borghi L (2003) Diferenciação das icnoéspecies e variantes de *Arthrophycus* e sua utilização problemática em icnoestratigrafia: O Resultado de Homoplasias comportamentais entre anelídeos e artrópodes. Rev Esp Paleontol 18:221–228 (Problems of arthrophycid ichnostratigraphy)

Seilacher A (2000) Ordovician and Silurian arthrophycid ichnostratigraphy. In: Sola MA, Worsley D (eds) Geological exploration in Murzuk Basin. Elsevier, Amsterdam, pp 237–258 (The fingerprint of *Arthrophycus* is compared to those of *Daedalus* and *Phycodes* and their stratigraphic and paleogeographic utility explored)

Plate 41: Teichichnid Burrows

Baldwin CT (1977) The stratigraphy and facies associations of trace fossils in some Cambrian and Ordovician rocks of northwestern Spain. In: Crimes TP, Harper JC (eds) Trace fossils 2. Geol J, Special Issue 9, pp 9–41 (*Teichichnus stellatus*)

Bland BH, Goldring R (1995) *Teichichnus* Seilacher 1955 and other trace fossils (Cambrian?) from the Charnian of central England. Neues Jahrb Geol P-A 195:5–23 (Cambrian *Teichichnus* in gravestones)

Chisholm JI (1970) *Teichichnus* and related trace-fossils in the Lower Carboniferous at St. Monance, Scotland. B Geol Surv GB (32):21–51

Frey RW, Bromley RG (1984) Ichnology of American chalks: The Selma Group (Upper Cretaceous), western Alabama. Can J Earth Sci 22:801–828 (Diagnosis of *Teichichnus zigzag*)

Jensen S (1997) Trace fossils from the Lower Cambrian Mickwitzia sandstone, south-central Sweden. Fossils and Strata 42:1–111 (Excellent description of *Syringomorpha nilssoni*)

Jensen S, Bergström J (1995) The trace fossil *Fucoides circinatus* Brongniart, 1828, from its type area, Västergotland, Sweden. Geol Foren Stock For 117:207–210 (The authors showed that *F. circinatus* is a vertically organized spreite burrow)

Mángano MG, Buatois LA (2004) Reconstructing early Phanerozoic intertidal ecosystems: Ichnology of the Cambrian Campanario Formation in northwest Argentina. In: Webby BD, Mángano MG, Buatois LA (eds) Trace fossils in evolutionary palaeoecology. Fossils and Strata 51:17–38 (*Syringomorpha* and its associated ichnofabrics)

Seilacher A (1955) Spuren und Fazies im Unterkambrium. In: Schindewolf O, Seilacher A (eds) Beiträge zur Kenntnis des Kambriums in der Salt Range (Pakistan). Akademie der Wissenschaften und der Literatur, Mainz, Abhandlungen der mathematisch-naturwissenschaftlichen Klasse, 10, pp 261–446 (Introduces *Teichichnus*)

Seilacher A (1957) An-aktualistisches Wattenmeer? Paläont Z 31:198–207 (Compares Cambrian *Teichichnus* with modern *Nereis* burrows)

Seilacher A, Hemleben C (1966) Beiträge zur Sedimentation und Fossilführung des Hunsrückschiefers, Teil 14, Spurenfauna und Bildungstiefe des Hunsrückschiefers. Hessisches Landesamt für Bodenforschung, Notizblatt 94, pp 40–53 (Description of *Heliochone hunsrueckiana.* See plate 8 for illustration)

Seilacher A, Buatois LA, Mángano MG (2005) Trace fossils in the Ediacaran-Cambrian transition: Behavioral diversification, ecological turnover and environmental shift. Palaeogeog Palaeoclim Palaeoecol 227:323–356 (Description of a Cambrian trace fossil similar to *Heliochone*)

Plate 42: Arthrophycids I

Desio A (1940) Vestigia problematiche paleozoiche della Libia. Pubblicazioni dell'Istituto di Geologia, Paleontologia e Geografia Fisica della R. Università di Milano, Serie P, Pubbli-cazione 2, 47–92 (Silurian *Arthrophycus alleghaniensis* described as *Harlania*)

Hall J (1852) Palaeontology of New York, 2. Containing descriptions of the lower Middle Devonian of the New York System (equivalent in part to the Middle Silurian rocks of Europe) Natural History of New York. C. van Benthuysen, Albany, 363 p (*Arthrophycus alleghaniensis*)

Harlan R (1831) Description of an extinct species of fossil vegetable, of the family *Fucoides.* J Acad Nat Sci Phila 1(6):289–295 (Diagnosis of *Fucoides* (now *Arthrophycus*) *allegha-niensis*)

Harlan R (1832) On a new extinct fossil vegetable of the family *Fucoides.* J Acad Nat Sci Phila 1(6):307–308 (Diagnosis of *A. brongniartii*)

Seilacher A, Cingolani C, Varela C (2003) Ichnostratigraphic correlation of early Paleozoic quartzites in central Argentina. In: Salem MJ, Oun KM, Seddig HM (eds) The geology of Northwest Libya. Earth Science Society of Libya 1, Tripoli, pp 275–292 (*Arthrophycus* from Silurian rocks in Argentina and North Africa)

Plate 43: Arthrophycids II

Del Valle A (1987) Nuevas trazas fosiles en la formacion Balcarce, Paleozoico inferior de las Sierras Septentrionales su significado cronologico y ambientla. Rev Mus La Plata (Nuer Ser) Secc Paleontol 9(52):19–41 (Fig. 5: *Arthrophycus alleghariensis.* Nevertheless considered Cambrian to Ordovician in age)

Rodriguez SG (1988) Trazas fósiles en sedimentitas del Paleozoico de las Sierras Australes Bonaerenses. 2° Jornadas Geológicas Bonaerenses, pp 117–129 (Description of giant *Daedalus verticalis* from Sierra de la Ventana, Argentina)

Seilacher A (1969) Sedimentary rhythms and trace fossils in Paleozoic sandstones of Libya. Petroleum Exploration Society of Libya, Guidebook 11th Annual Field Conference, Tripoli, pp 117–122 (Illustration of *Arthrophycus lateralis*)

Seilacher A (2000) Ordovician and Silurian arthrophycid ichno-stratigraphy. In: Sola MA, Worsley D (eds) Geological ex-ploration in Murzuk Basin. Elsevier, Amsterdam, pp 237–258 (Description of *Arthrophycus lateralis* and several ichnospecies of *Daedalus*)

Plate 44: Arthrophycids III

Rouault M (1883) Note sur le Grès armoricain – essai historique et Géologique sur *Vexillum desglandi.* (*Daedalus*)

Sarle CJ (1906) *Arthrophycus* and *Daedalus* of burrow origin. Proc Rochester Acad Sci 4:203–210 (Pioneer analysis of arthrophycid construction)

Seilacher A (1997) Fossil art. An exhibition of the Geologisches Institut Tübingen University. The Royal Tyrell Museum of Palaeontology, Drumheller, Alberta, Canada, 64 p (See p 40–41 for discussion of arthophycids)

Seilacher A (2000) Ordovician and Silurian arthrophycid ichno-stratigraphy. In: Sola MA, Worsley D (eds) Geological exploration in Murzuk Basin. Elsevier, Amsterdam, pp 237–258 (Description of several ichnospecies of *Daedalus*)

Plate 45: Arthrophycids IV

Mägdefrau K (1934) Über *Phycodes circinatum* Reinh. Richter aus dem thürungischen Ordovicium: Neues Jahrb Geol P-A 72:259–282

Osgood RG (1970) Trace fossils of the Cincinnati area. Paleontographica Americana 6:281–444 (Description of *Phycodes flabellum*)

Seilacher A (2000) Ordovician and Silurian arthrophycid ichno-stratigraphy. In: Sola MA, Worsley D (eds) Geological exploration in Murzuk Basin. Elsevier, Amsterdam, pp 237–258 (Description of *Daedalus multiplex* and several ichnospecies of *Phycodes*)

■ *Arthrophycus alleghaniensis* (L. Sil., SW Libya)

41

Plate 41
Teichichnid Burrows

In connection with tunnel systems made by crustaceans (Pl. 18), we have already encountered teichichnid constructions characterized by transversal backfill structures that are not evenly draped between the two shafts of a U-tube. Rather they were generated by J-tubes, very shallow U-tubes, or only in a limited stretch of an undefined tunnel system. Most of them have a *retrusive* backfill structure.

Teichichnus rectus from the Cambrian of Pakistan and elsewhere is not only the type species, but also the constructional prototype of the ichnogenus. Because of similar size (about the diameter of a finger) and simplicity, its relationship to *Trichophycus* (Pl. 17) is questionable due to the lack of scratch patterns.

Teichichnus palmatus increased mining efficiency by including multiple retrusive spreite bodies in fan-shaped bundles. Their arrangement is radial in *Teichichnus stellatus*, which is similar to *Brooksella* and *Dactyloidites* (Pl. 47). As all these forms occur in the Cambrian, they may be preservational variants of the same basic burrow architectures; but without more personalized fingerprints they are also hard to distinguish from similar, probably convergent forms in later periods.

Teichichnus spiralis from a Lower Jurassic sandstone in southern Germany should be mentioned, although it is represented only by a single loose specimen and has not yet been formally described. In median section it looks like two closely adjoined spreite bodies of standard *Teichichnus*, but in reality the gutter-like backfill lamellae are tight spirals! As such a structure could not have been produced by displacement of a corkscrew tunnel (e.g., *Gyrolithes*, Pl. 18), the generating tube is here reconstructed as a vertical shaft whose lower end was bent into a horizontal half coil and rotated during upward displacement. More specimens with known orientations would help in finding a more reasonable explanation.

While the previous examples were found in sandstones, the following two forms have been described from Upper Cretaceous chalks of North America. The name of the first one (*Teichichnus zigzag*) is descriptively appropriate, but in vertical section the spreite not only zigzags, but also changes width in the same rhythm. As the terminal tunnel is disproportionately narrow, one could assume that its upward course is meandering in cross-sectional view. Alternatively, the backfill could be partly radial (Pl. 37), if the animal was able to inflate its body.

The second Cretaceous ichnospecies (*Teichichnus sigmoidalis*) touches on the question of how spreite formation is triggered. In rhizocoralliid burrows (Pl. 19), the cue is the local curvature of the U as sensed by body

flexure. Alectorurid makers use the same signal or introduce their own sigmoidal subprogram (Pl. 38). In *Teichichnus rectus*, the whole tunnel appears to have been upwardly displaced. But how can spreite formation be induced in only a section of a long horizontal tunnel? The maker of *Teichichnus sigmoidalis* found the trick: plan an upward kink in the original tunnel and shave until it is evened out!

That similar trace patterns must not always relate to similar behavior is shown by sigmoidal variants of **Syringomorpha nilssoni** found in Cambrian sandstones of Sweden and Argentina. Its wing-shaped spreite, only a millimeter or so in thickness, is peculiar in several respects. (1) In cross section it is perfectly vertical to the bedding plane, but steeply inclined in lateral view; (2) it does not consist of gutter-shaped backfill lamellae, but of almost cylindrical rods of sand, so that fragments look like a minute pan-pipe (hence the Latin name); (3) nevertheless, indentations seen in cross section show clearly that the spreite was made in a retrusive mode, i.e. opposite to *Teichichnus sigmoidalis* (above) and *Daedalus* (Pl. 44). This means that the sigmoidal kink *resulted* from the spreite construction rather than inducing it. I formerly thought that the generating tube of *Syringomorpha* was U-shaped, but more likely it was dead-ended and its basal bend served mainly as a starter for the construction of the spreite.

Only because there is no better place for it, **Heliochone** is mentioned in the present context. It is equally remarkable for its size (up to 50 cm), its paleoenvironment (black shale facies of the Lower Devonian Hunsrück Slates of Germany), and its unique architecture. The bedding plane expression of this spectacular fossil (ring with rays) looks mysterious, except for a meniscate backfill structure in the radial sections. Only serial grinding (stereo pair of a **glass model**) revealed the three-dimensional structure shown in the reconstruction, which assumes a generating tube consisting of a basal ring with equally spaced vertical exits. The isometric enlargement and deepening of this tunnel system accounts for the observed backfill bodies (all protrusive): a cone left behind by the ring tube and radial spokes corresponding to the vertical shafts. Since the original description, more complete specimens of *Heliochone* have been found, but not serially sectioned. In one of them the central ring is open, so that some spirality may be involved. Speculating about the function of such an underground labyrinth, one is reminded of deepsea graphoglyptids (Pl. 54), whose complex horizontal tunnel systems probably serve as bacterial gardens, while the large number of exits allows for passive ventilation by diffusion. The Hunsrück Slate sediments were certainly not deposited in the deepsea, but they contained enough H_2S to make such a strategy feasible. A similar burrow has been found in a Lower Cambrian turbidite sequence in Argentina (Pl. 65).

Plate 41 · Teichichnid Burrows   121

Teichichnid Burrows

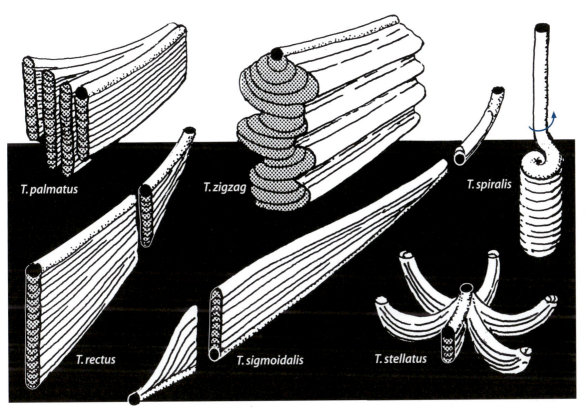

T. palmatus

T. zigzag

T. spiralis

T. rectus

T. sigmoidalis

T. stellatus

Teichichnus

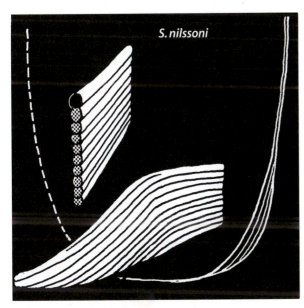

S. nilssoni

Syringomorpha

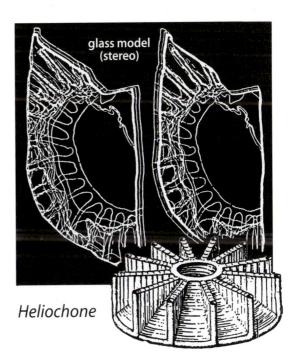

glass model
(stereo)

Heliochone

42

Plate 42
Arthrophycids I

After having explored similar burrow architectures made by unrelated kinds of animals, we can now focus on the proper arthrophycids, including the established ichnogenera *Arthrophycus*, *Daedalus* and *Phycodes*, that may be of stratigraphic use in otherwise unfossiliferous Ordovician and Silurian sandstones.

We begin with the simplest form, ***Arthrophycus linearis***. Such strings would normally be attributed to Repichnia (Pl. 31) or to head-on bulldozers. Yet, the typical arthrophycid fingerprints (see introduction) link them with much more complex burrow systems, which are here treated as different ichnospecies of the same ichnogenus (see below). Polished cross sections also reveal that hyporeliefs are neither open nor terminally backfilled tunnels, but result from an oblique displacement similar to that inferred for *Gyrochorte* (Pl. 35). But the teichichnoid structure seen in vertical section is *protrusive* (as predicted by the *Gyrochorte* model) only in forms from Jordan and Benin (note places where the filling of the terminal tunnel broke off). Both occur in sandstones that can be dated as Middle to Upper Ordovician by associated *Cruziana* ichnospecies (Pl. 68). In all other occurrences (including those in the Lower Ordovician of southern Algeria and the Kufra Basin, Libya) the linear *Arthrophycus* strings have a *retrusive* structure, suggesting that the animals burrowed along in a *head-down* position. In the hope that this difference will be stratigraphically useful and because the two variants may have given rise to different kinds of more complex feeding burrows, they are distinguished at the ichnosubspecies level. Large burrows of this type may show minute transverse wrinkles in addition to the typical arthrophycid segments (**upper right**, from Rochester, New York).

While the ichnosubspecies *A. linearis protrusiva* is strictly linear, *A. linearis linearis* occasionally splits bush-like at very small angles. These unilateral branches, however, look more like the fannings of *Gyrochorte* (Pl. 35) than intentional programs. This mode of branching became dominant in the Lower Silurian ***Arthrophycus alleghaniensis*** (also known as the junior synonym *Harlania harlani*), which is common in North Africa, but has also been found in Argentina and the eastern United States. Projecting from the sole faces of sandstone beds, these palmate burrow systems look like bundles of retrusive U-burrows, but in reality the generating tube was J-shaped as in other arthrophycids. Note also that – as in the palmate trilobite burrows of similar age and occurrence (Pl. 15) – the backfill was introduced from the sand bed on top, while the dugout mud must have been exported to the surface. In contrast to *Cruziana ancora*, however, not all branches radiate from one point, but tend to form smaller bundles within the larger ones.

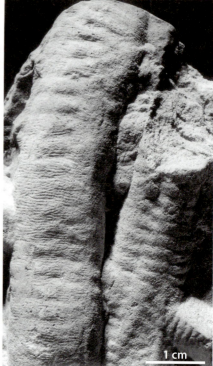

5 cm

1 cm

■ *Arthrophycus linearis* (L. Sil., Rochester, N.Y.)

Plate 42 · Arthrophycids I   123

Arthrophycids I

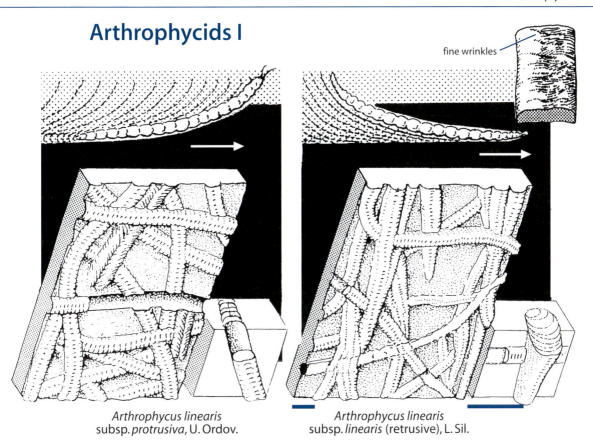

fine wrinkles

Arthrophycus linearis
subsp. *protrusiva*, U. Ordov.

Arthrophycus linearis
subsp. *linearis* (retrusive), L. Sil.

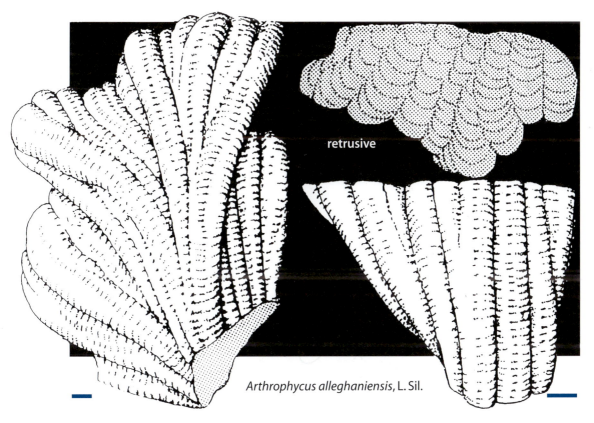

retrusive

Arthrophycus alleghaniensis, L. Sil.

43

Plate 43
Arthrophycids II

It is quite possible that different programs can in the future be distinguished among palmate variants of *Arthrophycus* that look similar at first glance. One such variant is **Arthrophycus lateralis**, which has so far been only recognized at a single locality (Tacharchuri Pass, southwestern Libya). Complete systems look almost like palmate *Arthrophycus alleghaniensis*, except that all branches bend to one side rather than radiating on two sides of a median axis. Cross sections reveal, however, a completely different burrowing program: (1) the stacks of crescentic backfill lamellae are *horizontal* rather than vertical; (2) relative to the curvature of the branches in plan view, the spreite structure is *protrusive* rather than retrusive; (3) as shown by intersecting relationships, multiple horizontal stacks were produced in an *upward succession*, possibly in response to sedimentation; (4) minor vertical excursions within one stack are expressed by *secondary lamellae* in a retrusive mode.

Unfortunately, the internal structure cannot always be seen as clearly as in the well weathered and iron-stained specimens of *A. lateralis*. Nevertheless it is this kind of evidence that may allow future distinction of different programs for a better stratigraphic resolution.

Daedalus. This ichnogenus, whose name refers to the wings of the father of Icarus in Greek mythology, is allied to *Arthrophycus* in various respects. It resembles *Arthrophycus* not only by similar size, with tube diameters ranging from those of a pencil to the thickness of a finger. *Daedalus* also has the same stratigraphic range (Ordovician-Silurian) and occurs in similar kinds of shallow-marine sandstones. Most importantly, however, there are the same transverse ridges as in *Arthrophycus*, even

though such fingerprints are less commonly recognizable on broken surfaces than at the interface between sand and mud.

The diagnostic differences from *Arthrophycus* are (1) a single *protrusive* spreite body made by a steeply inclined J-burrow; (2) the tendency (except in **Daedalus verticalis**) to proceed in a spiral.

We start with the non-spiral *Daedalus verticalis* from the Lower Silurian Medina Sandstone of New York, because it could be mistaken for the hairpin burrows of *Diplocraterion* (Pl. 18). Yet it differs by being perpendicular only in cross section, while the spreite lamellae are slightly inclined in lateral view. This relates to the J-shape of the generating tube, which is not always obvious on split surfaces. Only a highly polished horizontal section through the same specimen revealed the basic asymmetry: the spreite is widest on the steeper side and continuously tapers away from it. If burrowing was done by the head end, this would mean that the front part of the body tapered accordingly. A variant of *Daedalus verticalis* from the Sierra de la Ventana (Argentina) has giant spreite bodies that extend up to 40 cm horizontally and 70 cm vertically. In order to minimize intersection, the crowded burrows at this locality have the same azimuth, so that the sandstone breaks accordingly. Tube diameters are also large enough for the typical arthrophycid corrugation to be preserved.

In a large *Arthrophycus* slab from the Lower Silurian of New York (Yale Peabody Museum) a *Daedalus verticalis* reached the sole of the sandstone bed. But instead of continuing to burrow vertically, it followed the interface half a meter. In this unusual topology the typical arthrophycid fingerprints are clearly expressed. In addition, the now flat-lying spreite describes a curve, because the trace maker was no more guided by the gravitational compass, but by the bedding plane.

■
Arthrophycus alleghaniensis
(Sil., Chad)

5 cm

Plate 43 · Arthrophycids II   125

Arthrophycids II

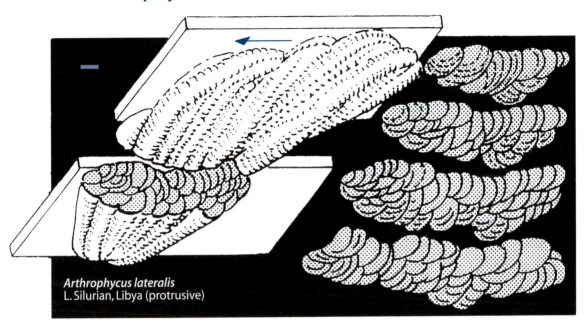

Arthrophycus lateralis
L. Silurian, Libya (protrusive)

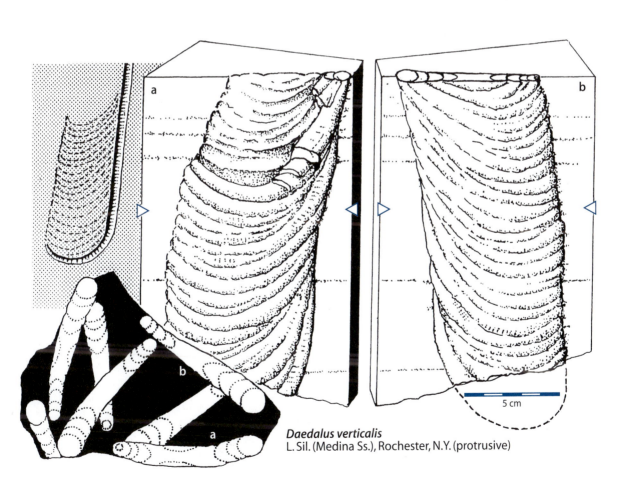

Daedalus verticalis
L. Sil. (Medina Ss.), Rochester, N.Y. (protrusive)

5 cm

44

Plate 44
Arthrophycids III

In this plate we discuss the more typical ichnospecies of *Daedalus*, in which the spreite forms a spiral cone. In order to understand the *kinematics* of these structures, we might imagine the trace of an elastic stick being moved through soft sediment around a point fixed at the surface. Its tip at depth would be automatically bent backwards and a stroboscopic X-ray picture of sequential positions would resemble the three-dimensional *Daedalus* structure. The actual worm, of course, cut its way through the sediment in a different manner – either by shoveling material from the ventral to the dorsal side of the body (as in *Gyrochorte*; Pl. 35) or, more probably, by constantly shaving away the convex side of an open tunnel (as in *Zoophycos*; Pl. 38). Either mode would result in the observed *protrusive* backfill structure. Yet, the lower part of the burrow did not lead in the transversal motion (as it does in *Dictyodora*; Pl. 29), but lagged passively behind. So we are again faced with the question whether the head or the tail end of the animal was located at the lower tip of the burrow?

Equally problematic is the biological *function* of these burrows. While the hypichnial *Arthrophycus* (Pl. 42) removed mud and replaced it by sand, the grain size of the *Daedalus* spreite appears to correspond to that of the sandy matrix. Moreover, the host rock is commonly a very clean quartzite. So one wonders: could this sand yield enough food to warrant such an expensive activity?

Three ichnospecies are distinguished in the Lower Ordovician ("Armorican") quartzites of France, Spain, Iraq, Oman and Antarctica. The spreite of **Daedalus halli** is only about 2 mm thick, but may penetrate half a meter deep. Its intersection with successive bedding planes looks like a worm trail that originally received a different name (**Humilis preservation**). The generating tube was also very steep and straighter than in **Daedalus labechei**, where the tube diameter is large enough to occasionally preserve the typical arthrophycid annulation (Oman Mountains near Muscat). **Daedalus desglandi** almost looks like a giant and very high-spired gastropod shell, because it is tightly coiled and the lower end of the generating tube – now itself slightly helicospiral – was bent *towards* the coiling axis (see diagram in Pl. 39). As seen in cross and longitudinal sections, the lower margin of the spreite is regularly intersected by the subsequent coil. This would have been fatal had there been a marginal tunnel as in a spiral rhizocoralliid (Pl. 20), but not with a J-tube.

A tightly coiled protrusive spreite and a helicoidal generating tube are also typical for **Daedalus archimedes**. It is common in thicker beds of the Medina sandstones (Lower Silurian of upstate New York), where *Arthrophy-*

cus linearis (Pl. 42) is found in thinner beds. Because the J-tube was not bent inward, the marginal spiral could change from steeply (left) to gently descending (middle), or even ascending (right), according to sedimentation rate. As seen in bottom views (middle level), younger coils of the spreite often intersect older ones, which would have cut off ventilation had there been a marginal tube. Note also that the bottom surfaces fail to show the arthrophycid corrugations. This is because the animal never touched or penetrated the interface with the underlying mud, which was essential in the preservation of *Arthrophycus*. Nevertheless, the typical fingerprints can occasionally be seen on the flanks of the spreite body.

2 cm

■ *Daedalus archimedes* (L. Sil., Rochester, N.Y.)

Plate 44 · Arthrophycids III   127

Arthrophycids III

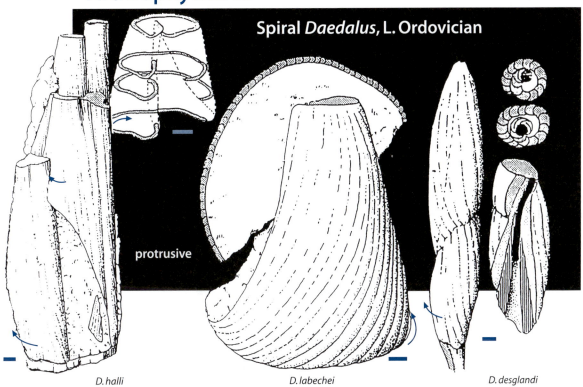

Spiral *Daedalus*, L. Ordovician

protrusive

D. halli

D. labechei

D. desglandi

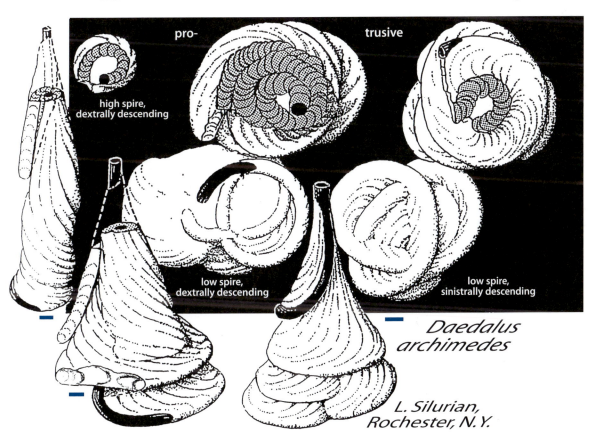

pro-   trusive

high spire,
dextrally descending

low spire,
dextrally descending

low spire,
sinistrally descending

Daedalus archimedes

L. Silurian, Rochester, N.Y.

45

Plate 45
Arthrophycids IV

While other *Daedalus* burrows consist of a single vertical or spiral spreite body, the maker of **Daedalus multiplex** from the Middle Ordovician introduced the new element of *branching* into its behavioral program. The three-dimensional reconstruction of a complete burrow system has been possible only because preservation differs in the two known occurrences.

In southwestern **Libya**, where the two photos were taken, the burrows are seen on vertical fractures through massive sandstone beds. They look like tufts of roots that radiate steeply down from a central stem. Closer examination, however, reveals vertical spreite bodies with a protrusive cross-sectional structure.

In Saudi Arabia, sandstones of similar age are weathered from the top, so that the burrow systems can be seen in horizontal section. In this aspect they look like a spiral nebula, in which up to seven curved branches radiate from a common center.

The program of *Daedalus multiplex*, being more complex than that of other *Daedalus* ichnospecies, allowed a given volume of sediment to be exploited without intersections. Yet it is uncertain whether the generating tube was a simple J that made one wing after another, or whether the radial branches remained open, so that all the wings could be deepened in concert.

Phycodes. In this ichnogenus we return to the morphospace and the preservational mode of *Arthrophycus* (Pls. 42, 43), i.e. the burrow systems are found as sand bodies at the bases of sandstone beds rather than within them. The distinction between the two ichnogenera is mainly based on *size*. While diameters of the generating tube are in the range of our fingers in *Arthrophycus*, they reach only a few millimeters in *Phycodes*. There is also a separation in time. While *Phycodes* appears to be restricted to the Ordovician, the behavioral radiation of *Arthrophycus* occurred in the Silurian.

In this context, a taxonomic clarification is in place. In earlier years I distinguished ichnogenera mainly on the basis of behavior. This means that heterogeneous forms of quite different ages were included in *Phycodes* just because they had similar modes of branching. Among these homeomorphs was "*Phycodes*" *pedum*, which later became important for defining the Precambrian/Cambrian boundary. It is now referred to *Treptichnus* (Pl. 64), while the affiliation of palmate probings without true spreite ("*Phycodes*" *palmatum*) remains in limbo.

We begin our discussion with the type ichnospecies, **Phycodes circinatum**, whose name is derived from the resemblance to seaweed and to the old-fashioned brooms on which witches ride through the air. It occurs in high relief on the soles of thin-bedded sands and silts in the Lower Ordovician of Germany, France, Iraq and North America – i.e. again on different paleocontinents. Cross sections reveal that the bundles of smoothly curved burrows seen in bottom view are actually the bases of closely packed vertical spreite bodies consisting of retrusive backfill lamellae. So *Phycodes circinatum* looks like a smaller version of *Arthrophycus alleghaniensis* (Pl. 42). Even the transversal lineation seen on the surface (enlarged small block) might be interpreted as an arthrophycid fingerprint. It is less clear whether the generating tube was J- or U-shaped.

This question becomes critical in the interpretation of **Phycodes fusiforme** from the upper part of the Lower Ordovician in Saudi Arabia. Here the branches are bundled at *both* ends of the system, so that it becomes difficult to distinguish a proximal and a distal side. Also, the spreite bodies radiate downward in cross section.

Phycodes parallelum is known from the Middle Ordovician of Utah and central Australia, i.e. from two different paleocontinents. It resembles *Phycodes fusiforme* by the lack of palmate fanning, but instead of being bundled at two defined ends, the retrusive spreite bodies are vertical and perfectly parallel to each other. They also form strands of indefinite lengths. Curved sets at the flanks suggest that these strands were made by progradation of teichichnoid spreite bodies that were themselves retrusive. So the burrowing program must have been more complex than in the Lower Ordovician ichnospecies.

The last representative, **Phycodes flabellum** from the Upper Ordovician of Cincinnati, looks very different. Instead of forming vertical spreite bodies that stick out from the base of the sandstone bed, its maker stripmined the nutrient-rich layer above the interface by *horizontal* displacement of a J-shaped generating tube. Therefore one sees the spreite lamellae only as faint impressions fanning from a base tunnel in a wing-like pattern. Being preserved as positive hyporeliefs at the sand/mud interface, these spreite lamellae also show the delicate transverse ornamentation characteristic of *Phycodes*.

In conclusion, the possibility to define arthrophycid burrows as a genetically coherent group allows us to view ichnospecies in an evolutionary context and to use them as guide fossils in otherwise unfossiliferous sandstones.

Plate 45 · Arthrophycids IV   129

Arthrophycids IV

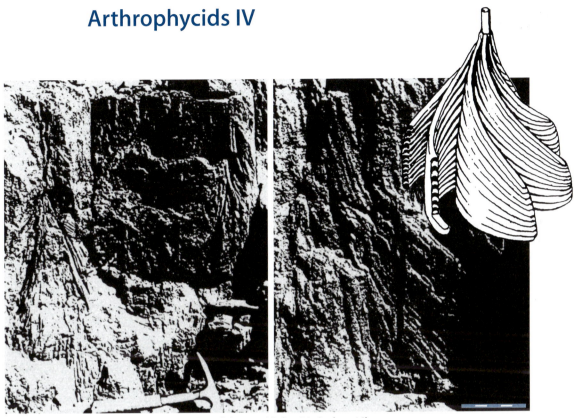

Daedalus multiplex, M. Ordov., Libya

Phycodes

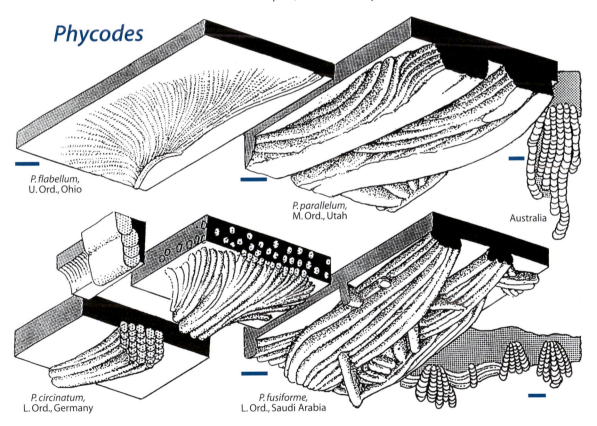

P. flabellum,
U. Ord., Ohio

P. parallelum,
M. Ord., Utah

Australia

P. circinatum,
L. Ord., Germany

P. fusiforme,
L. Ord., Saudi Arabia

Phycodes fusiforme
(M. Ordov., Saudi Arabia)

Phycodes parallelum
(L. Ord. Utah)

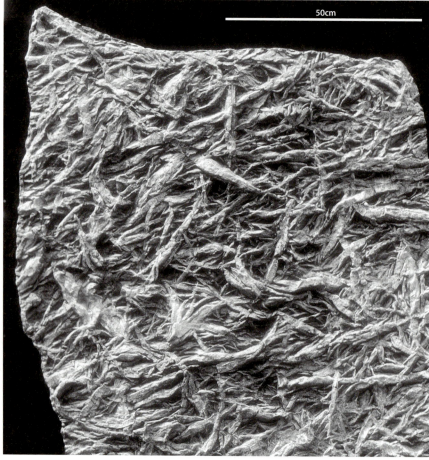

Probers

Arthrophycid burrows are either made in sand or – if they penetrate into mud – backfilled with sand. The heterogeneous burrows assembled in this chapter are mainly found in muddy sediments, where they contrast in color: white in dark bituminous shales, dark in light-colored micritic marls and limestones. The reason is that they are actively backfilled with faecal material different from the host sediment. We speak of probers because the backfilled branches of a system are in most cases made head-on and blind-ended. The tendency to intensively use a given area, or volume, of sediment with a minimum of intersections suggests that the makers were primarily sediment feeders. In the cases of pronounced distancing between branches we may also consider chemosymbiotic animals that pumped hydrogen sulfide or methane from the pore water. A third possibility is a sanitary function. In reality, however, a burrow system may have served more than one of these functions.

This group of burrows, commonly called fucoids because of their original interpretation as marine plants (*Fucus* is a modern kelp), has a long history of research. The oldest illustration (Pl. 50) is more than 400 years old. At that time, the Lower Jurassic Posidonia Shales were famous not for their fossils, but as a source of sulfurous mineral water; so families from the capital used Bad Boll near Holzmaden as a summer resort. For their entertainment, the naturalist Johannes Bauhin published a tourist guide in which the animals and fossils of the region were described. This was obviously a success, as shown by its three editions, of which the first two (1598 and 1600) were in Latin and the third (1602) in German. The chondritid he depicted is probably the earliest illustration of a trace fossil. Its legend reads: "A fissile stone through which run delicate veins of ash-colored clay in the process of becoming lithified. They resemble coral trees whose branches spread in different directions. The human figures have been mistakenly added either by the painter or the wood cutter". So we deal with the eternal problem arising when scientists have their illustrations done by artists whose perception of a given form may be very different from our own.

In this case we can be certain that the ichnospecies figured was *Phymatoderma granulosa* from the lower of the two burrow horizons. One of its characteristics is a ragged outline caused by ovoid faecal pellets, with which the burrows are backfilled. This feature reminded the 16th century artist of feathered wings in a heavenly combat. Later, the pellets were interpreted as sporangia of seaweeds ("*Algacites*" on the label written in the 19th century by F. A. Quenstedt).

The bituminous Posidonia Shales (Lower Jurassic) of nearby Holzmaden (southern Germany) are better known for their completely preserved vertebrate and crinoid skeletons. As usual in such *Konservat-Lagerstaetten*, the environment was largely hostile for benthic life. Nevertheless there are two horizons riddled with chondritid burrows that contrast sharply with the dark matrix by their lighter fill. Obviously these horizons correspond to short periods during which bottom waters were less anoxic, so that sulfide-tolerant species could exploit the nutrient-rich mud. Accordingly, each of these horizons preserves the original *tiering* with small chondritids in an upper tier and larger systems (plus thalassinoid crustacean burrows) in a lower tier (Pl. 73). Yet the chondritids in the two benthic horizons are not the same.

Bark Beetles

The burrow systems of modern bark beetles may serve as an analog to the fossil probings because they also exploit a strictly two-dimensional food source, the cambium separating wood and bark. The female of the species figured in Pl. 49 starts by excavating a straight tunnel parallel to the wood fibers – probably without being interested in wood as a food source. It then lays its eggs at equal distances along the sides of the tunnel. As the eggs hatch, the grubs immediately start to munch their path away from the maternal tunnel and across the supply lines of the tree, backfilling their own tunnels with faecal material. As the grubs and their tunnels widen, space becomes limiting, so that marginal individuals are forced to fan out their tunnels like spider legs. Others may end their young lives in competition with the siblings, lest they would violate the rule never to intrude into foreign territory. Finally the grubs metamorphose and start their second life as free beetles.

Literature

Plate 46: Asterosomids

Dahmer G (1937) Lebensspuren aus dem Taunusquarzit und den Siegener Schichten (Unterdevon). Jahrbuch der Preussisches Geologisch Landesanstalt für 1936 57:523–539 (First description of *Rosselia,* probably an eroded version of steeply inclined *Asterosoma*)

Jensen S (1997) Trace fossils from the Lower Cambrian Mickwitzia sandstone, south-central Sweden. Fossils and Strata 42:1–111 (Description of Cambrian *Halopoa.* See fig. 50 for illustration)

Książkiewicz M (1977) Trace fossils in the flysch of the Polish Carpathians. Paleontologia Polonica 36, 208 p (Description of flysch *Fucusopsis*)

Schlirf M (2000) Upper Jurassic trace fossils from the Boulonnais (northern France). Geologica et Palaeontologica 34:145–213 (Description of *Asterosoma ludwigae*)

Seilacher A (1997) Fossil art. An exhibition of the Geologisches Institut Tübingen University. The Royal Tyrell Museum of Palaeontology, Drumheller, Alberta, Canada, 64 p (See p 44–45 for discussion on asterosomids)

Uchman A (1998) Taxonomy and ethology of flysch trace fossils: Revision of the Marian Książkiewicz collection and studies of complementary material. Annales Societatis Geologorum Poloniae 68:105–218 (Re-evaluation of *Fucusopsis*)

Zacharov V (1972) 112:78–89, 1 pl. (*Arctichnus* resembles *Rosselia*)

Plate 47: Medusiform Burrows (Gyrophyllitids)

Adam KD (1953) Neue Funde von *Palaeosemaeostoma geryonoides* (Cnidaria) aus dem unteren Dogger Südwest-Deutschlands. Jber Mitt Oberrh Geol Ver 35:88–96 (Medusa interpretation)

Caster KE (1945) A new jellyfish (*Kirklandia texana* Caster) from the Lower Cretaceous of Texas. Paleontographica Americana 3:173–220 (Epirelief and concetionary preservations)

Fu S (1991) Funktion, Verhalten und Einteilung fucoider und lophocteniider Lebensspuren. Courier Forschungsinstitut Senckenberg 135, pp 1–79 (Review of gyrophyllitids. See Plate 40 for illustration)

Fuchs T (1901) Über *Medusina geryonoides* von Huene. Centralbl Mineral Geol Paläontologie B 1901:166–167 (Recognizes burrow origin)

Häntzschel W (1970) Star-like trace fossils. In: Crimes TP, Harper JC (eds) Trace fossils. Geol J, Special Issue 3, pp 201–214 (General review of gyrophyllitids)

Harrington HJ, Moore RC (1956) In: Moore RC (ed) Treatise on invertebrate paleontology, part F, F21–F23. Geological Society of America and University of Kansas Press, New York and Lawrence, Kansas (*Brooksella* as jellyfish)

Prantl F (1945) Two new problematic trails from the Ordovician of Bohemia. Bull Int Acad Sci 46(3):1–11 (Unnamed stellate burrows similar to *Dactyloidites* or *Teichichnus stellatus,* Pl. 41)

Uchman A (1998) Taxonomy and ethology of flysch trace fossils: Revision of the Marian Książkiewicz collection and studies of complementary material. Annales Societatis Geologorum Poloniae 68:105–218 (Updated review of gyrophyllitids)

Volk M (1960) *Bifasciculus radiatus* n.g., n.sp., eine Lebensspur aus dem Griffelschiefer des thüringischen Ordovizium. Geol Bl NO-Bayern 10(4):152–156 (Probings shaped like spiral nebula, otherwise similar to *Dactyloidites*)

Walcott CD (1898) Fossil medusae. US Geological Survey, Monograph 30, 201 p (Fossil jellyfishes from all occurrences then known, with an extensive discussion of *Brooksella,* which is now interpreted as a trace fossil)

Wetzel A, Uchman A (1997) Ichnology of deep-sea fan overbank deposits of the Ganei Slates (Eocene, Switzerland) – A classical flysch trace fossil locality studied first by Oswald Heer. Ichnos 5:30–162 (Description of *Gyrophyllites* from Eocene flysch deposits)

Plate 48: Various Fucoids

Aguirrezabala LM, Gibert JM de (2004) Paleodepth and paleoenvironment of *Dactyloidites ottoi* (Geinitz, 1849) from Lower Cretaceous deltaic deposits (Basque-Cantabrian Basin, west Pyrenees). Palaios 19:276–291 (Study of *Häntzschelina* within its paleoenvironmental framework)

Fischer-Ooster C von. (1858) Die fossilen Fucoiden der Schweizer Alpen, nebst Erörterungen über deren geologisches Alter. Huber, Bern, 73 p (Description of fucoids in a paleobotanic framework)

Fu S (1991) Funktion, Verhalten und Einteilung fucoider und lophocteniider Lebensspuren. Courier Forschungsinstitut Senckenberg 135, pp 1–79 (Discussion of various "fucoids". See Plate 40 for illustration)

Fuchs T (1895) Studien der Fucoiden und Hieroglyphen. Akademie der Wissenschaften zu Wien, mathematisch-naturwissenschaftliche Classe, Denkschriften 62, pp 369–448 (Pioneer work an re-interpretation of fucoids)

Fürsich FT, Bromley RG (1985) Behavioural interpretation of a rosetted spreite trace fossil: *Dactyloidites ottoi* (Geinitz) Lethaia 18:199–207 (*Häntzschelinia*)

Gregory MR (1969) Trace fossils from the turbidite facies of the Waitemata Group, Whangaparoa Peninsula, Auckland. J Roy Soc New Zeal, Earth Sci 7:1–20 (Cross section of *Cycloichnus*)

Häntzschel W (1934) Schraubenförmige und spiralige Grabgänge in turonen Sandsteinen des Zittauer Gebirges. Sternspuren, erzeugt von einer Muschel: *Scrobicularia plana* (da Costa). Senckenberg 16(4/6):313–330

Krejci-Graf K (1936) Zur Natur der Fukoiden. Senckenberg 18: 308–315

Lorenz von Liburnau JR (1900) Zur Deutung der fossilen Fucoiden-Gattungen *Taenidium* und *Gyrophyllites*. K. Akademie der Wissenschaften zu Wien, mathematisch-naturwissenschaftliche Klasse, Denkschriften 70, pp 523–583 (Algal interpretation)

Seilacher A (1997) Fossil art. An exhibition of the Geologisches Institut Tübingen University. The Royal Tyrell Museum of Palaeontology, Drumheller, Alberta, Canada, 64 p (See p 48–49 for discussion of *Haentzschelinia* from Japan)

Stepanek J, Geyer G (1989) Spurenfossilien aus dem Kulm (Unterkarbon) des Frankenwaldes. Beringeria 1:1–55 (Introduction of the ichnogenus *Taxichnites*)

Plate 49: Undermat Miners

Aceñolaza FG, Durand FR (1984) The trace fossil *Oldhamia*: Its interpretation and occurrence in the Lower Cambrian of Argentina. Neues Jahrb Geol P M 1984, pp 728–740 (*Oldhamia* from the Puncoviscana Formation, albeit interpreted as trilobite scratch marks)

Buatois LA, Mángano MG (2003) Early colonization of the deep sea: Ichnologic evidence of deep-marine benthic ecology from the Early Cambrian of northwest Argentina. Palaios 18:572–581 (Reinterpretation of *Oldhamia* from the Puncoviscana Formation and evaluation of its association with microbial mats)

Churkin M, Brabb EE (1965) Occurrence and stratigraphic significance of *Oldhamia*, a Cambrian trace fossil, in east-central Alaska. U.S. Geol., Survey Prof. Paper (525):120–124

Crimes TP (1976) Trace fossils from the Bray Group (Cambrian) at Howth, Co. Dublin. Bull Geol Soc Ireland 2:53–67 (Description of the type locality of *Oldhamia*)

Delgado JFN (1910) Terrains paléozoiques du Portugal. Étude sur les fossiles et des schistes à néreite de San Domingos et des schistes a néréite et à graptolites de Barrancos. Commission Service Geologia de Portugal 56, 68 p, 51 pls. (Illustration of *O. pinnata*)

Hofmann HJ, Cecile MP, Lane LS (1994) New occurrences of *Oldhamia* and other traces in the Cambrian of the Yukon and Ellesmere Island, arctic Canada. Can J Earth Sci 31:767–782 (Description of various ichnospecies of *Oldhamia*)

Lindholm RM, Casey JF (1990) The distribution and possible biostratigraphic significance of the ichnogenus *Oldhamia* in the shales of the Blow Me Down Brook Formation, western Newfoundland. Can J Earth Sci 27:1270–1287

Pfeiffer H (1968) Die Spurenfossilien des Kulms (Dinant) und Devons der Frankenwälder Querzone (Thüringen). Jahrb Geol 2:651–717 (Description of *O. fimbriata*)

Ruedemann R (1942) *Oldhamia* and the Rensselaer Grit problem. Bull N Y State Mus 327:5–12 (*Oldhamia* from the Cambrian of New York State)

Seilacher A (1999) Biomat-related lifestyles in the Precambrian. Palaios 14:86–93 (Interpretation of *Oldhamia* as an undermat miner)

Seilacher A, Buatois LA, Mángano MG (2005) Trace fossils in the Ediacaran-Cambrian transition: Behavioral diversification, ecological turnover and environmental shift. Palaeogeog Palaeoclim Palaeoecol 227:323–356 (General review of *Oldhamia* ichnospecies)

Sollas WJ (1900) *Ichnium Wattsii*, a worm track from the slate of Bray Head, with observations on the genus *Oldhamia*. Q J Geol Soc London 56:273–286 (Description of *Oldhamia* from its type locality in Ireland)

Plate 50: Chondritids

Bauhin Johann (Bauhini, Johannis) (1598) Historia novi et admirabilis fontis balneique Bollensis in Ducatu Wirtembergico ad acidulas Geopingenses. Adjunctur plurimae figurae novae variorum fossilium, stirpium et insectorum, quae in et circa hunc fontem reperiuntur. Montibelgardi, 291 p (Earliest known illustration of a trace fossil)

Bauhin Johann (Bauhini, Johannis) (1600) De lapidibus metallicisque miro naturae artificio in ipsis terrae visceribus figuratis: necnon de stiribus, insectis, avibus, aliisque animalibus, partim in fontis admirabilis Bollensis penetralibus, dum eius venas aquileges fodiendo perscrutabantur: partim in vicinia inuentis et observatis, quorum multa nunquam visa vivis eiconibus hic oculis pubiiciuntur. Montisbelgardi (König.), 222 p (Illustration of *Phymatoderma granulata*)

Bromley RG, Ekdale AA (1984) *Chondrites*: A trace fossil indicator of anoxia in sediments. Science 224:872–874 (*Chondrites* as an agrichnium)

Brongniart AT (1828–1838) Histoire des végétaux fossiles; ou, Recherches botaniques et géologiques sur les végétaux renfermés dans les diverses couches du globe. G. Dufour & E. d'Ocagne, Paris, 1, 488 p; 2, 72 p (One of the earliest works on fossil "seaweeds", most of which are trace fossils including several species of *Fucoides* later assigned to *Chondrites*. Nevertheless, later ichnospecies were all this well described and illustrated!)

Chiplonkar GW (1975) *Chondrites*, Sternberg, a trace fossil from the Dalmiapuram Formation, Trichinopoly District, S. India. Curr Sci 44(4):123–124

Chlupac I (1987) Ordovician ichnofossils in the metamorphic mantle of the Central Bohemian Pluton. Casopis pro Mineralogii a Geologii 32(3):249–260, 4 pls. (*Pragichnus* for bundles of steep branching burrows vagkely like *Chondrites*)

D'Alessandro A, Bromley RG (1987) Meniscate trace fossils and the *Muensteria-Taenidium* problem. Palaeontology 30(4):743–763 (Discuss unbranched and branched forms. *Cladichnus* n.igen.)

Fu S (1991) Funktion, Verhalten und Einteilung fucoider und lophocteniider Lebensspuren. Courier Forschungsinstitut Senckenberg 135, pp 1–79 (Review of chondritids. See Plate 40 for illustration)

Miller W III, Vokes EH (1998) Large *Phymatoderma* in Pliocene slope deposits, northwestern Ecuador: Associated ichnofauna, fabrication and behavioral ecology. Ichnos 6:23–45 (Spectacular examples of *Phymatoderma*)

Seilacher A (1955) Spuren und Fazies im Unterkambrium. In: Schindewolf O, Seilacher A (eds) Beiträge zur Kenntnis des Kambriums in der Salt Range (Pakistan). Akademie der Wissenschaften und der Literatur, Mainz, Abhandlungen der mathematisch-naturwissenschaftlichen Klasse, 10, pp 261–446 (Burrowing program of *Chondrites*)

Seilacher A (1990) Aberrations in bivalve evolution related to photo- and chemosymbiosis. Hist Biol 3:289–311 (Chemosymbiosis in bivalves. *Thyasira* makes chondritid probes with its worm-like foot)

Simpson S (1957) On the trace fossil *Chondrites*. Q J Geol Soc London 107:475–499 (Three-dimensional reconstruction of *Chondrites* as made by a sipunculid worm)

Uchman A, Wetzel A (1998) An aberrant, helicoidal trace fossil *Chondrites* Sternberg. Palaeogeog Palaeoclim Palaeoecol 146:165–169 (Spiral variant explained by nucleocave behaviour)

46

Plate 46
Asterosomids

The term asterosomids is again based on a certain burrowing technique rather than taxonomic relationships. It combines burrow systems in which waste material was stowed away in the form of *radial backfills* (Pl. 37). This means that the animal deposited such material in the wall of its tunnel and then pressed it radially out, so that the original tunnel diameter was maintained. In this process, the oldest and outermost layers of the backfill body became tensionally stressed and, depending on their mechanical properties, either became thinner or developed longitudinal cracks and microfaults parallel to the burrow axis.

Halopoa imbricata from Lower and Middle Cambrian sandstones of Sweden and Estonia and the Devonian *Lennea schmidti* form straight or branching burrows that are preserved as positive epi- or hyporeliefs. They have in common that (1) the diameter may vary along one burrow; (2) the surface is ornamented by irregular longitudinal rugosities; (3) the adjacent sand surface is deflected around the burrow. All these features relate to the deformations implied in the expansion of a radial backfill around the generating tube. The makers were either wormlike animals able to hydraulically expand their bodies (i.e. they had no tough cuticles), or crustacean that could press their smooth carapaces against the burrow wall.

Fucusopsis, found as post-turbidite burrows on the soles of flysch sandstones, expresses the radial expansion by downward deflection and longitudinal cracking of the sole surface (another case of a shadow trace). So the tunnels were made on top of the interface, but it is improbable that the observed deformation is only due to the initial wedging mode of penetration. In overcrossings the burrows appear to be superimposed, as in *Psammichnites* (Pl. 27), but in a hypichnial expression.

Radial backfill is most clearly expressed in *Asterosoma* (= *Astrophycus*), which occurs in shallowmarine sandstones of Devonian to Cretaceous ages. It differs from the previous forms in that the sandy backfill bodies are restricted to delimited sections, or branches, of the tunnel system. Therefore they have a *fusiform* shape and show longitudinal microfaults at the surface, whereas backfill laminae are concentric around the generating tunnel in cross section.

Two modes of preservation are responsible for the very different appearances of *Asterosoma* in the field. In the *Rosselia preservation* (not figured), steep backfill bodies cut through the upper interface of a sandstone bed. Thus we see their weathered cross sections with a seemingly conical arrangement of the laminae. In the *Asterosoma preservation*, backfill bodies have a more or less horizontal axis, so that their full fusiform shape can be seen.

Ichnospecies of *Asterosoma* (lower shaded area) can be distinguished by behavioral programs expressed in the arrangement of the backfill bulbs. In the type ichnospecies from the Upper Cretaceous of Germany (*Asterosoma radiciformis*), the wrinkled bulbs radiate atop of a vertical shaft. In a *A. coxii*

found in the Devonian of northern **Canada** and the Carboniferous of **Arkansas**, they branch off from a horizontal gallery either in an alternating fashion or only on the outer sides of a curved master tunnel. In **Jurassic** forms the backfill bodies appear like flower buds along tree twigs.

All these structures were made within the sediment. This is shown by the left-hand block from the **Middle Jurassic** of Germany, in which a series of *Asterosoma* bulbs emerges obliquely from a storm sand. Where the backfill body approached the interface from below, its expansion caused extensional cracks to propagate beyond the bulb onto the adjacent rippled sandstone surface. As this process also disrupted a preexisting *Gyrochorte* (Pl. 35), *Asterosoma* probably belonged to a deeper tier and shifted upward in response to sedimentation. In the thicker block to the right (from the same locality), bulbs radiate from a vertical, once open shaft of much smaller diameter. Now weathered out freely, the bulbs were once surrounded by black mud. This shows that the sandy backfill material was actively exported from the sand bed below.

From the size of *Asterosoma* it has been assumed that the makers were some kind of shrimp. This assumption could potentially be corroborated by scratch (rather than crack or microfault) patterns in parts of the tunnel that were not coated with backfill and passed into the underlying mud.

Finally, we add here a spectacular, but unnamed, trace fossil that stands out by its size, preservation, and incredible regularity. Found on a top surface in a Triassic limestone, a slab 7 m in length and 50 cm wide has been quarried and transported to the Geology Department of the Università la Sapienza in Rome (Italy), where it is now on exhibit. It is here informally called "*Tatzelwurm*" after a legend among Bavarian hunters who saw a burrow, but never the mystical animal that made it. Not surprisingly, this giant fossil was initially interpreted as the track of a large reptile. Several details, however, are in conflict with this interpretation.

1. The host rock is a marine pelagic limestone that deposited at a considerable depth.
2. The "footprints" are conical holes rather than resembling a toed foot.
3. The "trackway" ends suddenly at the distal side.
4. The supposed footsteps are connected by narrower series of globular chert nodules that form a racemose pattern. In every branch there are always five equally sized nodules between the branching point and the next "footprint".

At the moment, the most reasonable explanation for this enigmatic fossil is that it is a giant asterosomid burrow system whose rigid program resembled that of *Asterosoma ludwigae*. At the end of each branch there was a steeply inclined radial backfill body that has weathered out because it consists of different and diagenetically less cemented material. What induced the formation of exactly five chert nodules in every connecting tunnel during diagenesis is hard to understand. Probably there was some kind of biogenic prepattern.

Plate 46 · Asterosomids   135

Asterosomids

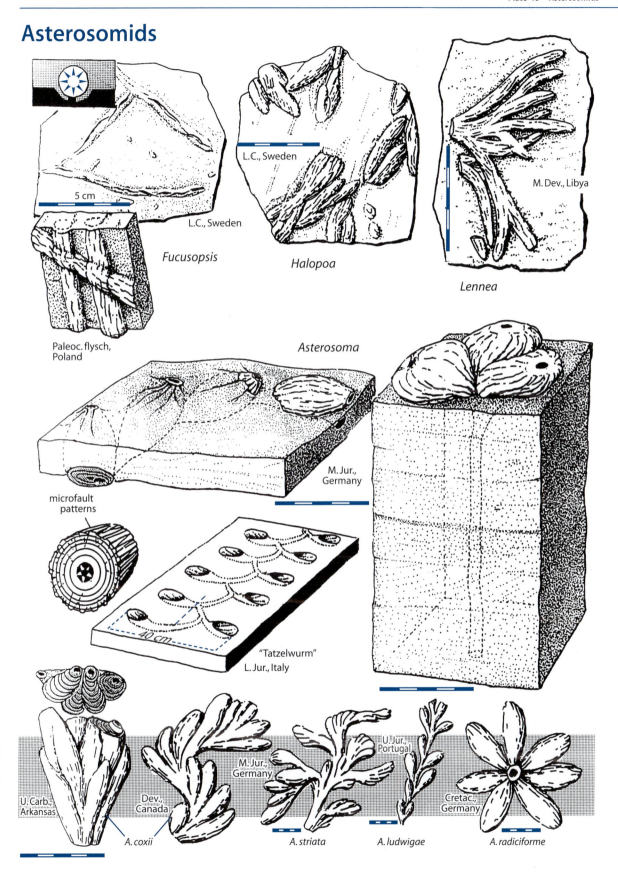

5 cm

L.C., Sweden

Fucusopsis

Paleoc. flysch, Poland

L.C., Sweden

Halopoa

M. Dev., Libya

Lennea

Asterosoma

M. Jur., Germany

microfault patterns

40 cm

"Tatzelwurm" L. Jur., Italy

U. Jur., Portugal

M. Jur., Germany

U. Carb., Arkansas

Dev., Canada

A. coxii

A. striata

A. ludwigae

Cretac., Germany

A. radiciforme

47

Plate 47
Medusiform Burrows (Gyrophyllitids)

Star-shaped trace fossils come in different versions. We have already discussed (Pl. 24) asterozoan resting burrows that reflect the body shape of the maker. Another kind is commonly observed on tidal flats: radial scratches, probings, or feeding traces reflecting the utilization of a superficial film of algae by tube dwellers around the openings of their domiciles. As they are made by different organs in bivalves (e.g., *Scrobicularia*), worms (*Nereis*), and crustaceans (*Uca, Ocypode, Corophium*; Pl. 36), their distinctive shapes and sizes can be used by marine biologists to map animal distributions within tidal flats. However, the low fossilization potential of these surface features reduces their interest for the paleoichnologist. A third group consists of radial teichichnoid burrows (Pls. 41 and 45), whose probes are backfilled with gutterlike lamellae – in contrast to the radial backfill in *Asterosoma radiata* (Pl. 46) as a fourth type.

In the present plate a fifth type of stellate burrows is illustrated, in which radial probings around a vertical shaft are stuffed with mud rather than sand and have a terminal rather than a transversal backfill structure (Pl. 37). Also, probings tend to broaden towards their distal ends.

Many of these burrows have in the past been interpreted as fossil medusae. We group them here as "gyrophyllitids" without claiming that the makers were taxonomically related. What unites them is a similar fabricational program and the ability to inflate parts of their body, e.g., an evertible proboscis.

Ethologically, gyrophyllitids are feeding burrows (fodinichnia), but this affiliation does not exclude the possibility that the preserved structures had a sanitary function, which is still part of the nutritional cycle.

The general appearance of gyrophyllitid burrows is very much influenced by preservation. In various occurrences, one or two out of three distinctive *preservational modes* can be observed:

1. *Negative epireliefs*. Because they correspond to muddy backfill bodies, outlines are very sharp. Also, the pressure exerted in the backfilling action is expressed by the bulging morphology of the impressions and faint compressional wrinkles contouring them on the surrounding bedding plane;

2. *Compressed silhouettes* on split bedding planes of mudstone, distinguished by the different color of the backfill;

3. Calcareous or sideritic *concretions* preserving the three-dimensional shape of the burrow system, but not the internal backfill structure.

As this is not the place for a formal revision, gyrophyllitids of different geologic ages and facies are presented under their original names (none of them created as ichnogenera). Many of these names can probably be synonymized in the future and distinguished at the ichnospecies level. Similar pseudofossils from the Precambrian will be discussed in another context (Pls. 58, 62).

In contrast to stellate teichichnoid burrows (Pls. 41, 45) all forms figured in the upper three rows have in the center the scar of a vertical shaft, whose diameter is much smaller than that of the radial probes. Note also that in multitiered rosettes (*Stelloglyphus*) the probes tend to become longer in the upper tiers. This suggests that they were made in upward succession in concert with sedimentation and opposed to spiral sequences (Pl. 48) that may have a similar appearance in bedding plane sections. *Kirklandia* and *Palaeosemaeostoma* are from shallow-marine units, all the others (*Stelloglyphus, Gyrophyllites, Atollites*) from flysch-like deposits.

In the Cambrian burrows of the lower row (*Brooksella, Dactyloidites*), also described originally as medusae, the central shaft comes from above. Thus they resemble *Teichichnus stellatus* (Pl. 41), except that their radial spreite structures are protrusive rather than retrusive. In *Brooksella*, a secondary concretionary overprint added to the bulky medusoid appearance. Such burrows should not be confounded with radial *pseudofossils*, which shall be discussed in Pls. 58 and 62.

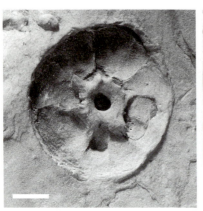

■
Gyrophyllites
(left: M. Jur., Germany; right: Eoc., Vienna, Austria)

Plate 47 · Medusiform Burrows (Gyrophyllitids)   137

Medusiform Gyrophyllitids

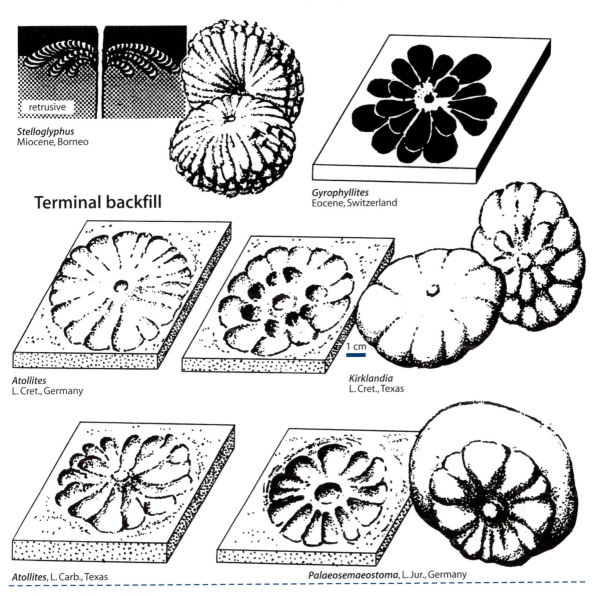

retrusive

Stelloglyphus
Miocene, Borneo

Gyrophyllites
Eocene, Switzerland

Terminal backfill

Atollites
L. Cret., Germany

Kirklandia
L. Cret., Texas

1 cm

Atollites, L. Carb., Texas

Palaeosemaeostoma, L. Jur., Germany

Lateral backfill

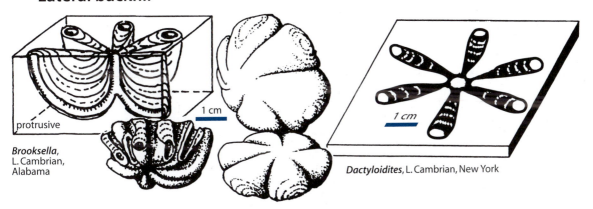

protrusive

1 cm

Brooksella,
L. Cambrian,
Alabama

1 cm

Dactyloidites, L. Cambrian, New York

48

Plate 48
Various Fucoids

The burrows assembled on this plate (except *Haentz-schelinia*) are unified mainly by their occurrence in post-turbidite assemblages of marly Upper Cretaceous and Tertiary flysch sequences. Compared to associated *Zoophycos* (Pl. 38) and *Scolicia* (Pl. 26), they are relatively small. All can be referred to wormlike sediment feeders backfilling their blind probes with terminal backfill that usually appears darker than the matrix. We shall discuss them in the order of their burrowing programs, which result in radiating, lyre-shaped, palmate or biserial arrangements of the probes around a central shaft connecting with the surface. In a broad functional sense, these fucoids are related to the chondritids (Pl. 50).

In line with the previous plate, we begin with the fairly large **Cladichnus fischeri**, whose probes radiate at different levels from a vertical shaft. In contrast to gyrophyllitids (Pl. 47), these probes are (1) more widely spaced; (2) subequal throughout in diameter, but unequal in length; (3) developing lateral branches at an angle of about 30°. In this case, the meniscate backfill structure within the compressed probes is clearly seen and sometimes produces a ragged outline, but its retrusive structure relative to the central shaft faces us with a problem that applies also to chondritids (Pl. 50). Assuming that the probes were first excavated to their full lengths, the backfill can be interpreted as reworked or digested host sediment only if its whole volume was stored within the gut and then left the body. This situation would be easier to explain in a sanitary burrow where probes were first made by wedging and then continuously filled with faecal material introduced from above.

In **Haentzschelinia** (formerly decribed as *Spongia ottoi* from Cretaceous shallowmarine sandstones of Germany), the probes are backfilled with sand and connected with the surface by an *oblique* shaft. Therefore, branches are arranged in a spreading palmate pattern comparable to *Cruziana ancora* (Pl. 15), *Oldhamia antiqua* (Pl. 49), *Chondrites recurvus* and *Phymatoderma alcicornis* (Pl. 50). While *Haentzschelinia* appears as positive epirelief, cf. **Taxichnites** is a *negative* epirelief on top of a vertical shaft. Its short secondary probelets may be made in opposite sequence on the two flanks of each radial probe, as in *Chondrites targionii* (Pl. 50). In contrast, the backfill bodies of **Polykampton**, from silty turbidites, form alternating wings along a horizontal trunk. As they appear feathered on broken surfaces, but with concentric scratches on hyporeliefs, they may be fabricationally compared to the "fannings" of *Gyrochorte* (Pl. 35). **Halimedides fuggeri** from a calcareous flysch has similar wings, but they are paired and terminally backfilled.

In horizontal section, **"Gyrophyllites" doblhoffi** resembles a minute medusoid burrow (Pl. 47). Its terminally backfilled lobes, however, are recurved, because they are spirally arranged around a vertical shaft. Also in contrast to *Stelloglyphus* (Pl. 47), lobes become larger at lower levels, suggesting a downward progression.

Lyrate bifurcation is also typical for **Hydrancylus** from **Italy**, but its probes are broad and leaflike, so that the whole burrow system looks like a minute version of a lobate *Zoophycos* (Pl. 38). The specimen from **Spain** shows that the lobes are one-sided and curve back as in *Phycosiphon* (Pl. 40), being made in similar succession. That the probes were actually tongue-shaped rather than compactionally compressed, is shown by much more complex burrow systems from the Miocene flysch of Borneo, here tentatively referred to **Cycloichnus**, which was originally described from horizontal sections in New Zealand. Preserved as ironstone concretions, they fail to show internal backfill structures, but radial progression is expressed as a delicate pattern of concentric scratches. The distinctiveness of *Cycloichnus* results from the close packing and upturning of *Hydrancylus*-like lobes into bodies that resemble a cabbage head. The enlarged detail shows an individual "leaf" from the outer and inner side with concentric scratch patterns.

One may also add burrow systems that look like twigs of a conifer (**Taxichnites**). Because they are so small, the underlying burrowing program is difficult to decipher. Lateral backfilling appears to have been involved and the generating shaft was probably a straight tunnel at a higher level, in which the maker retracted after having completed a branch, possibly in an outward succcession of bites on convex side and backwards on the other. This is indicated by asymmetric profiles of the "leaves" on opposite sides and their undulated connection along the midline.

The complex search pattern of **Criophycus** is so similar to that of *Lophoctenium* (Pl. 40) that it has at times been synonymized with this ichnogenus. Yet, probes and probelets can be made on either side and are in less intimate contact with one another. Occuring as positive hyporelief on soles of sandy turbidites, *Criophycus* definitely belongs to the postturbidite association. But it is unrelated to *Lophoctenium*, which did not survive the end-Permian extinction.

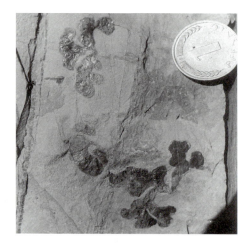

■ *Hydrancylus oosteri*. Holotype (U. Cret., Switzerland)

Plate 48 · Various Fucoids   139

Various Fucoids

Cladichnus fischeri,
U. Cret., Austria

Haentzschelinia,
Mioc., Japan

cf. Taxichnites,
U. Cret., Japan

Polykampton,
U. Cret., Austria

Halimedides,
U. Cret., Austria

Italy

1 cm

"Gyrophyllites"
doblhoffi,
U. Cret., Austria

Hydrancylus, U. Cret.

Spain

Cycloichnus,
Mioc., Borneo

Taxichnites wurmi,
L. Carb., Germany

T. minimus, Paleoc., Spain

Criophycus, Eoc., Austria

49

Plate 49
Undermat Miners

Most chondritid burrow systems (Pl. 50) exploit the sediment along bedding planes – at least in their distal probes. But they become oblique to bedding in proximal sections and were presumably centered by a vertical shaft that maintained an open connection with the surface.

In contrast, the burrows shown on this plate are completely confined to single bedding planes and probably originated only a few millimeters below the sediment surface. They represent a lifestyle that was particularly common in Precambrian and Cambrian times (Chap. XIII), when deep burrowing was neither feasible (because of low oxygen levels in the pore water) nor required (because of low predator pressure). Undermat mining survived, and became more elaborate, in the early Phanerozoic. At the same time it became restricted to extreme environments that were either hostile or unattractive for other bioturbators due to low oxygen levels or high salinity in bottom waters. Even present-day rain puddles fall into this category, because they are too short-lived for higher animals to complete their life cycles, while they may do for larval stages. We shall come back to matground biota in Chap. XIII. Here we are only interested in behavior patterns as compared to those discussed in the previous plates.

Eochondrites. Bark beetles come to mind when one first sees the figured trace fossils in the Upper Cambrian of Oman. While other beds in the section (mostly storm sands) teem with trilobite and other burrows (Pl. 14), this horizon of fine silt is perfectly laminated and contains no other trace fossils. So these sediments were deposited in a low-oxygen period, in which biomats could develop. Yet the similarity of *Eochondrites* (an informal name) with the excavations of **bark beetles** is misleading. Whereas the meniscate backfill lamellae of the grubs are convex towards the mother tunnel, those of the trace fossil curve in the opposite direction. Thus they present terminally backfilled feeding probes of the stem tunnel's inhabitant rather than products of its offspring. Nevertheless one can say that this animal was able to sharply turn by 90° when making the probes. Also note that the main tunnel is at a somewhat lower level than the probings, which makes sense in an animal that grazed on the decaying zone of a microbial mat from below and possibly used it as an oxygen mask.

Oldhamia. While *Eochondrites* was found in a shallow marine sequence, *Oldhamia* occurs in thick turbidite sequences usually interpreted as deepsea sediments. We shall come back to the biohistoric significance of Early Cambrian *Oldhamia* ichnospecies in discussing Pl. 62.

Besides the facies context, the figured burrow systems share (1) occurrence in finely laminated (i.e. otherwise unbioturbated) clays and silts; (2) restriction to one bedding plane with only a very short exit to the surface; (3) relatively small size and (4) terminal backfilling. The last feature may

not be obvious, because one usually observes only a bedding plane relief, but this relief may *reverse* on an adjacent lamina or on the same bedding plane, which is a proxy for backfill.

The earliest representative (*Oldhamia recta* from the Late Precambrian Albemarle Group of North Carolina) does not catch the eye, because it consists of irregular bundles of parallel burrow fills. The earlier comparison with *Syringomorpha* (Pl. 41) would in fact be appropriate if the orientational difference were disregarded.

In Lower Cambrian flysches, burrowing programs become much more elaborate and more ornamental. The type ichnospecies, *Oldhamia antiqua*, has the widest distribution. Its probes radiate from an oblique entrance like fireworks and in some cases several such centrifugal clusters are serially arranged; i.e. they were successively made by the same individual. *Oldhamia curvata* is similar except for the gap in front. This suggests that the probes were made in *centripetal* succession.

The Puncoviscana Formation of Argentina has yielded a number of additional ichnospecies showing a remarkable behavioral diversification. *Oldhamia flabellata* resembles the Precambrian form, but has a defined leaflike outline, because probes converge towards the distal as well as the proximal tips. In *Oldhamia radiata*, probes radiate from a vertical entrance and branch to fill the widening interspaces. In *Oldhamia alata* the probes are so closely spaced that they could, in a fabricational sense, be compared to *Zoophycos*. Note that in the figured specimen the second "wing" was made inward (retrusive), opposite to the others. The most unusual behavior is represented by *Oldhamia geniculata*. Here the entrance is again oblique and its gradual displacement forms a kind of central spreite. In concert, probes are made only on one side of the oblique master tunnel and in protrusive succession, as shown by kinks made to avoid collision with previous probes (other probes deviated into a higher or lower plane for the same purpose). Most unusual, however, is the way in which radial interspaces are utilized: instead of widening or branching, each probe turns back in a hairpin curve! This means that the coverage problem was solved by a preset program, rather than by local reaction. All these forms are restricted to particular horizons within the Puncoviscana Formation; i.e. only one of the five ichnospecies is found on a single bedding plane.

"Planning ahead" is also involved in the program of *Oldhamia pinnata* from the Ordovician flysch of Barrancos (southern Portugal). Its maker first produced a horizontal master tunnel, to which lateral probes were then added in a retrusive sequence. The program thus resembles that of *Chondrites bollensis* (Pl. 48), although the biological purpose was different. Close packing of probes in *Oldhamia pinnata* indicates sediment feeding, or rather grazing on the decaying zone underneath a biomat.

The latest representative, *Oldhamia fimbriata* from Lower Carboniferous flysches of Germany and Morocco makes again fanning probes, but in a less regular mode than earlier forms. It persists into Permian flysches in Sicily.

Plate 49 · Undermat Miners   141

Undermat Miners

Bark beetles

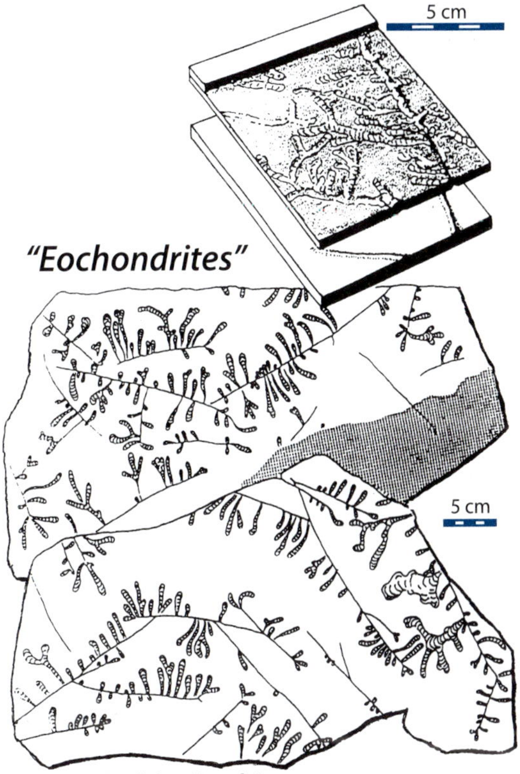

"Eochondrites"

5 cm

5 cm

Upper Cambrian, Huquf, Oman

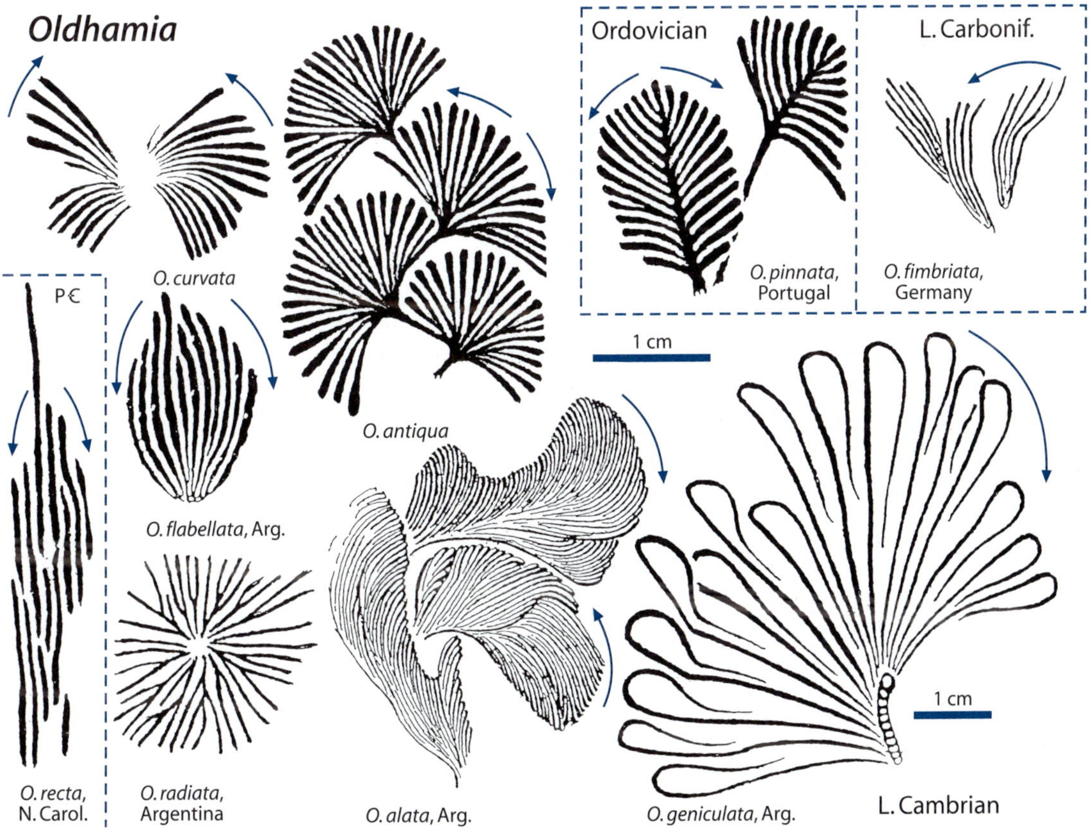

Oldhamia

PЄ

O. curvata

O. antiqua

O. flabellata, Arg.

O. radiata, Argentina

O. recta, N. Carol.

O. alata, Arg.

Ordovician

O. pinnata, Portugal

L. Carbonif.

O. fimbriata, Germany

1 cm

O. geniculata, Arg.

1 cm

L. Cambrian

50

Plate 50
Chondritids

This is one of the most common and distinctive groups of trace fossils. Ranging from the Ordovician to the Tertiary and to modern deepsea muds, chondritids are found in turbidite series as well as in shallowmarine shales and even in storm sands. So one may assume a heterogeneous origin. Yet the different forms share not only a plantlike kind of branching along bedding planes (hence the term "fucoids"), but also a preference for relatively quiet and low-oxygen environments. Like the other forms in this chapter, chondritids are clearly feeding burrows (fodinichia), but details of the probing, feeding and backfill processes are less uniform and must be analyzed separately in every case.

The pellet fillings of **Phymatoderma**, which reminded the ancient artist of angel's wings, are still vexing: consisting now of a material different from the matrix, have they been altered by digestion and differential diagenesis, or are they imported from a food source at the sediment surface? In the following, we bypass this problem by concentrating on the behavioral aspects of burrow *design*. In this respect, *Phymatoderma* differs from *Chondrites* mainly by the poor definition of burrow contours. This is not only due to pellets sticking out. Delicate *feathering* of the probes also implies a transversal component in the backfill (see enlargements). In addition, the angle of branching is rather narrow, allowing for a denser packing of probes than in *Chondrites*. Therefore it may be assumed that the maker of *Phymatoderma* was more strictly a sediment feeder that gathered its food directly, either at depth or at the sediment surface.

In the Posidonia Shales, the dominant form in the upper burrow horizon (Pl. 73; actually at the very top of the bituminous facies) is **Chondrites bollensis**. Partly due to its smaller size, faecal pellets or backfill structures cannot be recognized. On the other hand, the sharp boundaries of the burrows allow us to observe the detailed morphology of branching points.

The reconstructed *behavioral program* is in a way counterintuitive. In the "fucoid" model that still influences our perception, branches would grow as in a Christmas tree, i.e. in distalward (protrusive) succession. In other trace fossils, succession can be tested by interpenetrations between different burrows or parts of the same system. In chondritids, however, the primary rule is to *avoid* contact with any other burrow: "Stop short before a collision can happen". This "*phobotactic*" behavior has been the chief argument against a seaweed interpretation, but it can also be used as a criterion of succession within and between burrow systems.

Applying this criterion to the large (stippled) system in the figured specimen of *C. bollensis*, one discovers that branches must have been made in a *backward* (retrusive) succession: at every scale, proximal probes stop short of more distal ones. To make this possible, a second command must from time to time be introduced: "Penetrate as far as you can in a straight direction to open new areas for subsequent backward exploitation along this axis!"

Still this program is incomplete, because it does not specify the physiological *instrumentation*: (1) how did the animal sense the proximity of another burrow? (2) What limited the probing if there were no other burrows to stop the animal? (3) Why did the animal never return to earlier branches of the system? Chemical sensing is the most probable answer to the first question, while the ultimate length of probings may have been constrained by body length and ventilation. The answer to the third question is more difficult. The rule that earlier parts of the burrow system were never revisited is expressed not only by the consistently retrusive branching patterns, but also by the morphology of the *branching points* themselves: the sediment corner is always shunted towards the older burrow. This leads to the assumption that at any time there was only a single unbranched open tunnel, while the rest of the system (the historical part) was closed off by active terminal backfill (Pl. 37). So we must add the command: "Backfill the previous probe before making a new one!"

Elegant as it may appear, this model has its difficulty. In a *sediment feeder* that swallows the dug-out sediment for food, the volume of the backfill (and the length of each probe) would have been constrained by the storage capacity of the gut. Also, an ordinary worm would have had to turn around in order to deposit the faeces at the former eating place, unless these were regurgitates. Kotake's *sanitary* model avoids this problem, but fails to account for the complexity of the burrowing program.

Most favored at the moment is a *chemosymbiotic* model. It assumes that the *Chondrites* animal used the H_2S of the pore water to feed its bacterial endosymbionts and in addition collected more proteinaceous food from the surface. Supporting this model is not only the preference of *Chondrites* for low-oxygen environments. Pumping of pore water would also explain why the probes are more widely spaced than in stripmining sediment feeders. In addition, backfilling of the probes would provide a double advantage by: (1) getting rid of faecal material and (2) avoiding to divert the pumping energy to abandoned wells. In fact, modern lucinid bivalves (*Thyasira*) use their worm-shaped foot to pump H_2S-laden water for the bacterial symbionts housed in their gills. They thus produce dendritic canal systems, but nothing is known about their backfilling.

Before moving to other examples, we should note another interesting feature of the figured specimen, namely the interaction between different burrow systems. Burrow diameters being rather constant within any one system of *Chondrites bollensis*, the associated smaller burrows are probably the work of conspecific juveniles. In a regular community, these would be expected to occupy ▶

Plate 50 · Chondritids   143

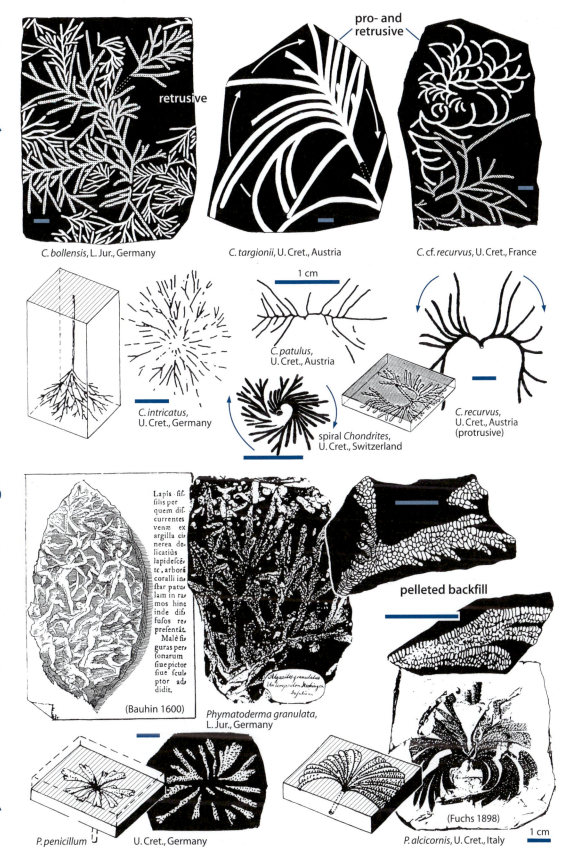

Chondrites (Chemosymbiosis)

C. bollensis, L. Jur., Germany

retrusive

pro- and retrusive

C. targionii, U. Cret., Austria

C. cf. recurvus, U. Cret., France

C. intricatus,
U. Cret., Germany

1 cm

C. patulus,
U. Cret., Austria

spiral *Chondrites*,
U. Cret., Switzerland

C. recurvus,
U. Cret., Austria
(protrusive)

Phymatoderma (Sediment Feeding)

Lapis fiſ-
ſilis per
quem diſ-
currentes
venæ ex
argilla ci-
nerea de-
licatiùs
lapideſcé-
te, arborè
coralli in-
ſtar patu-
lam in ra-
mos hinc
inde diſ-
fuſos re-
præſentát.
Malè fi-
guras per-
ſonarum
ſiue pictor
ſiue ſcul-
ptor ad-
didit.

(Bauhin 1600)

Phymatoderma granulata,
L. Jur., Germany

pelleted backfill

(Fuchs 1898)

P. penicillum   U. Cret., Germany

P. alcicornis, U. Cret., Italy

1 cm

the upper tier. Therefore they would be the first generation on a bedding plane on which an upward tier shift (induced by sedimentation) is recorded. In this case, however, the small burrows were made *after* the large ones; so we may be dealing with a minor *downward* shift due to erosion.

Examples from mud turbidites in **Cretaceous** flysch sequences illustrate modifications of the basic program. In *Chondrites targionii*, branches are very long with few secondary ramifications. The two sides of the system also differ. On the left side, branches curve back and stop short of more proximal ones, indicating a *protrusive* sequence, while branches on the right side are straight and their secondary branches abut distally. This indicates that the program did not begin with the excavation of a main axis to which branches were added on either side in retrusive sequence. Rather, all branches were made to the left in a protrusive succession and then became retrusive on the right side, as in *Taxichnites* (Pl. 48).

Chondrites cf. *recurvus* is smaller, but further complicates the *targionii* program. In the figured specimen, there is no straight axis. Instead, branches radiated lefthandedly and smoothly curved in *protrusive* succession. When the permissible distance was reached, the animal withdrew. On the way back it produced either no probe to the other side or only a few. These, however, curve towards the tip and are made in *retrusive* succession. By addition of new tips, the whole system developed into a dextral spiral that evenly covered a given surface with adequate drainage areas in between. This job took most of a lifetime, as shown by a gradual increase in the lengths and distances of the probes. Note also that this trace was respected by the burrow of another *Chondrites* (stippled) that followed the program of *C. bollensis*.

Other ichnospecies differed in their access to the nutritious zone within the sediment. The delicate **Chondrites intricatus** looks like an inverted tree. It first made a very long vertical shaft that reached below the tiers of associated larger burrows, such as *Zoophycos* and *Scolicia*. It "bloomed" only where the sediment was richest in H_2S, but still soft enough for burrowing in the mode of worms (Pl. 21). Otherwise the branching pattern resembles that of *Chondrites bollensis*, but as the deep layers suffered less compaction, it is preserved in three dimensions.

The next two forms approached the nutritious horizon by an oblique shaft and then bifurcated. In **Chondrites patulus**, which probably employed the *bollensis* program, the two wings are straight. In **Chondrites recurvus**, however, they became lyrate, because its maker subscribed to the *targionii* program. A form recently described from the flysch of **Switzerland** used the same program, but only to one side and without curving back, to produce a *helicospiral* pattern This is more likely than an induction of spirality by a (now invisible) *Gyrolithes* (Pl. 18).

Switching to the sediment feeders, the radial arrangement of **Phymatoderma penicillum** resulted from a vertical shaft and the lyrate pattern of **P. alcicornis** from an oblique one. But as the producers were sediment feeders, the patterns are not ramose with sharp outlines, but stripmined fans transversely stuffed with faecal pellets, as in **P. granulata**. Also, the shaft tunnel did not come from above, but from below.

Interestingly there is at least one other occurrence in which *Phymatoderma* is the dominant element: the Pliocene Esmeralda Shales on the coast of Ecuador, where it occurs in association with *Zoophycos*.

This is another case where the overall shape of the burrow should be given less taxonomic weight than structure, because it resulted from general rules of space utilization (note similarity with radiating trilobite burrows, Pl. 15; and arthrophycids, Pl. 42). True biologic relationships are probably better reflected in modes of backfilling and pattern execution.

■ *Lophoctenium* (Dev., Chios, Greece; see Pl. 40)

Deepsea Farmers

Environmental Conditions

Deepsea bottoms are a very special environment – not only because they cover more than half of this planet. In the absence of light there is also no primary production other than bacterial chemosynthesis, whose fuels (H_2S and CH_4) are partly derived from organic matter. Otherwise the main food source consists of the breadcrumbs sinking down from the photic zone. Even these have usually passed through other stomachs, before they reached the bottom. So food at the deepsea floor is at premium not only in a quantitative, but also in a qualitative sense. Under such extreme conditions, one might expect a lowly diverse fauna of generalists. But the opposite is the case: deepsea dredges reveal levels of within-habitat diversity and differentiation that are reminiscent of tropical reefs and forests! To account for this paradox, the marine biologist Howard Sanders proposed the *Time-Stability Hypothesis*. It claims that within-habitat diversity depends on the permanence of environmental conditions through long periods of time, more than on levels of nutrient supply. Over time, diversity may be enriched by species that immigrate from shallower environments, but probably *sympatric speciation* is an equally important factor.

Despite being based on behavioral rather than developmental differentiation, deepsea trace fossils corroborate this rule. Starting in the Precambrian (Pls. 49, 65), their diversity has gradually increased until it reached levels well above those of shallowmarine ichnocoenoses by Late Cretaceous and Early Tertiary times. Interestingly, this trend applies not only to the preturbidite background association, but also to post-turbidite communities (Pls. 26, 29, 33, 34, 38, 40, 48, 50). Obviously, turbidity events were too local to disrupt community evolution as a whole, but at the same time frequent enough to create a predictable and time-stable environment of its own. On modern deepsea bottoms, areas that have been covered by a turbidite maintain a distinctive fauna for centuries after the event.

Diversification, however, is always combined with *specialization*, which increases the vulnerability against unpredictable changes at a global scale. So we should expect a great impact of the event, or chain of events, at the end of the Cretaceous, which decimated "Golden Age" biota in shallowmarine and terrestrial realms. Thanks to orogenic events, deepsea sediments of this critical interval are preserved in many parts of the world. Yet, their rich and highly diverse ichnofauna is not at all affected by the end-Cretaceous mass extinction. In fact, we rely on planktonic microfossils to distinguish between Late Cretaceous and Tertiary flysches. Although being ecologically decoupled from other realms, deepsea biota are not necessarily immune to extinction, but the triggering events in this biotope are of a different kind, such as oceanic anoxia.

There appears also to be a difference with regard to the driving forces of evolutionary change. Morphological trends in the shells of shallowmarine organisms can be largely accounted for by increasing predator pressure. In deepsea ichnocoenoses one gets the impression that *competition* was more important and that it worked by accommodating new species rather than by the old ones being outcompeted. This tendency is expressed not only by increasing levels of diversity within communities, but also by a general *miniaturization* trend. Given a limited but rather constant food supply, reduction of body size allowed more individuals and species to share the budget.

Yet, miniaturization is not the only trend. Baited deepsea traps have caught isopod and amphipod crustaceans the size of a lobster. These *giants* depend not on the constant rain of microscopic food particles, but on a discontinuous food supply: large carcasses. To quickly reach them and to survive the long starvation between feasts, large body size is at a premium. Giant trace fossils in the post-turbidite association (see *Zoophycos*, Pl. 38) are also opportunistic by penetrating deeper into the turbidite. They tap a food sources that cannot be reached by associated smaller sediment feeders.

Burrow Patterns and Functions

Burrows in deepsea environments show more complex patterns than neritic ones. In addition to strip miners (Pls. 38, 40) and probers (Chap. X), one observes a great variety of guided spirals (Pl. 33) and *meanders*. Being different in a fabricational sense, they all cover a given surface evenly. The way in which food is recovered from this area, however, varies from group to group. In the post-turbidite association (Pl. 72), nereitids (Pl. 34) are intensive sediment feeders. So are the associated strip miners – only that they proceed continously instead of following a repetitive program. In contrast, meandering *graphoglyptids* (Pls. 52–54) carefully maintain a distance to neighboring burrows. In addition, their burrows did not become actively backfilled, but seem to have functioned in the way of open *drainage* systems. They are here united with radiating (Pl. 54) and net-like tunnel systems (Pl. 55) that met the same requirements through different fabricational pathways. All these tunnel systems served most probably for cultivating bacteria or other chemoautotrophs in the mucus linings of the tunnel walls. In this flower bed, they could use H_2S or CH_4 diffusing from the pore water and be occasionally harvested by the host.

The name "Graphoglyptids" was introduced by Theodor Fuchs as an informal term for a diverse group of ornamental trace fossils found as positive reliefs on the soles of flysch sandstones that we now interpret as turbidites. Although there is no ichnogenus of that name, the term is here used in the sense of an ichnofamily, whose most prominent representative is *Paleodictyon*. This is also the form that has been photographed and sampled on modern deeo-sea bottoms. Still, we have as yet no idea about the taxonomic identity of the tracemakers.

After these general considerations, we shall first discuss principles of meandering and then proceed to the fascinating world of graphoglyptids, which dominate the *preturbidite* ichnocoenoses.

Literature

Chapter XI

Azpeitia Moros F (1933) Datos para el estudio paleontólogico del Flysch de la Costa Cantábrica y de algunos otros puntos de España. Bol Inst Geol Min Esp 53:1–65 (First description of flysch trace fossils from Spain, including definition of new ichnotaxa)

Buatois LA, Mángano MG, Sylvester Z (2001) A diverse deep-marine ichnofauna from the Eocene Tarcau Sandstone of the eastern Carpathians, Romania. Ichnos 8:23–62 (Study of a highly diverse Eocene flysch ichnofauna from Romania)

Crimes TP (1974) Colonization of the early seafloor. Nature 48:328–330 (Pioneer paper on the colonization of deep marine environments during the Paleozoic)

Crimes TP, Fedonkin MA (1994) Evolution and dispersal of deepsea traces. Palaios 9:74–83 (Discussion of colonization of deep marine environments during the Paleozoic)

Fuchs T (1895) Studien der Fucoiden und Hieroglyphen. Akademie der Wissenschaften zu Wien, mathematisch-naturwissenschaftliche Classe, Denkschriften 62, pp 369–448 (Diagnosis of graphoglyptids)

Gómez de Llarena J (1946) Revision de algunos datos paleontologicos del Flysch Cretaceo y Numulitico de Guipuzcoa. Notas y Comunicaciones del Instituto Geológico Minero de España 15:113–165 (Additional descriptions of flysch trace fossils from Spain)

Książkiewicz M (1977) Trace fossils in the flysch of the Polish Carpathians. Paleontologia Polonica 36, 208 p (Classic monograph including morphologic classification of deepsea trace fossils)

Orr PJ (2001) Colonization of the deep-marine environment during the early Phanerozoic: the ichnofaunal record. Geol J 36:265–278 (Detailed analysis of colonization of deep marine environments during the Paleozoic, including *Oldhamia*)

Sanders HL (1968) Marine benthic diversity: A comparative study. Am Nat 102:243–281 (Time-stability hypothesis)

Seilacher A (1967) Fossil behavior. Sci Am 217(2):72–80 (Behavioral evolution of deep-sea trace fossils)

Seilacher A (1974) Flysch trace fossils: Evolution of behavioural diversity in the deep-sea. Neues Jahrb Geol P M 1974, pp 233–245 (Progressive colonization of deep marine environments during the Phanerozoic)

Seilacher A (1977a) Evolution of trace fossil communities. In: Hallam A (ed) Patterns of evolution, Elsevier, pp 359–376 (Ichnodiversity in of deep-marine environments through the Phanerozoic)

Seilacher A (1977b) Pattern analysis of *Paleodictyon* and related trace fossils. In: Crimes TP, Harper JC (eds) Trace fossils 2. Geol J, Special Issue 9, pp 289–334 (Morphologic analysis of different graphoglyptid patterns and their ethologic significance)

Seilacher A (1986) Evolution of behavior as expressed in marine trace fossils. In: Nitecki MHG, Kitchell JA (eds) Evolution of animal behavior: Palaeontological and field approaches. Oxford University Press, New York, pp 62–87 (Summary of the evolution of deep marine trace fossils)

Uchman A (1998) Taxonomy and ethology of flysch trace fossils: Revision of the Marian Książkiewicz collection and studies of complementary material. Annales Societatis Geologorum Poloniae 68:105–218 (Further development of morphologic classification of deepsea trace fossils including graphoglyptids)

Uchman A (2003) Trends in diversity, frequency and complexity of graphoglyptid trace fossils: Evolutionary and palaeoenvironmental aspects. Palaeogeog Palaeoclim Palaeoecol 192:123–142 (Analysis of graphoglyptid evolution through geologic time. Results do not support the time-stability hypothesis)

Uchman A (2004) Phanerozoic history of deep-sea trace fossils. In: McIlroy D (ed) The application of ichnology to palaeoenvironmental and stratigraphic analysis. Geological Society of London, Special Publication 228, pp 125–139 (Updated analysis of the evolutionary history of deep marine trace fossils through the Phanerozoic based on the study of 151 flysch formations)

Wetzel A, Uchman A (1999) Deep-sea benthic food content recorded by ichnofabrics: A conceptual model based on observations from Paleogene flysch, Carpathians, Poland. Palaios 13:533–546 (Discussion of food supply as a major controlling factor on deepsea trace fossils)

Plate 51: Meandering

Bourne DW, Heezen BC (1965) A wandering enteropneust from the abyssal Pacific, and the distribution of "spiral" tracks on the sea floor. Science 150:60–63 (Large meandering fecal strings on sediment surface)

Emery KO (1953) Some surface features of marine sediments made by animals. J Sediment Petrol 23(3):202–204 (Remote photography)

Ewing M, Davis RA (1967) Lebensspuren photographed on the ocean floor. Deep-sea photography: The Johns Hopkins oceanographic studies (3):259–294 (Includes *Spirorhaphe* seen at surface and many meandering traces)

Ewing M, Vine A, Worzel JL (1946) Photography of the ocean bottom. J Opt Soc Am 36(6):307–321

Frey RW, Seilacher A (1980) Uniformity in marine invertebrate ichnology. Lethaia 13:183–207 (Optimization of feeding strategies as revealed by meandering traces)

Minter NJ, Buatois LA, Lucas SG, Braddy SJ, Smith JA (2006) Spiral-shaped graphoglyptids from an Early Permian intertidal flat. Geology 34:1057–1060 (*Paraonis* spirals as epichnial grooves in the Robledo Mountain Fm., New Mexico)

Papentin F, Röder H (1975) Feeding patterns: The evolution of a problem and a problem of evolution. Neues Jahrb Geol P M 1975:184–191 (Relationships between Darwinian evolution and feeding optimization based on computer simulations)

Prescott TJ, Ibbotson C (1998) A robot trace maker: Modelling the fossil evidence of early invertebrate behavior. Artif Life 3:289–306

Raup DM, Seilacher A (1969) Fossil foraging behavior: Computer simulation. Science 166:994–995 (First computer models of meandering)

Seilacher A (1967a) Fossil behavior. Sci Am 217(2):72–80 (Feeding optimization as revealed by trace fossils)

Seilacher A (1967b) Vorzeitliche Mäanderspuren. In: Hediger H (ed) Die Strassen der Tiere. Verlag Vieweg, Braunschweig, pp 294–306 (Evolution of meandering)

Plate 52: Graphoglyptids I

Ekdale AA, Berger WH (1978) Deep-sea ichnofacies: Modern organism traces on and in pelagic carbonates on the western equatorial Pacific. Palaeogeog Palaeoclim Palaeoecol 23:167–168, 263–278 (Includes *Spirorhaphe*)

Leszczyński S, Seilacher A (1991) Ichnocoenoses of a turbidite sole. Ichnos 1:293–303 (Detailed analysis of Eocene trace fossils preserved at the base of a large slab from Spain)

Seilacher A (1977) Pattern analysis of *Paleodictyon* and related trace fossils. In: Crimes TP, Harper JC (eds) Trace fossils 2. Geol J, Special Issue 9, pp 289–334 (Discussion on *Spirorhaphe* and tiering)

Plate 53: Graphoglyptids II

Macsotay O (1967) Huellas problematicas y su valor paleoecologico en Venezuela. Universidad Central de Venezuela, Escuela de Geologia, Minas y Metalurgia, Geos 16:7–79 (Description of turbidite trace fossils from Venezuela)

Seilacher A (1989) *Spirocosmorhaphe*, a new graphoglyptid trace fossil. J Paleontol 63:116–117

Plate 54: Graphoglyptids III

Desio A (1941) Un nuovo reperto di *Lorenzinia carpatica* (Zuber) nel Flysch dell'Albania Settentrionale. Riv. Ital. Paleont. 47 (1–2):7–8

Seilacher A (1977) Pattern analysis of *Paleodictyon* and related trace fossils. In: Crimes TP, Harper JC (eds) Trace fossils 2. Geol J, Special Issue 9, pp 289–334 (Detailed analysis of graphoglyptid patterns, including biramous meanders and radiating traces)

Serpagli E (2005) First record of the ichnofossil *Atollites* from the Late Cretaceous of the northernApennines, Italy. Acta Palaeont Pol 50:403–408 (Diagnosis of *Atollites italicum*)

Vialov OS (1971) Rare Mesozoic problematica from the Pamir and Caucasus. Paleont Sbornik Izdatel Llov Univ Vyp Vtoroy 7:85–93 (In Russian)

Plate 55: *Paleodictyon*

Crimes TP, Crossley JD (1991) A diverse ichnofauna from Silurian flysch of the Aberystwyth Grits Formation, Wales. Geol J 26:27–64 (Study of a diverse Silurian ichnofauna from Wales, including descriptions of *Paleodictyon* and *Squamodictyon*)

Ekdale AA (1980) Graphoglyptid burrows in modern deep-sea sediment. Science 207:304–306 (*Paleodictyon* in modern deep sea cores)

Gaillard C (1991) Recent organism traces and ichnofabrics on the deep-sea floor off New Caledonia, southwestern Pacific. Palaios 6:302–315 (Bathymetric distribution of modern deep sea traces)

Książkiewicz M (1977) Trace fossils in the flysch of the Polish Carpathians. Paleontologia Polonica 36, 208 p (Classic monograph on the flysch trace fossils from the Polish Carpathians. See plate 53 for *Paleodictyon*)

Paczesna J (1985) Ichnogenus *Paleodictyon* Meneghini from the Lower Cambrian of Zbilutka (Gory Swietokrzyskie Mts.) (in Polish) Kwart Geol 29:589–596 (Earliest record, still from a shallow-marine environment)

Rona PA, Merrill GF (1978) A benthic invertebrate from the mid-Atlantic ridge. Bull Mar Sci 28:371–375 (Photographs of the deep-sea floor revealed regular patterns of holes, "living trace fossils" confirming Seilacher's 1977 reconstruction of *Paleodictyon nodosum*)

Sacco F (1939) *Paleodictyon*. R. Accademia delle Scienze di Torino, Memorie 69, pp 267–285

Vialov OS, Golev BT (1964) Printsipy podrazdeleniya *Paleodictyon*. Geologija i Razvedka, Izv Vyssh Uchebn Zaved 1964(1):37–48 (Ichnotaxonomy based on mesh size)

Wetzel A (2000) Giant *Paleodictyon* in Eocene flysch. Palaeogeog Palaeoclim Palaeoecol 160:171–178 (Description of giant specimen in Spain)

51

Plate 51
Meandering

We have already discussed the advantage of meandering over spiral programs for covering a given surface with a continuous trail (Pls. 33, 34). In the present plate, modern examples of *guided meanders* are compared with computer simulations in order to get a feel for the biological functions involved and the commands necessary to evolve and execute the program.

Modern Meanders. The first two meanders originate by superposition of locomotion with the swinging action of a terminal organ. In **limpets** and other gastropods grazing algal films on rock surfaces or an aquarium wall, this organ is the radula at the tip of the head (for fossil examples, see Pl. 63). In contrast to the sine-wave trace of the swinging siphon in *Psammichnites* (Pl. 27), the amplitude of molluscan bite series is much larger than the wavelength. This allows the direction of locomotion to be read from the arcuate shape of the turns. Note that locomotion of the snail must proceed incrementally; otherwise distances to previous turns would increase during the swings. Guidance is also expressed by younger turns contouring the tips of earlier ones as the animal changed direction. **Dragonflies** laying their eggs on the lower surfaces of water-lily leaves use a meander program to reach optimal spacing.

In all examples that follow, meandering trails, or burrows, are made by steering the whole body. In **bark beetle** species laying their eggs in isolation rather than in series (Pl. 49), the grubs can choose their course without being constrained by siblings. So they start making meanders across the grain of the cambium, with the burrow as well as the meander loops becoming wider with the growth of the grub. The turns are not significantly arcuate, meaning that the grain of the substrate is a stronger compass than guidance. Similarly, the larvae of **leaf moths** contour the veins as they mine the green tissue between leaf cuticles (the spreite in its original meaning).

Closer to trace fossils are meander trails of **acorn worms** photographed on the deepsea bottom. Although unlikely to become fossilized, they reflect a behavior that is also used by infaunal sediment feeders (Pl. 33): with a *starter spiral* as an initial reference, the guided meanders become arcuate. At the same time, a string of digested material is left behind.

While the function of *Paraonis* spirals in beach sands (**f**) has already been discussed (Pl. 33), a vexing air photo from the fifties brings us back to an analogous human behavior. A **farmer** in the American Midwest sprayed his fields to protect them against wind erosion, so he steered his tractor in guided meanders. Because the plowed pattern served as an additional compass, the turns did not become automatically arcuate, but widened at the ends due to a limited turning radius of the vehicle.

Computer Simulations. Back in the sixties, David M. Raup used his early computer skills to simulate meandering worms with only two commands: (1) to follow the previous trace at a certain distance "d" and (2) to turn back after a certain length "l" had been covered in one direction. By using random numbers to vary both parameters, and changing their proportions in each run, he produced patterns that a trace fossil expert would have affiliated with known ichnotaxa. Most interestingly, all simulated trails start with a spiral, although no specification for it was given in the program (Pl. 33). Also, wider d relative to l led to a more irregular course.

Later on, Röder and Papentin wondered how such a behavioral program could have evolved by *Darwinian selection*. For this purpose they created a virtual worm, ***Rectangulus rectus*** (*nomen nullum*, because there is no type specimen). It had the habit of creeping straight ahead, but could also sense the proximity of another trail and was able to turn right or left at a right angle. A constant number of individuals was allowed to enter the screen. After the first run, selection rewarded the champions (i.e. the ones that had lower numbers of intersections) with a higher reproduction rate for the second generation and so on. After about ten generations, patterns began to emerge and coverage became improved. At the eightieth generation, there were only a few intersections and the dominant pattern was a spiral contoured by meanders. From then on, patterns and coverage changed very little, because an equilibrium had been reached.

The same experiment with another virtual species (*Rectangulus vagus*) had an identical result; i.e. under the same selection pressure, *homeomorph* patterns evolved in originally very different lineages. There was also a chance effect: in some clades of either species the terminal patterns were not spiral, but *linear* meanders.

With this background we are now ready to tackle the group of trace fossils in which behavioral complexity and diversification reached an unparalleled peak.

Plate 51 · Meandering   149

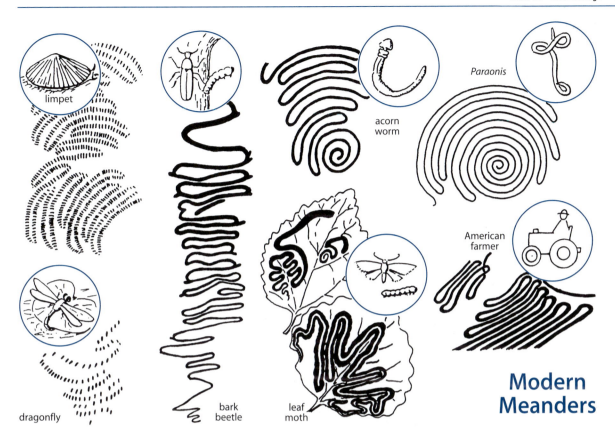

limpet

acorn worm

Paraonis

dragonfly

bark beetle

leaf moth

American farmer

Modern Meanders

Program test
(Raup)

Selection against overcrossing
(Röder and Papentin)

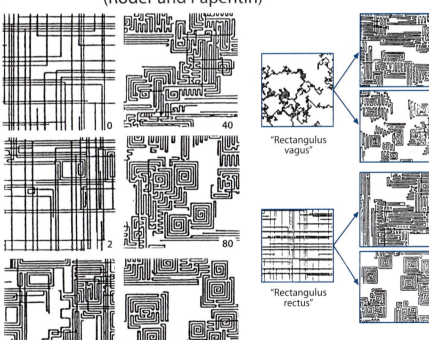

"Rectangulus vagus"

"Rectangulus rectus"

Computer Worms

52

Plate 52
Graphoglyptids I

Graphoglyptid Preservation. Besides their complex patterns, all graphoglyptids share a particular mode of **preservation**. Sharp boundaries and smooth burrow surfaces might suggest a post-turbidite origin. Yet one never sees them pass into hypichnial furrows (as *Granularia* does; Pl. 18), nor do vertical sections show any continuation into the overlying sand. But if they have been pre-depositional, how could the delicate impression in the mud survive the erosional phase of a turbidity current without being erased, or blurred, like other washed-out burrows (*Taphrhelminthopsis*, Pl. 26; *Spirophycus*, Pl. 33)? In order to understand this paradox, we have to consider the *sedimentary history* of a turbidite.

In all high-energy events, the depositional phase is preceded by an erosional one. Erosional features on turbidite soles include groove casts, flute casts and frondescent casts. They originated underneath a flow of suspended sand, which cast them immediately as the current slowed down. But graphoglyptids are never associated with such sedimentary structures. Still they may show more subtle signs of erosion, such as delicate flutings on the down-current side of tunnel casts (**c**). Locally the approach of the turbid water body would be felt as a sudden *shock wave*, as in front of a snow avalanche. As this means instant acceleration of the water, it would *suck* the unconsolidated surface mud into suspension; but only if the turbid cloud was already in its depositional phase, could it immediately cast the freshly exhumed burrow systems (**b**). Therefore one best searches for graphoglyptids in the distal zones of deepsea fans, where turbidites are relatively thin and where the biogenic structures exhumed by shock erosion were not secondarily erased.

Tractional erosion (as indicated by the sole marks mentioned above) is not only more destructive; it also reaches down to levels in which open tunnels have already disappeared by compactional collapse (**e,f**). Only extremely large graphoglyptids were exempt from this fate. Having been made at a more compacted level within the mud, their tunnels stayed open for a longer time (**h**); so they could become exhumed by frondescent rip-off and be sand-cast in their original geometry (**i**). Another possibility is for tunnels already compacted to be exposed by shock erosion. Unnaturally flattened casts (**f**) are the result.

Problems in Program Execution. While the regularity of patterns makes graphoglyptids ornamentally appealing, the meander in the upper left corner (*Helminthorhaphe reflecta* from an Upper Cretaceous flysch in Italy) cannot claim such quality. Instead of maintaining a proper distance (as in *Helminthorhaphe*, Pl. 53), meander turns start with a wide loop before they get in closer contact with the previous turn. In the Early Cambrian *Psammichnites saltensis* (Pl. 65) such behavior is general and can be explained by an underdeveloped

sensing of lobe distances. Here we probably deal with a personal problem that is unsuitable as an ichnotaxobase.

Shortcuts. A less conspicuous, but informative, irregularity consists of the *shortcuts* (asterisks) between the turns of this complex meander (*Cosmorhaphe sinuosa*) from the Eocene flysch near Zumaya, Spain. As in *Paraonis* meanders (Pl. 51), this tells us that graphoglyptid burrows were *open tunnels* that could be revisited on later occasions. Its also implies that the tunnel was reinforced by mucus, which explains the smoothness of the casts compared to washed-out backfill burrows such as *Taphrhelminthopsis* (Pl. 26) and *Spirophycus* (Pl. 33).

Function. As discussed in the introduction, graphoglyptids do not attempt to completely cover a given surface. Rather they subdivide it into areas of similar diameters, like a drainage system. This makes sense in **farming** burrows, in contrast to the **foraging** patterns of sediment feeders, such as the post-turbidite *Helminthoida* (Pl. 34).

Tiering. Turbidite erosion depicts bedding planes below the original sediment surface, on which different generations of graphoglyptids may be associated. In the figured specimen, the maker of the subsequent *Spirorhaphe* (broken line) was guided by a fellow spiral after having made its own. Both, however, disregarded the less well preserved smaller *Spirorhaphe*, as well as *Paleodictyon*, probably from an earlier generation. Assuming that tiers shifted upwards in response to background sedimentation, they may become telescoped into one level.

Multi-storey Meanders. In principle, graphoglyptid tunnel systems spread at one level. In the figured *Spirorhaphe involuta*, however, there is a *surplus whorl* (broken line). As in the previous example, another individual searching for a new estate could have become guided by the coils of a spiral already there, but it is very unlikely that an animal making the inward coil met a foreign turning loop right in the center, thus being forced to turn prematurely. Alternatively, this incident can be explained by a **multi-storey** system, in which the animal deviated from its proper plane when making the turn at the upper level. The multi-story model also has the advantage that the spiral can be derived from a meandering program. If turns in this three-dimensional meander were alternatingly made in a horizontal and in a vertical plane, guidance could switch accordingly; i.e. the animal could follow the outward spiral of the upper floor on the inward stretch and switch to a horizontal guidance on the way out. *Spirorhaphe azteca* has no turn in the center, but nevertheless shows surplus whorls (dotted). They can be explained by a spiral meander, in which all turns were vertical, but with a similar switch between horizontal and vertical guidance, and without adding a new whorl in the subsequent spiral.

Plate 52 · Graphoglyptids I   151

Graphoglyptids I

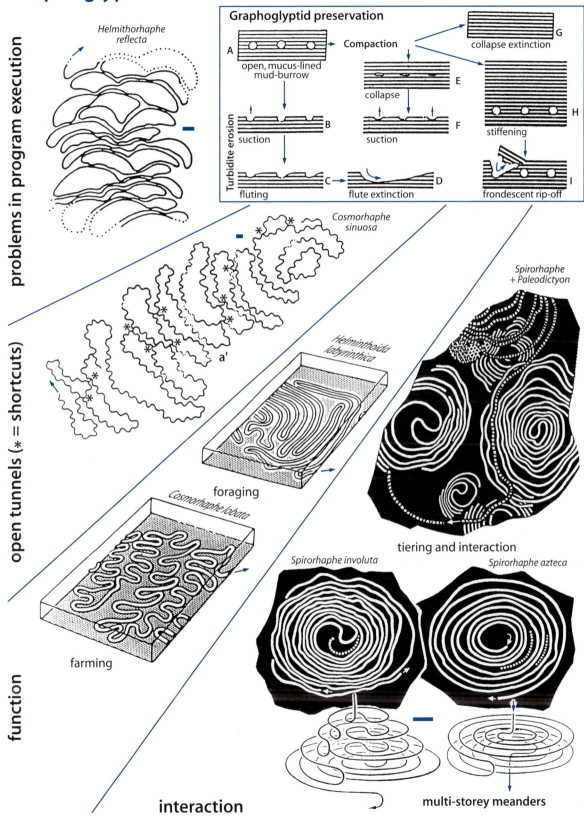

problems in program execution

Helmithorhaphe reflecta

Graphoglyptid preservation

A — open, mucus-lined mud-burrow → **Compaction** → G collapse extinction

Turbidite erosion

B — suction

C — fluting

E — collapse

F — suction

D — flute extinction

H — stiffening

I — frondescent rip-off

open tunnels (* = shortcuts)

Cosmorhaphe sinuosa

a'

Helminthoida labyrinthica

foraging

Spirorhaphe + Paleodictyon

tiering and interaction

function

Cosmorhaphe lobata

farming

Spirorhaphe involuta

Spirorhaphe azteca

interaction

multi-storey meanders

53

Plate 53
Graphoglyptids II

The notion that meandering graphoglyptid tunnels were bacterial gardens (*agrichnia*) rather than burrows of sediment feeders (*fodinichnia*; Pl. 31) is corroborated by the ways, in which the basic meander program became modified, and excess entrances introduced, in various ichnospecies.

Non-branched Meanders. Disregarding secondary shortcuts (Pl. 52), tunnels in this group do not branch and differences between ichnospecies refer to the small bends and wiggles along the large primary meanders. As the specimens shown were not collected in stratigraphic sequence, their arrangement in lineages is hypothetical and does not claim to depict the "tree" of behavioral evolution.

In Pl. 52, *Spirorhaphe involuta* was already discussed as a multi-storey meander with a central yin-yang turn. To make this design at a single level, the animal would have had no guidance for the inward spiral. In the three-dimensional model (derived from the surplus turn in Pl. 52), the reference problem is solved by *vertical* guidance on the way in and *horizontal* guidance on the way out. By adding an outward whorl at each successive level, spirals got automatically larger, with the number of inward whorls corresponding to the number of storeys above.

Spirorhaphe azteca had no such problem, because it presumably started from the center and made spirals alternatingly outward with horizontal, and inward with vertical guidance (Pl. 52). *Spirorhaphe graeca* from an Eocene flysch of Greece is too big (up to 80 cm!) to apply this model; still, the spaghetti shape and distancing of the hypichnial tunnel casts distinguish it from echinoid-made spirals (Pl. 26).

Helminthorhaphe did not require a starter spiral. Nevertheless, directionality is clear from the arcuation of the meanders. *H. crassa* is more loosely guided than *H. japonica*, whose turns widen as in the farmer's track (Pl. 51).

In *Cosmorhaphe* the evolutionary order becomes conjectural. Should we regard the more ornamental patterns as more advanced? In this view, *C. tremens* and *C. neglectens* would be derived from a kind of *Helminthorhaphe* with incipient smaller wiggles and lobes, while *C. parva*, *C. lobata*, and *C. involuta* (see also Pl. 52) would be more advanced. But in a fabricational sense, the latter forms could also be considered as more primitive, because they simply execute a set of nested meander programs. As the assumed function of even coverage improves in the more irregular forms, these have been taken as derived in *Cosmorhaphe* (secondary meanders only occasionally executed) and *Helicocosmorhaphe*. In the latter all or only some of the secondary meander lobes leave the bedding plane to perform a twisted three-dimensional looping. Note that in *H. sigmoidalis* (from Alaska) these lobes switch the sense of coiling at

every turn of the master meander, but irregularly in the much larger *H. helicoidea* (from Austria). (In this and the following plates scale bars are 1 cm unless stated otherwise).

Although a simple sinuous trace (*Cochlichnus*, Pl. 33) has here been figured for an "ancestor", this comparison is misleading. Graphoglyptid meanders are unrelated to sinuous locomotion, but represent intensional programs that can be not turned on and off in evolution and be modified according to the mood of the animal or the local situation. In *Cosmorhaphe lobata* directionality is expressed in the arching of the primary as well as the secondary meanders, because both seek guidance by the previous element. Which of the two nested meanders was primary in an evolutionary sense is another matter.

Uniramous Meanders. In order to oxidize H_2S or CH_4, chemosynthetic bacteria need oxygen, which may be limited in long tunnels not actively ventilated. To aerate the bacterial gardens by diffusion, additional shafts to the surface may be introduced. In the erosional reliefs preserved on turbidite soles, these shafts may not be obvious, but their presence can be inferred from the course of the meanders. In *Belorhaphe* ("lightning thread"), for instance, the pointed corners of the smaller meanders make sense only if a ventilation shaft branched off at every turn. These shafts become more visible in *Protopaleodictyon* because they followed the bedding plane before turning up.

In *Helicolithus*, the second-order meanders transform into corkscrew tunnels. As coiling direction changes at every turn of the first order meander, the bedding-plane expression of *H. sampelayoi* shows mushroom-shaped meander turns as in *Helicocosmorhaphe*. In *H. tortuosa*, the switch is also expressed in a change of sigmoidality (related to inclined ventilation shafts).

The dotted meanders of *Punctorhaphe* probably reflect a base tunnel undulating in the vertical plane with a shaft at every top turn. *Urohelminthoidea* modifies the program of *Belorhaphe* by lengthening the horizontal stretches of the second order meanders and inclining the ventilation shafts. Thereby the area covered became so large that broad meandering became unnecessary. The two ichnospecies (*U. dertonensis* and *U. appendiculata*) shown differ only in proportions.

Broad meanders were also reduced in *Dendrotichnium*. Instead its program emphasized the horizontal sections of the ventilation shafts while reducing the amplitude of the zigzagged basal tunnel. The resulting pattern resembles *Chondrites* (Pl. 50), but with branches made in a protrusive succession.

The two figured ichnospecies of *Hormosiroidea* (whose type species looks more like a string of beads and resembles graphoglyptids mainly by its mode of preservation) stick out by their giant size. Consequently they penetrated into more compacted zones of the mud and became exhumed by rip-off (Pl. 52i).

Plate 53 · Graphoglyptids II   153

Graphoglyptids II

U. Cret.–Eocene Flysch

(Scales = 1 cm, unless indicated otherwise)

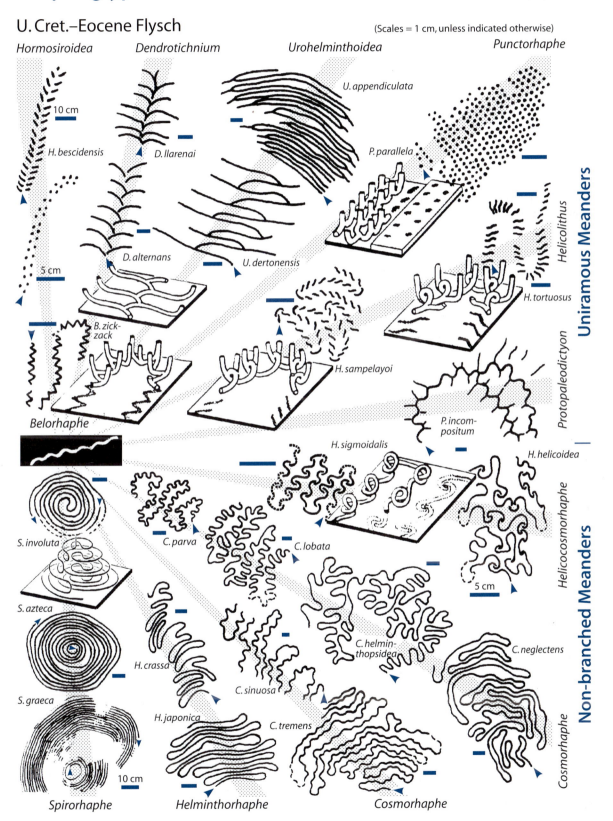

Hormosiroidea

10 cm

H. bescidensis

5 cm

Belorhaphe

S. involuta

S. azteca

S. graeca

10 cm

Spirorhaphe

Dendrotichnium

D. llarenai

D. alternans

B. zick-zack

C. parva

H. crassa

H. japonica

Helminthorhaphe

Urohelminthoidea

U. appendiculata

U. dertonensis

H. sampelayoi

H. sigmoidalis

C. lobata

C. sinuosa

C. tremens

C. helmin-thopsidea

Cosmorhaphe

Punctorhaphe

P. parallela

Helicolithus

H. tortuosus

Protopaleodictyon

P. incom-positum

H. helicoidea

5 cm

C. neglectens

Uniramous Meanders

Non-branched Meanders

Helicocosmorhaphe

Cosmorhaphe

54

Plate 54
Graphoglyptids III

Before going on to discuss other graphoglyptid programs, we should think about the evolutionary significance of all this behavioral diversification. In body fossils we have been trained to interpret every morphological change as improving *fitness*. This view can be extended to behavioral evolution, as we have done in previous chapters of this book. On the other hand, the within-habitat diversity of birdsong should remind us that diversification may also be driven by *character displacement*, if the signal reaches more than one population. If our interpretation of graphoglyptid burrows as bacteria farms is correct, the owner may have left them alone to return only for harvesting. In the meantime, usurpation by other species could become a problem. In contrast to the meanders of bark beetles (Pl. 51), graphoglyptid patterns maintain the same scale throughout the system. So an individual may have maintained several such gardens during its lifetime or even simultaneously. In this case, outgrown gardens could be used "secondhand" by younger members of the population.

In conclusion, the differences between the designs of graphoglyptid gardens should perhaps not be judged by ordinary fitness criteria, but by their distinctiveness – just as housekeys are different to allow access only to members of our kin.

Biramous Meanders. The housekey analogy is particularly useful to interpret minor ichnospecific differences in the biramous graphoglyptid meanders. Their basic scheme is best expressed by *Paleomeandron elegans*. Its "Greek" meanders are rectangular for the same reason that the second-order meanders of *Belorhaphe* appear zigzagged – except that there were *two* ventilation shafts per turn instead of only one. The transverse and the longitudinal connections between branching points could now be independently modified. On the other hand, the rectangular pattern is not as easily transformed into a three-dimensional corkscrew as in uniramous tunnels (Pl. 53).

In *Paleomeandron biseriale*, the basic program is modified by making the longitudinal stretches deeper, so that the transverse stretches are suppressed in the erosional hyporelief.

Paleomeandron transversum has the same arrangement, with a deeper longitudinal stretch connecting the two shafts into a transverse U-tube. In the following forms the transversal stretches have become extremely elongated. So, first-order meanders became impracticable. They again differ in the relative levels of the elements. The longitudinal connections in *Desmograpton ichthyforme* bend upward, but downward in *Desmograpton geometricum* and *D. inversum*. *Oscillorhaphe*, in contrast, lengthens the transversal stretches of the secondary meanders, which appear truncated by the bases of the U-shaped ventilation shafts.

Protopaleodictyon bicaudatum is again larger and more irregular than other forms in this group, but in contrast to *Protopaleodictyon impositum* (Pl. 53) there are regularly two branches per turn. In a future revision, all biramous ichnospecies could probably be united in a single ichnogenus.

Radiating Graphoglyptids. Here we return to patterns known from probers (Chap. X), with the difference that the present forms have a graphoglyptid mode of preservation (Pl. 52) and that probes were not backfilled blind tubes, but open tunnels with a ventilation shaft at the end. Otherwise the rules of pattern formation are similar: *oblique* entrances produce bilaterally symmetrical patterns, while arrangements become radially symmetrical around *vertical* entrances.

The graphoglyptid character is most clearly expressed in *Tuapseichnium*. *T. simplex* has no probe extending in the median plane. Rather, two U-shaped segments are arranged on either side. *Tuapseichnium cervicorne* starts in the same fashion, but adds more such modules to each of the initial ventilation shafts, plus a secondary radial branch on the outside of each of the two initial branches. Note that the two initial branches run strictly parallel to the midline, while the subsequent branches curve *back* towards them in order to reduce unused sectors – opposite to what we have seen in the palmate burrows of sediment feeders (Pls. 42, 48, 49, 50). *Tuapseichnium ramosum* from the Paleocene of Daghestan follows the same principle, as shown by the bilateral symmetry and the median seam. Yet the curved segments are so perfectly and closely knit that the actual performance is hard to reconstruct (If you manage to do it, let me know!).

Yakutatia from the Upper Cretaceous of Alaska appears to be also bilaterally symmetrical. The two branches presumably radiate from an inclined entrance, but in contrast to *Tuapseichnium* they curve outward into a spiral, with secondary branches taking off with an opposite curvature. More material would be required to fully understand this program.

Another line of radiating graphoglyptids is represented by *Glockeria*. Its bilateral forms (*G. dichotoma*, and *G. alata*, of very different sizes) can be derived from the design of *Tuapseichnium simplex* by adding bifurcations and reclining the ventilation shafts. The shift from a bilateral to a radially symmetrical arrangement of such branches in *Glockeria glockeri* is no big step either. As in *Oldhamia radiata* (Pl. 49), it was probably induced by the change from an inclined to a vertical entrance.

Resemblance to actively backfilled burrows has muddled the interpretation of other radiating graphoglyptids. The dendroid *Chondrorhaphe* (not figured), looks like *Chondrites* (Pl. 48). It is not only preserved like other graphoglyptids, but also has a dichotomous mode of branching.

Similarly misleading is the resemblance of *Lorenzinia* to backfilled stellate burrow systems (Pl. 47). As the erosional hyporeliefs never show a central shaft, branching must have occurred at a higher level. Also, the faint crosslinks in *Lorenzinia apenninica* suggest a branching mode like in *Glockeria*. *L. moreae* (formerly *Bassaenia*), on the other hand, may be the cross-sectional expression of very regular radial U-tunnels, because numbers are the same in both circles.

Plate 54 · Graphoglyptids III   155

Graphoglyptids III

biramous meanders

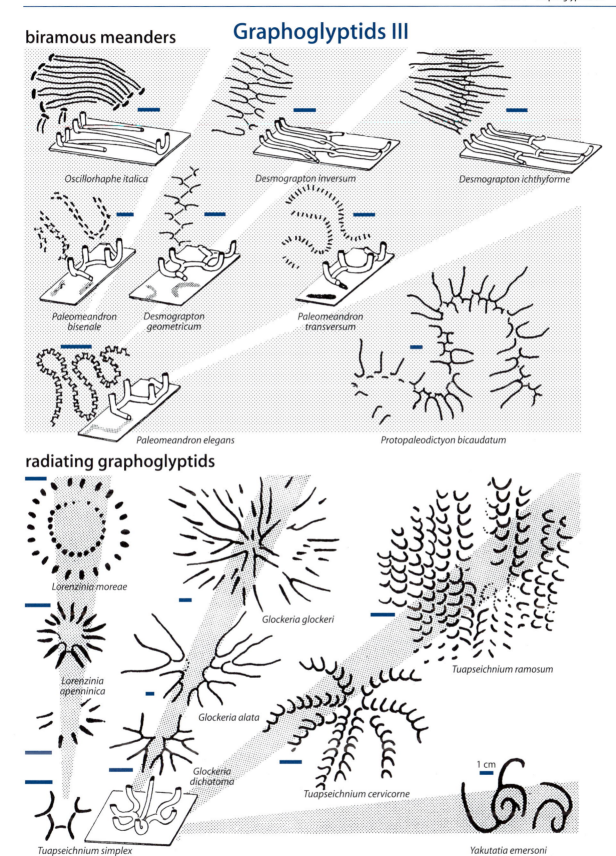

Oscillorhaphe italica

Desmograpton inversum

Desmograpton ichthyforme

Paleomeandron bisenale

Desmograpton geometricum

Paleomeandron transversum

Paleomeandron elegans

Protopaleodictyon bicaudatum

radiating graphoglyptids

Lorenzinia moreae

Glockeria glockeri

Tuapseichnium ramosum

Lorenzinia apenninica

Glockeria alata

Glockeria dichotoma

Tuapseichnium cervicorne

Tuapseichnium simplex

1 cm

Yakutatia emersoni

Plate 55
Paleodictyon

By its regular meshwork, *Paleodictyon* is the most conspicuous of all flysch trace fossils and therefore most likely to appear in field reports. Found in shallowmarine environments in the Cambrian, and in deepsea settings from the Ordovician to the present day, it has also the longest record of all graphoglyptids, with which it shares the distinctive mode of preservation (Pl. 52) and the tendency to circumscribe parcels of unused sediment like an irrigation system.

On the other hand, *Paleodictyon* is also most difficult to explain in terms of a simple behavioral program. First of all, a reticulate tunnel system cannot be fabricated without breaking a general taboo in graphoglyptid morphogenesis: "Avoid collision with previous tunnels!" (secondary shortcuts are another matter; see Pl. 52). Rather, links with previous parts of the tunnel system have to be made on a regular basis. Also, a certain amount of double coverage (which should be reduced if the maker had been a sediment feeder) cannot be avoided. Though similar hexagonal patterns result automatically from close packing of soap bubbles, eggs, corals (Pl. 57) and honeycomb cells, it turns out that "weaving" them is a more difficult task (compare the fabrication of hexagonal chicken wire!).

Theoretical Pathways. The unshaded parts of the diagram show potential modes of fabrication derived from a meandric, a spiral, or a radiating master plan. In the shaded area, these modes are translated into a *Squamodictyon*, whose scale-like meshes reveal the succession in which they were made.

Comparison with actual fossils shows that only the spiral mode has been used. Notably the simple method of forming rectangular meshes by superimposing guided meanders at a right angle has never been employed – possibly because it would have required overcrossing and the ability to move around sharp corners in harvesting.

Fossils through Time. Though the earliest deepsea *Paleodictyon* was found in an Ordovician flysch of Iraq, one of the richest records is available in the Silurian Aberystwyth Grits (Wales), which are exposed in beautiful coastal cliffs. In this occurrence, nets reach diameters of half a meter and mesh sizes of several centimeters. The dominant form there is *P. (Squamodictyon) petaloideum*. Its mesh size ranges from less than 1 to more than 3 cm, but remains uniform within each system. This underscores that an animal produced many nets during its lifetime. The scale-shaped meshes characterizing this ichnosubgenus are clearly made from the center outwards and probably in a spiral succession. With *Paleodictyon (Squamodictyon) tectiforme*, smaller versions of the same fabricational style persist into the Upper Cretaceous.

Associated in the Aberystwyth Grits is a form of similarly large size, but with angular meshes. The name *Paleodictyon (Glenodictyum) imperfectum* refers to this angularity as well as the fact that the hexagons are less regular than in Cretaceous and Tertiary ichnospecies such as *P. (Glenodictyum) strozzii*. In the still smaller *P. (Glenodictyum) minimum* the net tends to be eroded as a whole. The exhumed outline of the whole system is hexagonal with straight and smooth edges. Obviously, new meshes were added in a hexagonal spiral (see above), which together with a rigid linkage program guaranteed the uniform mesh size.

Most revealing is the third ichnosubgenus, *Ramodictyon*, whose name refers to the multiple *ventilation shafts* ascending from specified points of the net tunnels to the surface. In parts that were incompletely exhumed, these shafts are preserved as hypichnial knobs whose relationship to the net pattern can be inferred from more deeply eroded parts of the same system. In a very large form from an Upper Cretaceous flysch in Italy, *Paleodictyon (Ramodictyon) tripatens*, the shafts are arranged at more or less equal distances (although one direction appears to be more pronounced in the figured pattern, possibly due to tectonic deformation), but these distances correspond to the diameters of the meshes rather than those between branching points of the net. This means that there were only three ventilation shafts around each mesh, instead of the six openings one would expect. In the model, these shafts are placed at corners of the hexagon. This is surprising, because positioning in the centers of the connecting tunnels would have maintained the triple-junction style of other graphoglyptid traffic systems. In any case, the maker had to count turns in order to produce shafts only in every other element of the net.

In the smaller *Paleodictyon (Ramodictyon) nodosum* it was unnecessary to count, because every connecting tunnel of the hexagonal net had its own ventilation shaft. Accordingly, the pattern of the shafts in planview resembles the hexagonal pattern of the net underneath, but rotated by thirty degrees.

Modern *Paleodictyon*. The tunnel architecture reconstructed for *P. (Ramodictyon) nodosum* provided the explanation for enigmatic designs that were discovered by Peter A. Rona in photographs of modern deepsea bottoms in 1978. In size, outline, and pattern they are identical to the fossil ichnospecies – or rather to its reconstructed surface expression in the form of shaft exits. Later, these patterns could be directly observed from deepsea submersibles, which also allowed experimental exhumation of the underlying reticulate tunnel system. This we did in ALVIN by inverting a pump designed for sucking-in animals. To our surprise, the gentle blast did not produce a cloud of mud blocking the view. Rather, the mud between the openings peeled off in the form of small tablets, indicating ▶

Plate 55 · *Paleodictyon*   157

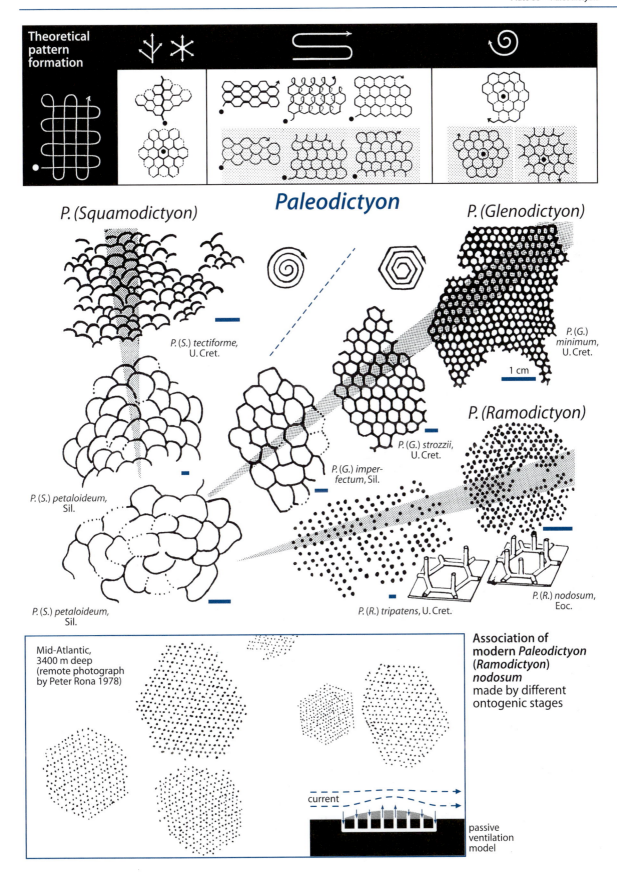

Theoretical pattern formation

Paleodictyon

P. (Squamodictyon)

P. (Glenodictyon)

P. (Ramodictyon)

P. (S.) tectiforme, U. Cret.

P. (S.) petaloideum, Sil.

P. (S.) petaloideum, Sil.

P. (G.) minimum, U. Cret.

1 cm

P. (G.) strozzii, U. Cret.

P. (G.) imperfectum, Sil.

P. (R.) nodosum, Eoc.

P. (R.) tripatens, U. Cret.

Mid-Atlantic, 3400 m deep (remote photograph by Peter Rona 1978)

Association of modern *Paleodictyon* (*Ramodictyon*) *nodosum* made by different ontogenic stages

current

passive ventilation model

some kind of (microbial?) binding. Underneath, the meshes of the horizontal tunnels could be seen as clearly as on a turbidite sole. Other horizontal tunnel systems came also into view, but unfortunately there was not enough time left to continue. By thus treating larger surfaces, one could certainly discover other graphoglyptid species that are not as easy to spot by surface openings.

As another surprise, the host sediment was light-colored and slightly reddish, rather than black as required for an H$_2$S-based symbiosis. Yet it is still possible that the farmed bacteria use methane, which might well be available in an area where vent activity has only recently ceased.

A third observation relates to *ventilation*. Pictures taken at an angle show that the fields of *Paleodictyon* are slightly elevated above the surroundings. Fabricationally, this could be the result of excavated sediment being dumped on the surface. Functionally, however, it means that water will automatically be sucked out of the more central shafts and be replaced from the margins – an ideal arrangement for passive ventilation in the absence of the owner!

One day even the maker may be caught and identified – provided it happens to harvest its garden at the time of sampling.

For the time being, we must be content with the videos taken for the IMAX film *Volcanoes of the Deep Sea* at a depth of 3 400 m in the Mid-Atlantic rift and continue to search the fossil sea bottoms. The sole faces of ancient turbidites remain a much a more convenient and a more promising place to study the biology and behavioral evolution of these most artistic, but anonymous, tracemakers.

■ *Paleodictyon nodosum*. **a** Atlantic, 3 400 m (Stephen Low, IMAX film); **b** Eocene flysch, Vienna; **c** *P. strozzii* and *P. minimum* (U. Cret. Flysch, Zumaya)

Pseudo-Traces

In old paleontology textbooks, trace fossils were treated as an appendix together with "Problematica". Although this situation has changed, the mesalliance persists. As trace fossils are themselves sedimentary structures, their distinction from non-biogenic structures is more difficult than in body fossils. Therefore every practicing ichnologist must be aware of pseudofossils that owe their origin to a variety of physical processes. The last edition of the *Treatise on Invertebrate Paleontology*, part W, lists 245 ichnogenera and no less than 76 "genera" of pseudofossils (not counting synonyms). Most of the latter come from the Proterozoic for two reasons. (1) When biohistorians work in the Precambrian, they are particularly eager to recognize – and name – anything that could potentially be the earliest documents of animal life. (2) The Precambrian was a world of microbes. In the absence of bioturbation they could form continuous films or *biomats* at the sea floor. Microbial mats not only produced the laminated build-ups called stromatolites; they also reinforced the top layer of otherwise soft sands and muds. Therefore, clastic Precambrian rocks contain a lot of "anactualistic" sedimentary structures that are rare or absent in later deposits. Accordingly, the present chapter has been placed before the one on Precambrian trace fossils. This is not the place to cover all *sedimentary structures*. They represent a field of their own that, like paleoichnology, still requires more observational and experimental research. Many eye-catching sedimentary structures (such as cross bedding, ripples, or desiccation cracks) would never be mistaken for trace fossils. Nevertheless, the present selection should make it also clear that a given structure does not lose scientific interest after having been identified as a pseudofossil. It remains a historical document that records sedimentary, diagenetic, or tectonic processes and environmental conditions.

The same is true for traces that turn out to be *body fossils*. Most misleading in this respect are fossils, in which sedimentary particles have been used to reinforce the shell (*agglutination*) or to build an internal skeleton (*inglutination*). A familiar example are the tubes of caddisfly larvae. If fossilized, should they be treated as body fossils or as a special kind of trace fossils with separate names? Inglutinated skeletons are known from passively implanted actinians (Psammocorallia) and sponges as an early kind of skeletonization. The former are difficult to distinguish from *algal sand balls* and could also be mistaken for the resting traces of actively burrowing actinians or sea pens (Pl. 25). *Xenophyophoran* protists (Pl. 56) agglutinate coarser particles in their outer wall and use finer ones for the inglutinated fill skeleton. They thus resemble mantle burrows with a fecal backfill (Pls. 32 and 34).

Another marginal case is *Shuichenyichnus spiralis* from Permian coal measures in China. While being described as a burrow, it is more probably the intestinal cast (*cololite*) of a tetrapod, as evidenced by occurrences in identical deposits of later times. Depending on definitions (Pl. 31), it might range as a trace fossil in either interpretation.

Other kinds of pseudofossils are produced by *modern* animals. Some insects build their nests within the soil and older sediments, so that the difference in age is not obvious. Scorpions often hide in rock clefts, as do some polychaetes in the intertidal zone. With their claws and bristles scorpions produce characteristic scratch patterns ("*Jiyuanichnus*") that can easily be mistaken for trace fossils if they happen to follow bedding planes. Similarly, the radular scratches left by gastropods grazing on algal films look like trace fossils when dry; but if you clean the rock surface, they disappear. As a last example, certain aquatic insect larvae produce deep elongate impressions on cobbles, probably while feeding on the algal films. Such *Furchensteine* are popular collectibles in Alpine lakes.

This shows that ichnologists should also be familiar with the local fauna when collecting.

Nor should they forget that postdepositional processes not only create pseudofossils, but also contribute to either the enhancement, or the elimination, of true traces. Enhancement is exemplified by burrows of a deep tier whose active or passive fill was more porous than the already compacted host sediment. Thus they became selectively lithified and weather out three-dimensionally. This process is particularly common in crustacean burrows of Cretaceous chalks, where it has also been observed in *Zoophycos*.

On the other hand, traces made on interfaces between lithologically similar layers may be altered by *stylolitization*. A notable example is the *Muschelkalk* (Middle Triassic, Germany), where bed surfaces preserve corroded burrows (*Rhizocorallium*, Pl. 18; *Solemyatuba*, Pl. 36) that were originally endichnial and became projected on the bedding plane by stepwise pressure dissolution. Similarly, such reliefs (e.g., *Asteriacites*, Pl. 24; or *Gyrochorte*, Pl. 35) are hardly recognizable any more in the Rhaetic Sandstone (Upper Triassic) of the same region, while they are perfectly preserved on bedding surfaces of environmentally equivalent Jurassic sandstones a few meters higher up in the section. The reason is a kind of early silification in which the quartz cement was derived not from grain-to-grain dissolution, but from the stylolitized bedding planes.

In the following discussion, pseudotraces are grouped according to their presumable origin into body fossils and syndepositional, biomat-related, diagenetic, tectonic, or weathering phenomena.

Literature

Chapter XII

Bromley RG, Ekdale AA (1984) Trace fossil preservation in flint in the European chalk. J Paleontol 58:298–311 (Preservational analysis of Cretaceous chalk trace fossils, with discussion on diagenetic enhancement)

Häntzschel W (1975) Trace fossils and problematica. In: Teichert C (ed) Treatise on invertebrate paleontology, Part W, Supplement 1. Geological Society of America and Univerity of Kansas, W1–W269 (Contains alphabetical lists of pseudofossils, trace fossils, *incertae sedis,* and unrecognized and unrecognizable "genera", some of which have since been recognized)

Jensen S (1997) Trace fossils from the Lower Cambrian Mickwitzia sandstone, south-central Sweden. Fossils and Strata 42:1–111 (Contains section on pseudofossils)

Lebesconte P (1887) Constitution générale du Massif breton comparée à celle du Finisterre. B Soc Geol Fr 14:776–820 (Introduction of "*Neantia*")

Seilacher A (1999) Biomat-related lifestyles in the Precambrian. Palaios 14:86–93 (Role of microbial mats in producing physical sedimentary structures)

Shrock RR (1948) Sequence in layered rocks. McGraw-Hill, New York, 507 p (Chapters 4 and 5 contain references and illustrations relevant to this chapter)

Wu X (1985) Trace fossils and their significance in non-marine turbidite deposits of Mesozoic coal and oil bearing sequences from Yima-Jiyuan Basin, western Henan, China. Acta Sediment Sin 3:23–31 (In Chinese with English abstract) (Original description of "*Jiyuanichnus*" made by modern scorpions)

Yang S, Zhang J, Yang M (2004) Trace fossils of China. Science Press, Beijing, pp 29–263 (*Shuichengichnus*. and other pseudofossils)

Plate 56: Trace-like Body Fossils: Xenophyophoria

Fedonkin MA (1985) Paleoichnology of Vendian metazoa. In: Sokolov BS, Ivanovskiy MA (eds) The Vendian system: Historic-geological and palaeontological basis, 1. pp 112–116 (In Russian; English translation, Springer-Verlag 1990) (Interpretation of *Palaeopascichnus, Neonereites, Intrites* and *Yelovichnus* as trace fossils)

Glaessner, MF (1984) The dawn of animal life: A biohistorical study. Cambridge University Press, Cambridge, 244 p (*Palaeopascichnus* interpreted as a grazing trace)

Seilacher A, Grazhdankin D, Legouta A (2003) Ediacaran biota: The dawn of animal life in the shadow of giant protists. Paleontol Res 7:43–54 (Reinterpretation of many Ediacaran trace fossils as body fossils of *Xenophyophoria*)

Seilacher A, Buatois LA, Mángano MG (2005) Trace fossils in the Ediacaran-Cambrian transition: Behavioral diversification, ecological turnover and environmental shift. Palaeogeog Palaeoclim Palaeoecol 227:323–356 (Reinterpretation of many Ediacaran trace fossils as body fossils of Xenophyophoria or pseudofossils)

Tendal OS (1972) A monograph of the Xenophyophoria (Rhizopodea, Protozoa). Galathea Reports 12:7–99 (Review of modern deep-sea Xenophyophoria)

Plate 57: Tool Marks

Linck O (1949) Lebens-Spuren aus dem Schilfsandstein (mittl. Keuper km 2) NW-Württembergs und ihre Bedeutung für die Bildungsgeschichte der Stufe. Verein für Vaterländische Naturkunde in Württemberg, Jahreshefte 97–101, 1–100 (Introduction of "*Ichnyspica*")

Lucas SGPIT, Lerner AJ (2001) Reappraisal of *Oklahomaichnus*, a supposed amphibian trackway from the Pennsylvanian of Oklahoma. Ichnos 8:251–253 (Reinterpretation as arthropod undertrack)

Nathorst AG (1881) Om spår af nagra everterbrerade djur m. m. och deras palæontologiska betydelse (Mémoire sur quelques traces d'animaux sans vertèbres etc. et de leur portée paléontologique) Konglinga Svenska Vetenskapsakademien, Handlingar (2) 18 (1880), 1–60 (Swedish), 61–104 (abridged, French) (*Eophyton* as tool mark)

Osgood RG (1970) Trace fossils of the Cincinnati area. Paleontographica Americana 6:281–444 (See Plate 19: "*Chloephycus*" and roll mark of tabulate)

Pavoni N (1960) Rollmarken von Fischwirbeln aus den oligozänen Flyschschiefern von Engi-Matt (Kt. Glarus). Eclogae Geol Helv 52:941–949 (Recognizes Peyer's Tunicates as rollmarks of fish vertebrae)

Peyer B (1958) Über bisher als Fährten gedeutete problematische Bildungen aus den oligozänen Fischschiefern des Sernftales. Schweizerische Palaeontologische Abhandlungen 4 (1957), 34 p (Roll marks of fish vertebrae interpreted as tunicates)

Sarjeant WAS (1967) Track of a small amphibian from the Pennsylvanian of Oklahoma. Tex J Sci 27:107–112 (Diagnosis of *Oklahomaichnus*)

Seilacher A (1963) Umlagerung und Rolltransport von Cephalopoden-Gehäusen. Neues Jahrb Geol P M 11:593–615 (Rollmarks of ammonite shells)

Seilacher A (1982) Distinctive features of sandy tempestites. In: Einsele G, Seilacher A (eds) Cyclic and event stratification. Springer-Verlag, Berlin, pp 333–349

Plate 58: Synsedimentary Structures

Bland BH (1984) *Arumberia* Glaessner and Walter, a review of its potential for correlation in the region of the Precambrian-Cambrian boundary. Geol Mag 121:625–633

Bloos G (1976) Untersuchungen über Bau und Entstehung der feinkörnigen Sandsteine des schwarzen Jura alpha (Hettangium und tiefstes Sinemurium) im schwäbischen Sedimentationsbereich. Arbeiten des Instituts für Geologie und Paläontologie der Universität Stuttgart 71, 270 p ("*Kinneyia*" and microripples)

Crowell JC (1958) Sole markings of graded graywacke beds: A discussion. J Geol 66(3):333–335

Hagadorn JW, Bottjer DJ (1997) Wrinkle structures: Microbially mediated sedimentary structures common in subtidal siliciclastic settings at the Proterozoic-Phanerozoic transition. Geology 25:1047–1050 (Discussion of wrinkle marks and elephantskin structures)

Hagadorn JW, Bottjer DJ (1999) Restriction of a late Neoproterozoic biotope: Suspect-microbial structures and trace fossils at the Vendian-Cambrian transition. Palaios 14:73–85 (Wrinkle marks and elephantskin structures)

Janicke V (1969) Untersuchungen über den Biotop der Solnhofener Plattenkalke. Mitt Bayer Staatssamml Paläontol Hist Geol 9:117–181 (Solnhofen elephantskin structures)

Książkiewicz M (1977) Trace fossils in the flysch of the Polish Carpathians. Paleontologia Polonica 36, 208 p (Mop structures held for *Lophoctenium*. See pl. 5, fig. 9)

Miller SA, Dyer CB (1878a) Contributions to paleontology, no. 1. J Cincinnati Soc Nat Hist 1:24–39 ("*Blastophycus*")

Miller SA, Dyer CB (1878b) Contributions to paleontology, no. 2, 11 p, privately published, Cincinnati, Ohio ("*Aristophycus*")

Müller AH (1955) Über die Lebensspur *Isopodichnus* aus dem Oberen Buntsandstein (Unt. Röt) von Göschwitz bei Jena und Abdrücke ihres mutmaßlichen Erzeugers. Geologie 4(5):481–489 (Figured hyporeliefs belong to *Lockeia*; associated epireliefs of pseudofossil *Aristophycus* are interpreted as gill impressions)

Osgood RG (1970) Trace fossils of the Cincinnati area. Paleontographica Americana 6:281–444 (Origin of "*Blastophycus*")

Pflüger F (1995) Morphodynamik, Aktualismus und Sedimentstrukturen. Neues Jahrb Geol P-A 195:75–83 (Origin of *Kinneyia*)

Pflüger F (1999) Matground structures and redox facies. Palaios 14:25–39 (Discussion of "*Kinneyia*" and "*Manchuriophycus*")

Seilacher A (1982) Distinctive features of sandy tempestites. In: Einsele G, Seilacher A (eds) Cyclic and event stratification. Springer-Verlag, Berlin, pp 333–349 (Fig. 2c "*Manchuriophycus*", Fig. 4a "*Kinneyia*")

Seilacher A (1997) Fossil art. An exhibition of the Geologisches Institut Tübingen University. The Royal Tyrell Museum of Palaeontology, Drumheller, Alberta, Canada, 64 p (See Introduction p 13 "*Manchuriophycus*", p 19–20, "*Astropolithon*")

Stanley DSA, Pickerill RK (1994) *Planolites constriannulatus* isp. nov. from the Late Ordovician Geogian Bay Formation of southern Ontario, eastern Canada. Ichnos 3:119–123 (see Pl. 32, "*Aristophycus*" interpreted as trace fossil *Walcottia rugosa*)

Plate 59: Diagenetic Structures

Astin TR (1986) Septarian crack formation in carbonate concretions from shales and mudstones. Clay Miner 21:617–631

Breyer JA, Busbey AB, Hanson RE, Roy EC III (1995) Possible new evidence for the origin of metazoans prior to 1 Ga: Sediment-filled tubes from the Mesoproterozoic Allamoore Formation, trans-Pecos Texas. Geology 23:269–272 (Allamoore structures)

Brown RW (1954) How does cone-in-cone material become emplaced? Am J Sci 252:327–376

Mayer D, Kronberg P (1989) Klüftung in Sedimentgesteinen. 148 p (Fig. 7: crack patterns on cleavage surfaces)

Müller AH (1956) "Parkettierende" Lebensspuren aus dem unteren Buntsandstein Thüringens. Geologie 5:411–412 (Plate 5 shows Liesegang rings compared to *Helminthoida*)

Müller AH (1962) Fossil oder pseudofossil? Geologie 11:1204–1213 (Discoid cone-in-cone concretion)

Pratt B (1998) Syneresis cracks: Subaqueous shrinkage in argillaceous sediments caused by earthquake-induced dewatering. Sediment Geol 117:1–10 ("*Manchuriophycus*" and "*Rhysonetron*" reinterpreted as synaeresis cracks)

Pratt B (2001) Septarian concretions: Internal cracking caused by synsedimentary earthquakes. Sedimentology 48:189–213 (*Septaria* interpreted as earthquake-induced cracking of under-compacted center)

Seilacher A (2001) Concretion morphologies reflecting diagenetic and epigenetic pathways. Sediment Geol 143:41–57

Yabe H (1939) Note on a Pre-Cambrian fossil from Lyoto (Liautung) Peninsula. Jap J Geol Geogr 16:205–207, 2 pls. (*Manchuriophycus* Endo 1933)

Yang S, Zhang J, Yang M (2004) Trace fossils of China. Science Press, Beijing, pp 29–263 (Pl. 55, Fig. 1: rectangular counter-septaria)

Yao P (1984) A new trace fossil genus from Jurassic Longzhaogou in eastern Heiongngjiang, China. Regional Geology of China, pp 117–120 ("*Mishanichnus*")

Plate 60: Tectograms

Cloud PE, Wright J, Glover L (1976) Traces of animal life from 620-million-year-old rocks in North Carolina. Am Sci 64:396–406 (Interpretation of *Vermiforma* as a body fossil)

Seilacher A, Meschede M, Bolton EW, Luginsland H (2000) Precambrian 'fossil' *Vermiforma* is a tectograph. Geology 28:235–238 (Reinterpretation of "*Vermiforma*" as a pseudofossil)

56

Plate 56
Trace-like Body Fossils: Xenophyophoria

The name **Palaeopascichnus** indicates that this structure from shallow marine Ediacaran sandstones was originally compared to guided meanders of sediment feeders and thus taken as an example of another onshore-to-offshore shift through geologic time. In fact, the transverse stretches are equidistant and tend to turn back against earlier ones as in nereitids (Pl. 34). If one traces them by pencil, however, the turns on the two sides never alternate as they would in a true meander. Also, the guidance between loops appears overly perfect for such early trace makers. Better preserved hyporelief specimens from Australia, Newfoundland and Russia show that we deal with a series of laterally attached sausage-shaped chambers, whose agglutinated sandy walls leave a meander-like pattern on the bed surface when broken off. This model can also be applied to associated structures in similar preservation, whose chambers are bubbleshaped and align either uniserially or biserially, like the fecal backfill pellets in *Neonereites* (Pl. 34). The new interpretation equally fits similar patterns of darker color reported by Peter Haines from the pre-Ediacaran Patooga Sandstone of South Australia. All these systems fill the given surface without intersecting. They also *bifurcate*, which is impossible in a continuous meander.

Fortunately, organisms survive in the modern deep sea that fit our reconstruction in size as well as patterns: **Xenophyophoria**. As revealed by the studies of Ole Tendal, these rhizopods transcend the size limit of unicells in a way similar to larger foraminifera and the quilted Vendobionta, namely by *compartmentalization* of the multi-nucleate (plasmodial) protoplasm by a tough outer wall. In contrast to the rigid foraminiferan tests, the wall of modern xenophyophorans is flexible, but reinforced by agglutinated sand, or by shells of planktonic foraminifera. If we accept the identification of the Precambrian fossils as Xenophyphoria, their difference to the modern representatives relates only to habitat. In Precambrian times they could still exist on well-lit *shallow marine* bottoms, where sand was available for agglutination and where biomats provided a substrate, into which these protists were embedded. In building their walls, they introduced coarser grains into the finer sediment below the mat, so that they now look like trace fossils in positive-hyporelief preservation. After having retreated to the muddy deepsea bottoms, forms like **Stannophyllum** grew upright, attached only by rootlets along the straight lower edge (other species may still be immersed in the sediment). Nevertheless the limitations of growth remained the same. Chambers are either *globular* and arranged in branching twigs. Alternatively, *sausage-shaped* chambers allow growth into *leaflike* structures that bifurcate as soon as the next chamber would exceed a permissible length.

Neoproterozoic xenophyphores and vendobionts (whose wall was not agglutinated and could therefore expand) follow the principle of allometric compartmentalization; but compared to the chamberlets of large Foraminifera, their chamber (or quilt) diameters are too large. The *inglutinated* skeleton (*stercomare*) of modern xenophyophores resolved this problem: it consists of smaller particles taken up with the food that subdivide the protoplasm inside the chambers into strands of permissible diameters.

The illustrations, taken from an unpublished study of White Sea material by Anton Legouta, show the modifications of the Neoproterozoic xenophyophores. They range from lobate with transverse chambers (**Palaeopascichnus**) to catenate and foamlike with globular chambers ("**Neonereites**") and star-shaped with radial tubes (**Eoporpita**). In **Hiemalora** the rays around a central body could be the impressions of naked pseudopodia, similar to a questionable form (cf. **Psammetia**) photographed at a depth of 7 000 m.

Assuming that *Paleodictyon* (Pl. 55) is not a xenophyophore but a graphoglyptid burrow, why is there hardly any fossil record of xenophyophores after the late Neoproterozoic? One reason may be that the group emigrated into the deepsea after the Cambrian Substrate Revolution (Pl. 65). At the same time, an important *taphonomic window* closed. Probably most of the forms figured here lived embedded in biomats, which protected them against uprooting and transport and make them visible as positive hyporeliefs on the former base of the mat. As bioturbation transformed matgrounds into mixgrounds in hospitable environments, this preservation became extinct. A possible exception is a large lobate specimen (collection of Renata Neto, Unisinos) from laminated Devonian silts in Brazil. It would be interesting to study its paleoenvironmental setting.

■ Chambered xenophyophore ("*Neonereites*") (Vendian, White Sea)

Plate 56 · Trace-like Body Fossils: Xenophyophoria   163

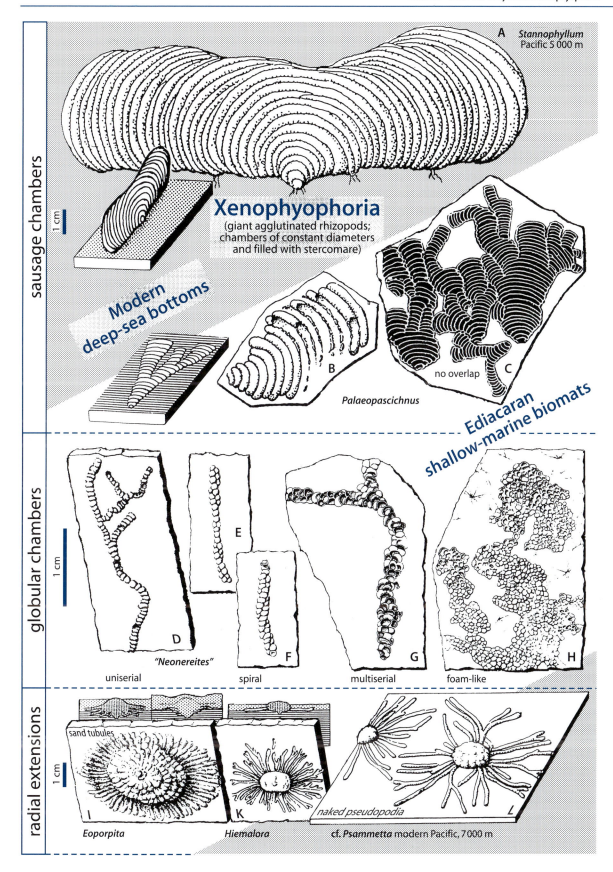

sausage chambers

A *Stannophyllum*
Pacific 5 000 m

1 cm

Xenophyophoria
(giant agglutinated rhizopods;
chambers of constant diameters
and filled with stercomare)

Modern
deep-sea bottoms

B

Palaeopascichnus

no overlap   C

Ediacaran
shallow-marine biomats

globular chambers

1 cm

E

D

"Neonereites"

F

G

H

uniserial   spiral   multiserial   foam-like

radial extensions

1 cm

sand tubules

I

Eoporpita

K

Hiemalora

naked pseudopodia   L

cf. *Psammetta* modern Pacific, 7 000 m

57

Plate 57
Tool Marks

Not only animals move over the bottom. Non-living objects may be passively transported by waves, wind, and current. They leave *tool marks* that can easily be mistaken for animal traces.

Good examples are the roll marks of dead ammonite shells found on particular layers of the famous lithographic limestones of **Solnhofen**, Germany. Probably driven by a muddy turbidity current, the flat shells of *Perisphinctes* rolled along like wheels, provided they had the right orientation (when they wheeled in the opposite sense, the aperture hit the bottom and made them tumble). Their tire tracks are necessarily discontinuous, because every time the shell rolled over the aperture, it made a jump and bounced back when touching bottom again (**A**). The result is a series of longer and shorter impressions in regular succession. Since they also bear the imprints of the shell ribs, it is not surprising that they were originally referred to fish – or to whole schools of them – swimming over the bottom, because all marks run in the same direction.

The basic roll pattern (**A**) became modified by wobbling (**B, C**) but also by the increasing wear on the rolling shell. Damage started with a dent in the ventral margin of the aperture (**B**), from which breakage proceeded, leaving two ridges of the lost part (**D**). Later, another breakage occurred where the shell hit the ground after the apertural jump (**E**). The resulting ruins could not roll like a wheel any more. Rather, they rolled sideways and left the most vexing patterns, but still with a regular repetition (**F, G**).

The bullate shells of *Aspidoceras* did not roll like a wheel. They always rolled sideways, so that the spines of the right and left flanks left their stamps in alternation. Such patterns (**H–K**) have at times been referred to a squid with spiral suckered arms moving sideways, while variants produced at higher speeds (**L–N**) were interpreted

as swimming tracks of a turtle with strong claws, if two shells happened to roll side by side.

A rare kind of repetitive markings associated with ordinary *Perisphinctes* roll marks was probably produced by a shell that still retained some buoyancy and therefore could not roll. A pendulum movement (**O**) is more likely the cause than an ammonite using its arms to crawl sideways (**P**).

Other biological objects of appropriate shapes also leave roll marks that were once interpreted as trace fossils, such as round tabulate corals (**Q**), fish vertebrae (**R, S**) and segments of large cattail stems (**T**). Regular tool marks not related to rolling include circles made by seaweed swinging around under wave action (**U**) or ruffled groove casts (**V**).

Since all these markings originated at the sediment surface, the presence of biomats was probably necessary to preserve them.

The study of syndepositional sedimentary structures, such as tool marks, is akin to that of trace fossils. Both are based on deformations of the original bedding planes. Also there is a majority of unspecific impressions, called *impact casts*, that at least indicate paleocurrent directions. The more complex "elite" structures, moreover, can be analyzed with the same forensic approach as used in ichnology. As roll marks are made by biologically standardized, but *dead* objects, there is also the advantage that the hypothesized processes can be experimentally simulated. For instance, a fossil ammonite rolling down an inclined board coated with modeling clay produces exactly the same intermittent impressions as found on Solnhofen bedding planes. Similarly, Pavoni discovered the roll marks of fish vertebrae on butter when preparing a sardine sandwich at a well in the Swiss Alps. They immediately reminded him of impressions that had been described as tunicate colonies from the Glarus Shales. Equally, "*Chloephycus*" (**V**) can be produced with a fork on the scum of whole milk. One does not need sophisticated instrumentation to make discoveries in this field!

■ Artificial roll marks of *Perisphinctes* (Tübingen Museum)

Plate 57 · Tool Marks   165

Tool Marks

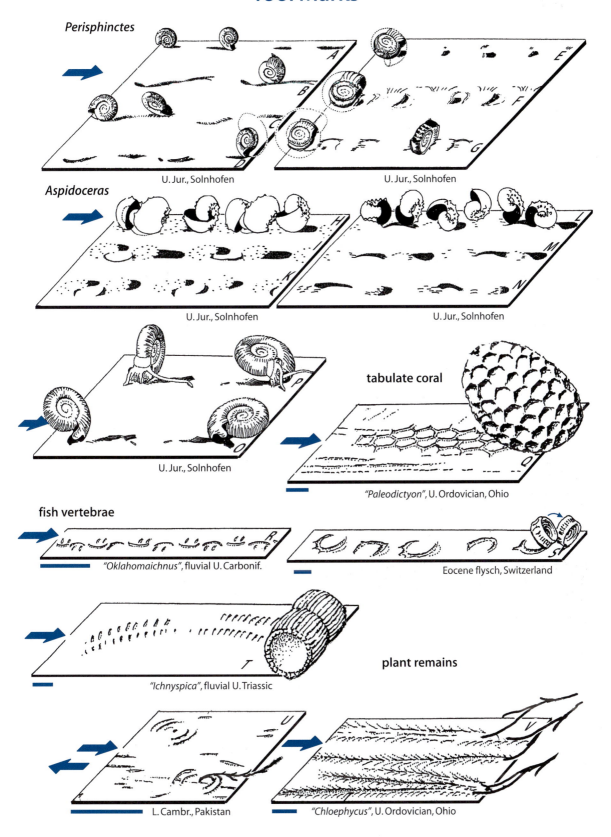

Perisphinctes

U. Jur., Solnhofen

U. Jur., Solnhofen

Aspidoceras

U. Jur., Solnhofen

U. Jur., Solnhofen

U. Jur., Solnhofen

tabulate coral

"*Paleodictyon*", U. Ordovician, Ohio

fish vertebrae

"*Oklahomaichnus*", fluvial U. Carbonif.

Eocene flysch, Switzerland

"*Ichnyspica*", fluvial U. Triassic

plant remains

L. Cambr., Pakistan

"*Chloephycus*", U. Ordovician, Ohio

58

Plate 58
Synsedimentary Structures

The repetitive patterns of ordinary depositional structures, such as ripple marks in sand and sun cracks in mud, are too familiar from modern environments to be mistaken for fossils in the rock record. This is not the case with *crypto-actualistic* structures that originate *below* the sediment/water interface. Among these are erosional features (flute casts; frondescent casts; scratched tool marks). They form at the peak of the event under suspended sand and are immediately covered after transition to the depositional phase. Therefore they are beautifully preserved on the soles of sandy event beds, but remain invisible in modern environments.

Other structures are *anactualistic*, because they are related to biomats that could not form on oxygenated sea bottoms after the Cambrian bioturbational revolution (Pl. 65).

The bilobate "**Blastophycus**" from the Upper Ordovician of Cincinnati could be mistaken for a rusophyciform trilobite burrow. In reality it is a bilobed scour behind an enrolled trilobite, which was certainly dead and probably reworked from a former mud as a heavy particle.

Mop structures form in turbidites during the initial state of the depositional phase by small avalanches of sand gliding down the lee slopes of erosional ripples (contoured by isohypses of eroded mud laminae). Feathering out in current direction and coalescing into a "moraine" on the downcurrent front, they resemble a household mop. A giant version in Ordovician turbidites of New York State is locally known as "Coxsackie dinosaur skin", while a minute version from **Poland** has been mistaken for the trace fossil *Lophoctenium*. Such structures also occur (and vex visitors) in the Ediacaran deposits of Newfoundland.

Elephant-skin structures form underneath biomats that are somewhat coarser-grained than the sediment below. They are common in Precambrian clastic sediments, but also occur in the toxic environments of the Solnhofen lithographic limestones. In principle we deal with a kind of *load casts*, but at a smaller scale than usual.

Upper surfaces of Precambrian sandstone beds are often coarsely striated. These structures have variously been called "*Neantia*" or "*Arumberia*", particularly if they converge in local depressions and therefore have a more biogenic aspect. They probably originated as slide marks underneath tough *biomats* that were exposed to tractional currents loaded with sediment. In the figured specimen from the Vendian of India, the overlying wrinkled mat (black) is still preserved.

The name "*Kinneyia*" (junior synonym of "Rivularites"; "Peanut blisters" is a handier term) stands for negative epireliefs that look like raindrop impressions. The restriction to flattened ripple tops, as well as experiments, led F. Pflüger to the interpretation as gas bubbles trapped underneath a biomat. Note that in the figured specimens (including latex cast on top) bubbles merge into chains, particularly along the boundary with a smooth ripple trough. A stick-like body fossil (graptolite?) is surrounded by a smooth halo, probably because the sediment around it had been altered into an incipient concretion. On the other hand, the burrow in the lower block was superimposed after "Kinneyia" had formed, possibly triggered by an earthquake. When this happened, the biomat was still in the reach of burrowers. Degassing or dewatering (and possibly biomats) may also have been involved in the concentric microfault systems that have been called "*Protospiralichnus*".

A dewatering structure called "*Aristophycus*" (= "*Lobichnus*") is always found as positive epirelief and is located on the crests if the surface is rippled. Although there is a similarity to *viscosity figures* (forming when the cover glass of a thin section is lifted from the molten resin), we probably deal instead with an *erosional* feature showing a similar fractal behavior. As pore water escaped during compaction from the sand, it was stopped by the overlying clay, or biomat, and dissipated in a form resembling a distributary river system. "Aristophycus" commonly radiates from burrows that served as a conduit, leading to the misinterpretation as a gill impression of the trace maker. It also intersects epireliefs of *Gyrochorte*, which indicates an intra-sedimentary origin below the surface.

Our last example, "*Astropolithon*", is much larger. Discovered in the jungles of Northern Australia, these structures were originally described as jellyfish. More properly they should be referred to fluidized sand that could not readily escape. Being held up by mucous horizons of former biomats, the fluidized sand first formed a dome. The mats then cracked radially before the fluidized sediment could erupt as sand volcanoes. Note crescentic laminae in the neck of the largest form, indicating lateral displacement during eruption. This might also have been the origin of the celebrated *Brooksella canyonensis* from the Neoproterozoic of the Grand Canyon (though not of the Cambrian forms of *Brooksella*; Pl. 47). The model can be also applied to the blunt-ended radii of the Ediacaran "*Mawsonites*" (Pl. 62) which as well has a crescentic structure in the central shaft. Burrow-like vertical shafts with radial tributary canals around the base have been observed in a thick tsunami deposit related to the asteroid impact at the K/T boundary. Such dewatering structures should be expected in sediments that were instantly dumped at this event.

Plate 58 · Synsedimentary Structures   167

Synsedimentary Structures

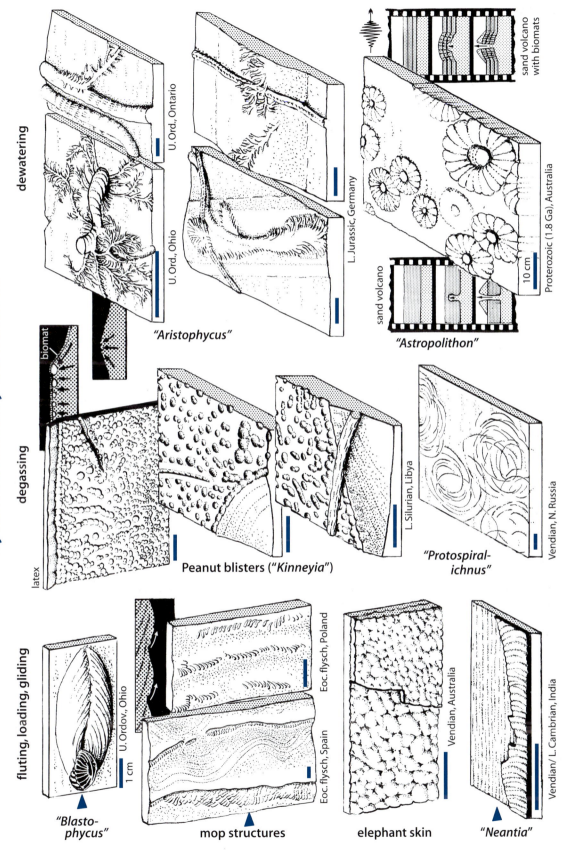

dewatering

U. Ord., Ontario

U. Ord., Ohio

L. Jurassic, Germany

"Aristophycus"

sand volcano with biomats

sand volcano

"Astropolithon"

Proterozoic (1.8 Ga), Australia

10 cm

degassing

biomat

latex

Peanut blisters ("Kinneyia")

L. Silurian, Libya

"Protospiral-ichnus"

Vendian, N. Russia

fluting, loading, gliding

U. Ordov., Ohio

1 cm

"Blasto-phycus"

Eoc. flysch, Poland

Eoc. flysch, Spain

mop structures

elephant skin

Vendian, Australia

"Neantia"

Vendian/ L. Cambrian, India

59

Plate 59
Diagenetic Structures

Early diagenesis is dominated by two processes, compaction and cementation. Mineralization may enhance the original stratification by differentially cementing more porous layers; but it may also be localized, using burrows, carcasses, or shell accumulations for nucleation. We shall first discuss rhythmic patterns caused by concretionary processes and then those related to crack propagation.

Dissipative Mineralization. In glacial silts of Norway, elongate concretions commonly formed around fish carcasses; but instead of forming an envelope all around, accretionary growth proceeded as pancake layers that expanded *transversally* from a point of origin. So there was a critical distance limiting radial expansion. This is a common property of *dissipative systems*. Another striking feature is the *symmetry* of such concretions relative to the bedding plane: every pancake and every "epaulette" has its counterpart on the opposite side! While there is no scientific term for such concretions, Norwegian children found one: "Marleik" (a doll one can play with).

For completeness, we add **siderite** and **chert** mummies to show that the dissipative process is not restricted to a particular mineral. In both cases the critical distance is not only expressed in the thickness of the encrusting envelope, but also in its segmentation into concentric sausages. It may also be significant that both examples come from arid environments, where iron and silica tend to migrate with meteoric water (Cretaceous chert nodules of northern Europe never have such encrustations).

Encrustations of a similar kind, but on a much larger scale, are the **Westerstetten structures**. They consist of micritic calcite and cover upper surfaces of blocks of Jurassic limestone floating in the residual clay of a karst fissure. They probably formed when the area was an arid land surface during Cretaceous and Tertiary times. Fine internal lamination also shows that the sausages accreted at their ends. This links them with the colorful **Liesegang** rings that decorate the sandstone walls of Petra (Jordan). Similar patterns have been described as meandering trace fossils (cf. *Helminthoida*) in Triassic sandstones of Germany.

Dissipative silicification may also explain the **Allamoore structures** reported as meandering burrows from Proterozoic limestones in Texas, although they are discoid rather than globular and lack turns at the ends.

Crack Propagation. Frondescent casts are diagenetic only in the sense that they formed after initial compaction. When in the peak of a storm or turbidity current erosion reaches an already compacted layer of mud, erosion no longer proceeds grain by grain. Instead, large chunks of stiff mud are plucked off at once (Pl. 52). Their scars, which become immediately buried under the settling sand, show a very characteristic fractal pattern. The reason is that in isotropic materials cracks tend to propagate in leaflike patterns. In the figured example, the similarity with plant leaves is enhanced by preexisting burrows, which served as conduits for the pressurized water.

Syneresis stands for shrinkage caused by the dewatering of gel-like sediments. As the mucus of biomats enhances this quality, it is not surprising that syneresis cracks of many kinds are most common in Precambrian rocks, or in low-oxygen environments of later times. If such cracks are filled with sand, they may easily be mistaken for trace fossils, such as incipient cracks resembling bivalve burrows (cf. *Lockeia*; Pl. 23) or sinusoidal cracks ("*Manchuriophycus*") that follow ripple troughs. They probably opened *within* the sediment in buried biomats.

Septaria are calcareous concretions with a boxwork of shrinkage cracks inside. Cut and polished, they make good conversation pieces, particularly if the cracks are healed with white calcite. In order to shrink, the concretion still had to be unlithified inside its hard outer shell, because each layer conserved the compactional state by the time of its formation. It is also interesting to speculate what released the dewatering. Was it an earthquake, as in other instabilities hidden in sediments? In contrast, the calcareous **cone-in-cone** concretion resembles "pyrite suns" by its discoid and centripetal growth along the median plane.

While in septaria it was the soft heart that shrank, **counter-septarias** show the crack pattern on the *outside* of the concretion or concretionary bed. The solution to the **Pliocene** specimen (courtesy of Sue Kidwell) came from a photograph made in the tailings of a **Surinam** bauxite mine, where toxic muds had developed an indurated crust under the tropical sun. When turned over, fragments revealed patterns that could have passed as artifacts of an ancient culture. Yet they are simply the products of *sequential cracking*. The first generation of cracks formed a polygonal pattern, the second generation was constrained by the first, forming perfect spirals, and the third was conchoidal within the spiral frame. So the final picture resembles the section through a *Nautilus* shell, complete to the vaulted septa. The basis for this miracle of self-organization is the gradational transition from a hard to a shrinkable matrix, so that fragments could not defoliate – just like the heart of a septaria, but with an opposite gradient. Not figured are other examples. In a variant from the Upper Ordovician of China (compared with *Squamodictyon*; Pl. 55), cracks of the first generation are subparallel and those of the second one form straight crossbars at smaller distances, just like the "septa" in the spiral version. On a small spherical concretion described by the late Alfred Eisenack there was room for only one generation of counter-septarian cracks; they follow the pattern of a perfect dodecahedron.

The last example is a block of **Jurassic oil shale** cracked by blasting. Because of its elastic properties, the cracks propagated in a rhythmically conchoidal fashion, as observed along the edges of a flint hand ax. What might the Neanderthal man have thought when he made it? Certainly not the same as the paleontologist that created the name "Mishanichnus"!

Plate 59 · Diagenetic Structures   169

Diagenetic Structures

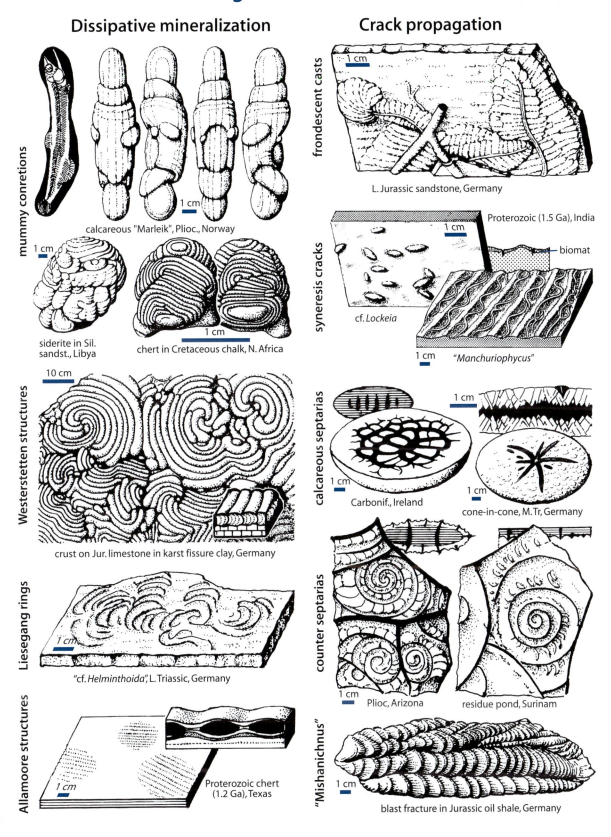

Dissipative mineralization

mummy concretions

calcareous "Marleik", Plioc., Norway

siderite in Sil. sandst., Libya

chert in Cretaceous chalk, N. Africa

Westerstetten structures

crust on Jur. limestone in karst fissure clay, Germany

Liesegang rings

"cf. *Helminthoida*", L. Triassic, Germany

Allamoore structures

Proterozoic chert (1.2 Ga), Texas

Crack propagation

frondescent casts

L. Jurassic sandstone, Germany

syneresis cracks

Proterozoic (1.5 Ga), India

biomat

cf. *Lockeia*

"*Manchuriophycus*"

calcareous septarias

Carbonif., Ireland

cone-in-cone, M. Tr, Germany

counter septarias

Plioc, Arizona

residue pond, Surinam

"Mishanichnus"

blast fracture in Jurassic oil shale, Germany

60

Plate 60
Tectograms

Diagenesis is not the last stage in which fossil-like structures can be added to the rock record. This is shown by the following examples.

"Vermiforma". Under the name *Vermiforma antiqua*, Preston Cloud described what he considered to be the USA's earliest animal fossils. Found on a bedding plane beneath a 620 million years old turbidite, the impressions looked to him like worms that were anchored by their tail ends in the mud. Later workers (including myself) reinterpreted them as trace fossils, because what Cloud had interpreted as scales resembled meniscoid terminal backfill structures (Pl. 37). The true nature of these impressions was revealed by the strange fact that a dozen of them exposed on the large slab (now in the Smithsonian Institution, Washington) all show an identical pattern: a meat hook, a kink, and a pretzel. Neither this congruence, nor the identical orientation of all "signatures", can be explained by spaghetti-like bodies being swept and curled by currents or by a complex burrowing behavior. There must have been a complex movement of two rock layers relative to each other parallel to the bedding plane.

As the signatures also vary in proportions, the sand grains producing them were not fixed in one bed; rather they rolled relative to either bed, as Hans Luginsland simulated with balls of modeling clay between moving glass plates. The experimental tracks show the same variation in the proportions of an otherwise identical pattern. It results from irregularities in grain shape controling friction on the upper or lower bed. The simulation also produced the "backfill" structures, because rolling of irregular balls is a differential process in which meniscoid contours are expressed behind the object. In the original rock slab the meniscoid structure is preserved not only as relief, but also by shading. This could mean that the tectonic shearing was slow enough to cause pressure solution at every station of the rolled grain.

The new interpretation of "Vermiforma" removed one of the oldest records of assumed metazoans. At the same time it demonstrates that in folded rocks the movements along bedding planes did not always follow a straight course perpendicular to the fold axis.

Other Tectograms (photo). Repetitive scratch patterns have also been found on bedding planes of the Permian *Kupferschiefer* in northern Germany. They do not either show the unidirectionality of normal slickensiding; nevertheless their repetiveness excludes comparison with trace fossils. It is also interesting that in this case the associated patterns *are* congruent. This means that the sand grains did not roll, but remained firmly anchored in the overlying bed.

Trace Fossils Indicating Lateral Deformation. In this context, it should be mentioned that regularly shaped trace fossils may also record the plastic deformation of the host sediment before it became lithified. In the Upper Cretaceous to Eocene turbidites, originally circular graphoglyptids (e.g., *Lorenzinia*, Pl. 54) are commonly *elongated* in the direction of the paleocurrent. This could be partly due to *lateral compaction* on a paleoslope exposed to contour currents. A large slab (cast in the Tübingen Geology department and in FOSSIL ART) shows several *Paleodictyon* systems; but only in one of them are the regular hexagons compressed in the direction of the current. In all these cases (including truly tectonic deformation), trace fossils are more reliable gauges than deformed body fossils, because they had the same consistency as the host sediment.

Another useful strain gauge are the vertical tubes of *Skolithos* (=*Tigillites*). While they are morphologically undifferentiated and stratigraphically useless, their mass occurrence in high-energy sands makes them good facies indicators (Pl. 71). In folded sequences, such "pipe rocks" reflect the shear parallel to bedding planes by their obliquity in vertical sections, particularly in layers that were less competent by the time of deformation. As seen on bedding planes, the circular cross sections of the tubes are also deformed, although this would not be predicted by a pure shear model (think of a pile of coins). In some localities, the directions of inclination and cross-sectional deformation do not coincide. So the deformation happened when the sediment was still compactable, with a down-slope vector. Shear deformation, however, could happen as long as there was a mechanical anisotropy parallel to bedding.

■ Spiked tectogram (Permian Kupferschiefer, Germany)

Plate 60 · Tectograms   171

Signatures of the Mountain Builder

Interpretations

① **body or trace fossil**
- individuals: must vary in shapes, sizes, orientations
- counterpart: mirror image

② **spike scratch**
- individuals: identical
- counterpart: none

③ **rolling grain**
- individual: may vary in proportions, but not in orientation
- counterpart: inverted image (point symmetry)

"Vermiforma antiqua"
Cloud 1976
Neoproterozoic
(620 m.y.)
North
Carolina

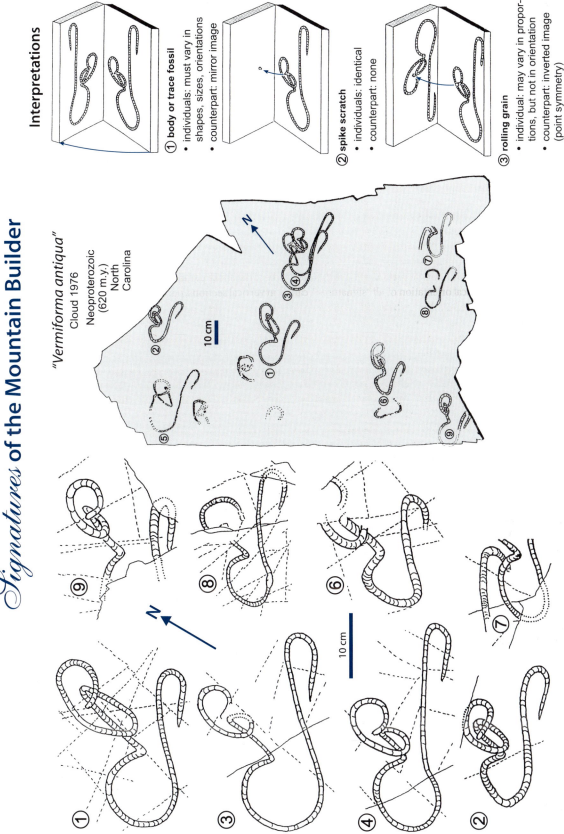

10 cm

■ Westerstetten structures

Earliest Trace Fossils

Trace fossils of the Precambrian and the Proterozoic/Cambrian transition deserve a chapter of their own, because this deep in time (1) the questions to be asked change considerably, and (2) environmental conditions were possibly different from later periods. In Phanerozoic rocks we can afford to pick the most telling kinds of trace fossils and neglect the poorly defined or dubious ones. In the Precambrian, however, even the poorest trace becomes important as a testimony for the presence and activities of a multicellular animal. On the other hand it is essential to eliminate pseudo-trace fossils (Chap. XII) that owe their origin to merely physical, microbial, or geochemical processes.

As we have seen in the previous chapter, the key feature of *Precambrian* sea bottoms was their microbial seal. In the absence of bioturbation, the mucous sheaths of cyanobacteria did not only form a coherent organic film, but also glued together the upper millimeters of sand. The former presence of such *biomats* and *biofilms* can be inferred from particular sedimentary structures. Among these are *palimpsest* wave ripples, in which the tops of earlier event layers have been overlain by ripple sets of the next storm (often with a different orientation) without becoming eroded and amalgamated. Biomats have also been involved in a variety of vexing pseudo-tracefossils (Pl. 58).

The ubiquity of biomats also had important *ecological* consequences. For instance, the microbial seal reduced the recycling of organics, which increased source rock potentials compared to later sediments of equivalent facies. Also, animals crawling over the sediment surface failed to leave trails – except for the raspings of radular teeth (Pl. 63) and resting traces of mobile vendobionts (Pl. 62). On the other hand, the indurated surface of otherwise soft sands allowed epifaunal organisms to employ lifestyles that were later restricted to rockgrounds.

Microbes are also responsible for the unique preservation of Ediacaran *body fossils*. As pointed out by Jim Gehling, carcasses became immediately coated by bacterial films. Their mineralization led to a kind of "*death masks*" that preserve external morphologies of fully softbodied organisms in all details. As this happened in sands and silts, the impressions are not even flattened by compaction. Nevertheless the taxonomic affilation of Ediacaran fossils remains a matter of debate. In my own interpretation (Vendobionta Hypothesis), the vast majority represents giant unicellular rhizopods that compartmentalized their multinuclear protoplasm by regular quilting of a flexible and expandible outer wall and a sandy fill skeleton.

The beginning of the Phanerozoic is commonly referred to as the "*Cambrian Explosion*": vendobionts disappeared and nearly all animal phyla with hard skeletons appeared within a few million years after the end of the Proterozoic. In the earliest phase, their preservation was favored by another time-specific preservational window: the tendency of small mineralized skeletons to become diagenetically phosphatized. This makes it possible to etch "*small shelly fossils*" from the carbonate matrix with weak acids. Shells of early molluscs, brachiopods and other phyla are thus known in considerable detail. Somewhat later in the Early Cambrian, *lagerstaetten* of the Burgess Shale type make their appearance (Chengjiang, China; Sirius Pass, Greenland). Along with the mineralized skeletons of trilobites and brachiopods, they preserve the flattened cuticles of various arthropods, worms, early fish and perhaps even jellyfish.

On this background, one might suspect that the Cambrian Explosion is largely a *preservational artifact*. The acquisition of mineralized skeletons – and thereby increased preservability – has undoubtedly been a key innovation, particularly since it also allowed for novel bauplans and new lifestyles, including predation. So the Early Cambrian marks the end of the peaceful "Garden of Ediacara" and the beginning of the "arms race" between predator and prey species, which has ever since remained a major driving force for evolutionary transformations.

The transition from matgrounds to mixgrounds was due to bioturbation of the sediment by burrowing organisms. Before this "*Agronomic Revolution*" (Pl. 65), burrowing animals had been rare and restricted to the zone just below the biomats (undermat mining, Pl. 49). Neither did Precambrian burrowers employ the elaborate behavioral programs that account for the distinctiveness of later burrow systems. Nevertheless, Latin names have been generously given to traces that would have passed as "nondescript" in Phanerozoic rocks; therefore lists of published names tend to exaggerate diversity compared to later ichnocoenoses. However, despite their low morphologic resolution, Precambrian trace fossils are uniquely important as witnesses of earliest metazoan life – provided that a nonbiological or non-metazoan origin can be excluded.

Literature

Chapter XIII

Bergström J (1990) Precambrian trace fossils and the rise of bilaterian animals. Ichnos 1:3–13 (Cambrian explosion interpreted as a true evolutionary event due to a shift from acoelomate-pseudocoelomate faunas to coelomate-dominated faunas)

Conway Morris S, Grazhdankin D (2005) Enigmatic worm-like organisms from the Upper Devonian of New York: An apparent example of Ediacaran-like preservation. Palaeontology 48:395–410

Crimes TP (1994) The period of early evolutionary failure and the dawn of evolutionary success: The record of biotic changes across the Precambrian-Cambrian boundary. In: Donovan SK (ed) The palaeobiology of trace fossils. Johns Hopkins Press, Baltimore, pp 105–133 (Review of the ichnology with extensive bibliography)

Gehling JG (1999) Microbial mats in terminal Proterozoic siliciclastics: Ediacaran death masks. Palaios 14:40–57 (Proposal of the Death Mask model to explain preservation of Ediacaran biota)

Glaessner, MF (1984) The dawn of animal life: A biohistorical study. Cambridge University Press, Cambridge, 244 p (The standard text on Precambrian paleontology)

Narbonne GM (2004) Modular construction of early Ediacaran complex life forms. Science 305:1141–1144 (Description of three-dimensionally preserved Ediacaran fossils supporting the unique nature of the Ediacaran biota)

Seilacher A (1956) Der Beginn des Kambriums als biologische Wende. Neues Jahrb Geol P-A 103:155–180 (Cambrian Explosion in trace fossils)

Seilacher A (1992) Vendobionta and Psammocorallia: Lost constructions of Precambrian evolution. J Geol Soc London 149:607–613 (Proposal of the Vendobiont hypothesis)

Seilacher A (1994) Early multicellular life: Late Proterozoic fossils and the Cambrian explosion. In: Bengtson S (ed) Early life on Earth. Nobel Symposium 84. Columbia University Press, New York, pp 389–400 (Trace metazoans)

Seilacher A (1999) Biomat-related lifestyles in the Precambrian. Palaios 14:86–93 (Discussion of the agronomic revolution and the role of microbial mats in Ediacaran ecosystems)

Sokolov BS, Iwanowski AB (eds) (1990) The Vendian System, 1, Paleontology. Springer-Verlag, Berlin Heidelberg, 383 p (Compilation of review papers on the Vendian System in Russia)

Plate 61: Pre-Ediacaran Dubiostructures

Budd GE, Jensen S (2003) The limitations of the fossil record and the dating of the origin of the Bilateria. In: Donoghue PCJ, Smith MP (eds) Telling the evolutionary time: Molecular clocks and the fossil record. CRC Press, Boca Raton, Florida, pp 166–189 (Critical evaluation of the pre-Ediacaran trace fossil record)

Dawes PR, Bromley RG (1974) Late Precambrian trace fossils from the Thule Group, western Northern Greenland. Report of Acivities (75):38–42

Fedonkin MA, Yochelson EL (2002) Middle Proterozoic (1.5 Ga) *Horodyskia moniliformis* Yochelson and Fedonkin, the oldest known tissue-grade colonial eucaryote. Smithson Contrib Paleobiol 94:1–29 (*Horodyskia* interpreted as a colonial eucaryote)

Rasmussen B, Bengtson S, Fletcher IR, McNaughton NJ (2002) Discoidal impressions and trace-like fossils more than 1 200 million years old. Science 296:1112–1115 (Description of supposed Mesoproterozoic discoidal body fossils and trace fossils from southwestern Australia)

Seilacher A, Bose PK, Pflüger F (1998) Triploblastic animals more than 1 billion years ago: Trace fossil evidence from India. Science 282:80–83 (Description and interpretation of the Chorhat structures as trace fossils)

Plate 62: Ediacaran Sole Features

Fedonkin MA (1985) Paleoichnology of Vendian metazoa. In: Sokolov BS, Ivanovskiy MA (eds) The Vendian system: Historic-geological and palaeontological basis, 1. pp 132–137 (In Russian; English translation, Springer-Verlag 1990) (Description of several Ediacaran ichnotaxa such as *Aulichnites, Bilinichus, Planolites* and *Nenoxites*)

Fedonkin, MA (2003) The origin of the metazoans in the light of the Proterozoic fossil record. Paleontol Res 7:9–41 (Documentation of *Yorgia* and *Dickinsonia* resting traces)

Webby BD (1970) Late Precambrian trace fossils from New South Wales. Lethaia 3:79–109 (Including diagnosis of the ichnogenus *Torrowangea*. However, the age may be Early Cambrian)

Webby BD (1984) Precambrian-Cambrian trace fossils from western New South Wales. Aust J Earth Sci 31:427–437 (Trace fossils from the Farnell Group of Australia. The age is poorly constrained)

Plate 63: Traces of Early Molluscs

Caron JB, Schelterma A, Schander C, Rudkin (2006) A soft-bodied mollusc with radula from the Middle Cambrian Burgess Shale. Nature 442:159–163 (*Odontogriphus* interpreted as a relative of *Kimberella*)

Fedonkin, MA (2003) The origin of the metazoan in the light of the Proterozoic fossil record. Paleontol Res 7:9–41 (Documentation of *Kimberella* radular marks. See Fig. 16 for *Kimberella* preserved at apex of fan-shaped scratch pattern)

Fedonkin MA, Waggoner BM (1997) The Late Precambrian fossil *Kimberella* is a mollusc-like bilaterian organism. Nature 388:868–871 (Interpretation of *Kimberella* as a primitive mollusc)

Seilacher A (1977) Evolution of trace fossil communities. In: Hallam A (ed) Patterns of evolution. Elsevier, pp 359–376 (Giant *Radulichnus* from the Cambrian of Saudi Arabia)

Seilacher A (1999) Biomat-related lifestyles in the Precambrian. Palaios 14:86–93 (Interpretation of *Kimberella* as a mat scratcher)

Wade M (1972) Hydrozoa and Scyphozoa and other medusoids from the Precambrian Ediacara fauna, South Australia. Palaeontology 15:197–225 (*Kimberella* described as cubomedusa)

Yochelson EL, Fedonkin MA (1993) Paleobiology of *Climactichnites*, an enigmatic Late Cambrian fossil. Smithson Contrib Paleobiol 74:1–34 (Summary and reanalysis of the large mollusk trail *Climactichnites* in the Potsdam Sandstone)

Plate 64: *Treptichnus pedum*

Fedonkin M, Liñan E, Perejon A (1983) Icnofósiles de las rocas precámbrico-cámbricas de la Sierra de Córdoba, Espana. Bol R Soc Esp Hist Nat Geol 81(1–2):125–138 (*Treptichnus pedum*)

Gehling JG, Jensen S, Droser ML, Myrow PM, Narbonne GM (2001) Burrowing below the basal Cambrian GSSP, Fortune Head, Newfoundland. Geol Mag 138:213–218 (*Treptichnus pedum* just below the Precambrian-Cambrian boundary)

Germs GJB (1972) Trace fossils from the Nama Group, Southwest Africa. J Paleontol 46:864–870 (*Treptichnus pedum* from the Ediacaran of Namibia)

Jensen S (1997) Trace fossils from the Lower Cambrian Mickwitzia sandstone, south-central Sweden. Fossils and Strata 42:1–111 (Description of *Treptichnus pedum*)

Jensen S, Saylor BZ, Gehling JG, Germs GJB (2000) Complex trace fossils from the terminal Proterozoic of Namibia. Geology 28:143–146 (Description of relatively complex, branched burrow systems from the Ediacaran of Namibia)

Landing E, Narbonne GM, Myrow O (1988) Trace fossils, small shelly fossils and the Precambrian-Cambrian boundary. Bull N Y State Mus 83:1–76 (Review of the ichnology of the Precambrian-Cambrian transition at the global stratotype section in Newfoundland)

Lane AA, Braddy SJ, Briggs DEG, Elliot DK (2003) A new trace fossil from the middle Cambrian of the Grand Canyon, Arizona, USA. Palaeontology 46:9987–9997 (Cambrian *Treptichnus*-like systems interpreted as arthropod trackways and called *Bicavichnites*)

Seilacher A (1955) Spuren und Fazies im Unterkambrium. In: Schindewolf O, Seilacher A (eds) Beiträge zur Kenntnis des Kambriums in der Salt Range (Pakistan). Akademie der Wissenschaften und der Literatur, Mainz, Abhandlungen der mathematisch-naturwissenschaftlichen Klasse, 10, pp 261–446 (Diagnosis of *T. pedum* under the name *Phycodes*)

Plate 65: The Cambrian Revolution

Seilacher A (1997) Fossil art. An exhibition of the Geologisches Institut Tübingen University. The Royal Tyrell Museum of Palaeontology, Drumheller, Alberta, Canada, 64 p (See Introduction, p 29, 38, 39, 91)

Seilacher A (1999) Biomat-related lifestyles in the Precambrian. Palaios 14:86–93 (Discussion of the agronomic revolution and the role of microbial mats in Ediacaran ecosystems)

Seilacher A, Pflüger F (1994) From biomats to benthic agriculture: a biohistoric revolution. In: Krumbein WE, Paterson DM, Stal LJ (eds) Biostabilization of sediments. Bibliotheks und Informationssystem der Carl von Ossietzky Universität Oldenburg, pp 97–105 (Proposal of the agronomic revolution)

Seilacher A, Buatois LA, Mángano MG (2005) Trace fossils in the Ediacaran-Cambrian transition: Behavioral diversification, ecological turnover and environmental shift. Palaeogeog Palaeoclim Palaeoecol 227:323–356 (Analysis of the Cambrian explosion based on ichnologic evidence)

Chorhat structures ("worm burrows"; 1.5 Ga, India)

61

Plate 61
Pre-Ediacaran Dubiostructures

According to a general consensus, metazoans made their appearance no earlier than 600 to 800 million years ago, probably only after the Minoan *snowball earth* event. Accordingly, one must be extremely critical towards trace fossils older than that. The long list of potential animal traces, described from the Neoproterozoic in various parts of the world, is in urgent need of revision. In such an effort, some claims will inevitably turn out to be physical structures; others may have been wrongly identified (for instance by disregarding preservational aspects) with much later forms, while still others may be algal body fossils.

In this and the following plate we shall discuss only a few examples of Precambrian sedimentary structures whose biogenic (metazoan) interpretation is in conflict with the radiometric ages of the host rocks.

Chorhat "Worm Burrows". The report on these structures in 1998 raised a fierce discussion, because the Chorhat Sandstone of central India, in which they occur, had been dated at 1 100 (and 1 500 million years by subsequent workers) – much older than the assumed origin of the metazoans. Thus the claim that they were made by worm-like (and presumably triploblastic) animals in sediments almost three times the age of the Phanerozoic called for rigorous testing. The following questions came up.

(1) Could these structures be *pseudofossils*? As shown in the previous chapter, syneresis can produce structures that may easily be mistaken for worm burrows. "Manchuriophycus" (Pl. 58), for example, is very common on rippled tops of the Chorhat Sandstone; but it fundamentally differs from the alleged burrows with regard to preservation as well as morphospace. Chorhat structures are not keeled epichnial ridges following ripple troughs, but epichnial *grooves* with a rounded profile and without a regular relationship to the relief of the bed surface. Second, they are neither smoothly curved, nor pinched where they abut against another element, but wiggle irregularly and merge broadly at branching points.

Gas bubbles may also produce burrow-like structures; but gas escapes are usually vertical, mimicking *Skolithos*. If trapped underneath a biomat (the model proposed for "Kinneyia", Pl. 58), bubbles could move parallel to the bedding plane, but always uphill on a rippled surface. Nor could bubble tracks branch at nearly right angles. There was no physical process known that could produce the observed patterns. So the Chorhat burrows were, by exclusion, considered as biogenic.

(2) Could these burrows have been produced by *modern animals*? In fact, a modern origin – either by termites or plant roots – was initially suspected for the

Chorhat structures, because they are only found on bed surfaces that had been exposed to weathering under a thin cover of soil – never on fresh rock. Plant roots, however, can etch calcareous rocks but not a hard sandstone, which would also be difficult for insects to scratch. In one place (arrow in block diagram) the burrow is still covered by a little piece of original sandstone. So the burrows are as old as the rock; but slight weathering is necessary to expose them. This would fit the lifestyle of an undermat miner (Pl. 49).

(3) New *experiments* (Bons, Plüger and Seilacher, in preparation) suggest a solution. At a critical diameter, films of pure water form between adjacent bubbles. They have less surface energy than walls of equal size containing sediment. Approach to this threshold by *growing* bubbles leads to "Kinneyia" (Pl. 58; now called "*peanut blisters*" because the type specimen of *Kinneyia* shows microripples rather than gas bubbles), whereas larger but shrinking bubbles may end up as Chorhat structures. Their foam menisci may also shift within bubble strings and thus produce a burrowlike structure without gravitational control.

In conclusion, Chorhat structures would pass as worm burrows in later rocks. Only because of the radiometric ages a non-biological origin must be considered.

Sterling Biota. The still older *hairpin structures* from the Sterling Quartzite of western Australia defy a physical origin. Preserved on bed tops, they have been explained as trails of minute flatworms with borders of displaced sediment (see *Climactichnites*, Pl. 63 for a much larger analog). This, however, does not explain why the bordering ridges are always connected only at one end and why the enclosed areas do not differ from the surface around. Alternatively one could interpret the string itself as an agglutinated microorganism. But why should it have the shape of a hairpin?

Horodyskia. These "strings of beads" have been found in rocks of similar age of North America and Australia. They are certainly biogenic, but difficult to explain. The drawing was made from a rubber cast, so the relief is reversed and will be referred to as if it were a hyporelief. The strings are aligned, presumably along a current coming from the lower side of the picture. This excludes a trace-fossil interpretation. The rows of shallow impressions across current direction could be minute flute casts. Similar scours connect the beads, which are sharply impressed in opposite relief. But because there are no such scours at either end of the string, scouring must have been induced by connecting structures rather than in the shadow of the beads.

In conclusion, no undoubted metazoan trace fossils are known before the last Neoproterozoic glaciation (snowball earth).

Plate 61 · Pre-Ediacaran Dubiostructures   177

Pre-Ediacaran Dubiofossils

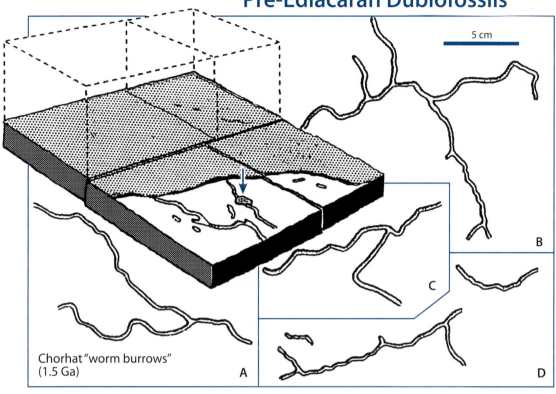

Chorhat "worm burrows"
(1.5 Ga)   A

5 cm

B

C

D

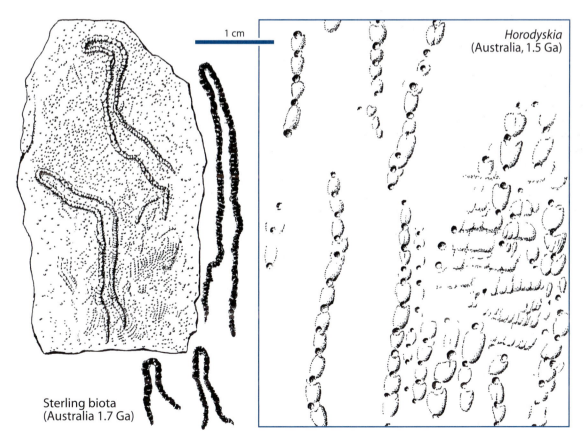

1 cm

Sterling biota
(Australia 1.7 Ga)

Horodyskia
(Australia, 1.5 Ga)

62

Plate 62
Ediacaran Sole Features

Precambrian trace fossils almost exclusively reflect horizontal motion, while vertical bioturbation was virtually absent. This does not mean that all trace makers were epifaunal. In fact, the general rule that surface traces have a very low preservation potential was at that time modified by the ubiquity of *biomats*: except for scratches made by hard organs, soft-bodied epibenthic animals did not leave tracks to begin with! Consequently we could assume that trace fossils other than *Radulichnus* (undertraces Pl. 63) were made by *infaunal* undermat miners burrowing along bedding planes.

Why did these animals not penetrate deeper into the sediment? An obvious answer is that there was no need to escape predators. But there still was the menace of *physical* perturbations. Other explanations would be the steep oxygen gradient in matgrounds and the unability of early burrowers to produce and ventilate open tunnels. In fact, U-burrows have not yet been found in Precambrian rocks. Instead, most burrows shown on this plate suggest undermat miners. Backfilled burrows (*Aulichnites*, Pl. 29; *Nenoxites*, Pl. 32) support this view.

Aulozoon. Everybody knows the reciprocal preservation potential of body versus trace fossils: a sandstone series may contain thousands of trilobite burrows and no trilobite, while a contemporaneous limestone yields an abundance of trilobite carapaces, but not a single burrow. In Precambrian rocks, however, there is no such divergence. Because animals had no mineralized hard parts, body as well as trace fossils are mainly found as impressions on bedding planes covered by biomats.

On the figured sole face of a **South Australian** tempestite (found by Jim Gehling) we see a rich assemblage of quilted *vendobionts*. Are they washed together by a storm or buried in life positions? First one notes **elephant skin** structures indicative of microbial mats (Pl. 58). The sharp *positive* hyporeliefs of the elongate *Phyllozoon* (their uniform sizes suggesting a single generation) are not aligned or accumulated by current; rather they show a tendency to "hug" one another. In contrast, associated *Dickinsonia* lacks respect for other organisms; it is preserved only as phantoms, suggesting that its stiff septa (but not the margin) were pressed through from a higher level. All this speaks for a smothered matground, in which *Phyllozoon* grew below the biomat and *Dickinsonia* lived on top.

The third fossil, *Aulozoon* (informal name), clearly belongs to the undermat tier. Instead of being quilted in the style of vendobionts, it looks like a smooth sand-filled sausage that has become flattened by compaction. Sand, however, is hardly compactable (remember the fillings of mud cracks, Pl. 2). So the cross section was probably flat from the beginning. One also notes that *Aulozoon* does not behave like a hydrostatic sausage, whose outline should be smoothly curved and have sharply defined endings. Instead

it tends to wiggle and to contour *Phyllozoon* when meeting it obliquely, but passes under or over it in a head-on approach. Such behavior would agree with the trace of an undermat miner that bulldozed along instead of repeatedly probing. Nor did it penetrate *Phyllozoon* bodies in its way. Flatworms would fit this description. On the surface they move by spreading their own mucus carpet, against which the ciliary epithelium can act. Translated into an infaunal lifestyle, this kind of locomotion would produce a flattened mucus sausage, into which the processed sand is backfilled. Obviously this model is still hypothetical; but as flatworms are still living today, it is more testable than the idea of another strange organism.

Radulichnus. This trace will be discussed in more detail in Pl. 63. The slab from Australia (after a photograph by Jim Gehling) shows the trace maker's (*Kimberella*) smooth *dorsal* death mask and its neglect for a next-door vendobiont (*Dickinsonia*).

Palaeophycus. These burrows resemble the Chorhat structures (Pl. 61), except that they are preserved in *negative* hyporelief. Their age also fits wormlike undermat miners that avoided the nearby *Tribrachidium*, probably a sponge-like organism attached to the top of the biomat.

Mawsonites. The crescentic backfill structure in the central shaft of this positive hyporelief is in conflict with the original interpretation as a jellyfish. The fossil also resembles radial burrow systems (Pl. 47); but as such deep penetration and complex behavior program would be unusual for Ediacaran times, comparison with the pseudofossil "Astropolithon" (Pl. 58) is more adequate. Alignment of the "probes" along radial cracks speaks for a non-biological origin.

Dickinsoniid Resting Traces. The discovery of *Yorgia* (preserved as negative hyporelief) at the end of a series of fitting **resting traces** (with opposite relief) changed our view of vendobionts. Impressions of the trace maker's quilting (arrow; clearer in other specimens) leave no doubt about authorship. Later, the same kind of traces was found in Australia, together with the closely related *Dickinsonia*. Obviously, these epifaunal vendobionts were not firmly attached to the biomat (elephant skin structures).

Yet there are important differences from other resting traces (Chap. V): (1) As shown by the broken outline, the author *contracted* upon being smothered. (2) The traces show *no scratches*. Instead, the trace has a bumpy surface reminiscent of load casts. This would fit *digestion* of the biomats by the minute pseudopodia of a giant protist. (3) There is no trail connecting the resting traces, i.e. the weight of the organism was not sufficient to deform the sediment.

In conclusion, the biomat was essential not only for forming a dorsal death mask of these vendobionts, but also for the origin and preservational history of their resting traces.

Plate 62 · Ediacaran Sole Features   179

Ediacaran Sole Features

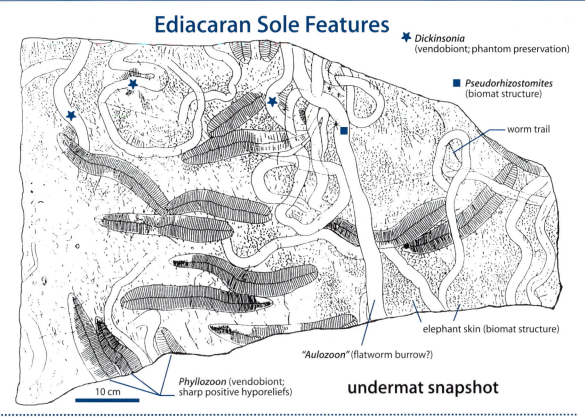

★ *Dickinsonia*
(vendobiont; phantom preservation)

■ *Pseudorhizostomites*
(biomat structure)

— worm trail

elephant skin (biomat structure)

"Aulozoon" (flatworm burrow?)

Phyllozoon (vendobiont;
sharp positive hyporeliefs)

10 cm

undermat snapshot

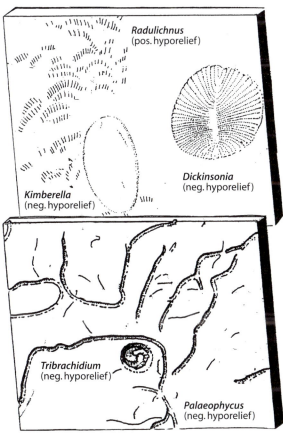

Radulichnus
(pos. hyporelief)

Dickinsonia
(neg. hyporelief)

Kimberella
(neg. hyporelief)

Tribrachidium
(neg. hyporelief)

Palaeophycus
(neg. hyporelief)

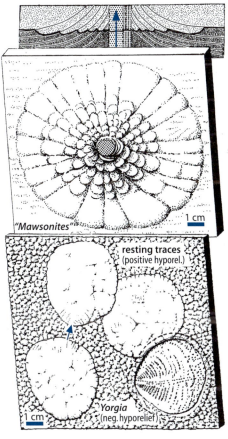

"Mawsonites"

1 cm

resting traces
(positive hyporel.)

1 cm

Yorgia
(neg. hyporelief)

63

Plate 63
Traces of Early Molluscs

Thanks to the presence of matgrounds, Ediacaran biota provide us with an unusual record of stem-group molluscs and their grazing behavior.

Trace Fossils. Let us start with specimens from **Saudi Arabia**. In this case, the hypichnial radula scratches are associated with desiccation cracks and with trilobite burrows (*Cruziana* sp.). They make it clear that we deal with post-Proterozoic intertidal sandstones of **Cambrian** age. The scratches are bifid (inset), somewhat deeper on their distal sides, and connect into continous *meanders*. Except for their giant size (up to 12 cm wide), they resemble the grazing produced by modern gastropods on the algal film of an aquarium wall. This pattern results from the combination of a pendular swinging of the head with forward motion of the whole body. In contrast to *Psammichnites gigas* with its swinging snorkel, however (Pl. 27), locomotion was discontinuous: the animal moved one step ahead at the end of each swing, so that meander loops retain a constant separation.

Meandering radular bites (*Radulichnus*) have been described before from surfaces of fossil *shells*. Like modern radular traces, they are much smaller than the ones from Arabia, whose giant size also implies a preservational divergence: on soft mud, the delicate scratches should properly have been erased when the heavy body bulldozed over them. That this did not happen can only be explained by the original presence of a *biomat* that made the sediment surface strong enough to carry the body without being deformed, while it could be scratched by the hard radular teeth.

This model can be directly applied to the upper Proterozoic (**Ediacaran**) *Radulichnus* from the White Sea, where the presence of a biomat is indicated by elephant skin structures (Pl. 58). Yet, there remains the basic difference in the trace pattern: why should the swing regularly widen away from the origin like a fan, instead of forming a continuous meander band? Here the bulldozer model must be replaced by that of a *stationary* crane, whose swing becomes wider with increasing distance of the shovel from the fixed base. Being equipped with a long proboscis, the animal could remain stationary while grazing. Nevertheless it left no trace when moving to a new site, because the biomat was strong enough to carry its body.

Body Fossils. In reconstructing the Ediacaran trace maker, another preservational effect of sandy biomats comes handy: by inducing mineralization immediately after burial, the microbes produced what Jim Gehling has appropriately called "death masks" of otherwise unpreservable soft bodies. This effect accounts for the negative hyporelief of most Ediacaran vendobionts (Pl. 62); but it may also apply to other soft-bodied organisms. In the *Radulichnus* from S. Australia (Pl. 62), the associated death mask is smooth, because it

shows the upper side of the carcass. This explains why it was originally described as a cubomedusa under the name *Kimberella*. The White Sea death masks, however, show the more differentiated ventral sides of the same organisms. Their outer rims correspond to the margin of a dorsal hood, which was not yet a rigid shell, because its outline is sometimes softly indented. Still following the limpet model, an inner crenulated ring may correspond to gills or the margin of a flat foot whose circular muscles shrank segmentally upon death, while the weaker central sole became secondarily pressed upward into the intestinal cavity during later stages of necrolysis. Where this did not happen (uppermost specimen), one can still see its original transversal wrinkling.

A stumbling block in the correlation between *Kimberella* and the associated *Radulichnus* are the different size ranges. While *Kimberella* measures from a few millimeters to 8 cm, associated scratch patterns fit only the largest body fossils. This can be explained as an undertrace effect: smaller scratches simply did not penetrate deep enough into the sandy biomat to be recorded at its base.

Accordingly, *Kimberella* can be reconstructed as a kind of "soft limpets". Today, this lifestyle in represented by certain gastropods, polyplacophorans and holothurians, but only on rockgrounds. Yet, Precambrian sandy biomats have probably been sufficiently tough to allow a similar mode of life, in which the broad sole with its stronger rim could act as a sucker in case of emergency. With the hood being pressed tightly to the substrate, the animal could also become smothered in place during a storm, without sand entering the gill slit around the foot – a prerequisite for the formation of the ventral death mask. With this evidence, the affiliation of *Kimberella* can be narrowed down. Of all known Early Cambrian molluscs, the strange halkieriids come closest to our reconstruction of *Kimberella* and its radular traces.

Climactichnites. In the case of the **Cambrian** *Radulichnus* there are no death masks of the trace maker. In their stead, trails of equivalent size (*Climactichnites*) have been found in Cambrian sandstones of similar age in North America. They resemble motorbike tracks on the rippled sand of a modern beach. *Climactichnites* has not yet been found outside North America, but it corresponds to the Arabian radula scratches not only in size. Its V-shaped corrugations (opening in the direction of movement) and marginal piles of sand also suggest a gastropod-like molluscan foot that pushed the animal ahead by waves of muscular contraction. In other cases, *Climactichnites* is preserved in the form of positive hyporeliefs (specimen from **New York**). Some of these trails start abruptly, as if the animal had been dropped from the water column; but in an undertrack scenario it might have simply dug to the interface before plowing along by muscular waves. Near the starting point, one also observes a much finer *concentric lineation*, which probably reflects and the ciliary progression of the rear edge of the foot and thus fits the outline of a *Kimberella*-like halkieriid.

Plate 63 · Traces of Early Molluscs   181

5 cm

5 cm

Climactichnites,
Upper Cambrian,
New York

Radulichnus,
Cambrian, Saudi Arabia

1 cm

Early Molluscs:
Death Masks and Traces

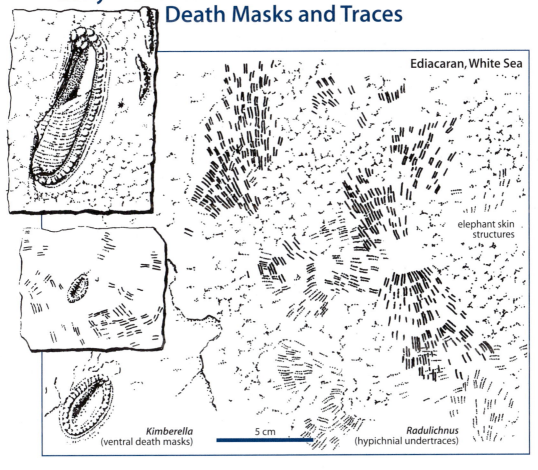

Ediacaran, White Sea

elephant skin
structures

Kimberella
(ventral death masks)

5 cm

Radulichnus
(hypichnial undertraces)

64

Plate 64
Treptichnus pedum

As we have seen throughout this book, most "worm" burrows are rather useless for stratigraphic correlation, because they are either too featureless, or because similar behavior patterns may have evolved independently in different clades. Yet, the trace fossil shown here should be familiar to every stratigrapher, because its first appearance has been chosen to define the boundary between the Proterozoic and the Phanerozoic eras. In order to convey a feeling for the variability, biological purpose and ichnotaxonomic affiliation, we devote a whole plate to it.

Hypichnial Preservation. The feature that distinguishes *Treptichnus pedum* from ordinary worm burrows is its subdivision into modular segments. They look like buds along a twig, while the burrow as a whole does never branch, but may follow a straight, sinusoidal, or coiled course.

Although backfill lamination has not been verified inside the hypichnial ridges, we probably deal with the rhythmical probing and backfilling action of a wormlike animal. We may also assume that it was an undermat miner, because its burrow as a whole strictly follows the bedding plane without ever probing into the underlying mud layer. The main activity was in the thin sand layer on top, where the animal probably emerged to the surface in every segment. So the trace may be described as a series of small teichichnoid backfill bodies with U-shaped contours, arranged along a continuous path that did not necessarily remain open.

The affilation of this ichnospecies with *Treptichnus* is arguable. Because the probing somewhat resembles that of *Phycodes* (Pl. 45), *T. pedum* was originally affiliated with this ichnogenus, while the species name refers to the similarity of coiled forms with the bishop's crosier (*pedum*) of Christian tradition. Examination of more extensive material (for instance from the Upper Vendian or lowermost Cambrian of **South Africa**) shows a considerable variability. Among the bedding planes (cast on a local farmer's terrace), some show *disjunct* impressions, because only the deepest parts of the U-shaped segments reached the interface.

The straight variant (**A**) appears to be current-induced, because burrows are aligned and run in the same direction (upward in the drawing), perpendicular to the ripples pressed-through from a higher level. On another bedding plane at the same locality (**B**), burrows show sinusoidal patterns. As they are found in different sizes, each individual must have produced many such systems throughout life.

Occurrences at higher levels of the Lower Cambrian (**C, D**), the Middle Cambrian (**E**), and the Lower Ordovician (**F**) show a similar variability, although a particular pattern may prevail at any one site. One end member resembles the "feather stitch" pattern of typical *Treptichnus*, while coiling becomes extremely tight in the Grand Canyon (**E**).

Because no clear trend can be observed through geologic time, it is probably wisest to maintain one name (*Treptichnus pedum*) and perhaps to distinguish the vari-

ants by informal terms or as ichnosubspecies. Nevertheless, this trace is sufficiently distinctive, common and widely distributed to use its first occurrence as a marker for the Precambrian/Cambrian boundary – even though it does not yet represent the new, truly bioturbational style of burrowing that led to the high diversity of Cambrian ichnocoenoses.

Epichnial Preservation. *Treptichnus pedum* is usually found in *positive hyporelief* preservation, where the cavities made in the underlying mud have been actively backfilled with sand from the layer above. On sandstone *tops*, one should expect corresponding negative epireliefs, in which mud from above was actively introduced into the sand layer. Apart from this reversal, one should also expect that the upper parts of the burrow system (i.e. the distal branches) are more pronounced than the axial parts, which normally remain hidden in the sand.

Such structures actually occur in the Middle Cambrian of the Grand Canyon under the name *Bicavichnites*. They were tentatively interpreted as trackways of early lobopods. Given that we mostly deal here with two rows of alternating impressions, this view is understandable. On the other hand, footprints in epichnial arthropod undertracks are usually narrow slots rather than broad furrows, because sand closes in again after the impression has been pierced. The figured specimen, however, also preserves the sandy fills of a central shaft and some proximal branches. These structures indicate a long, worm-like tracemaker. Therefore *Bicavichnites* likely represents straight burrows with alternating branches and is here considered as a preservational variant of *T. pedum*.

Stratigraphic Implications. Stratigraphic boundaries must be defined by *stratotypes* that can be referred to in other sections. Working groups of the International Correlation Programs democratically decide, in which country and at what level to put the "Golden Spike". Nevertheless such decisions have a subjective (if not political) element.

Of the many changes that took place in the transition from the Proterozoic to the Phanerozoic Era, the onset of vertical bioturbation is just one. *Treptichnus pedum* has the advantage that it is distinctive, common and can be recognized in a variety of facies. On the other hand, it could be argued whether this burrow really represents vertical bioturbation or just a more sophisticated mode of undermat mining.

Thus it is not surprising that the first appearance does not exactly coincide with other bioevents, such as the disappearance of the formerly ruling vendobionts, the radiation of "small shelly fossils", and eventually of the trilobites. In fact, *Treptichnus pedum* has lately been found in association with the last vendobionts in Australia and Namibia. Thus it is formally true to say that these strange organisms survived into the Early Cambrian, even though their extinction, and with it the demise of the Ediacaran ecosystem, was certainly more decisive. Possibly *geochemical* or *paleomagnetic* excursions would in this case be better chronometers than guide fossils; but unfortunately they cannot be recognized right in the field!

Plate 64 · *Treptichnus pedum*   183

Treptichnus pedum

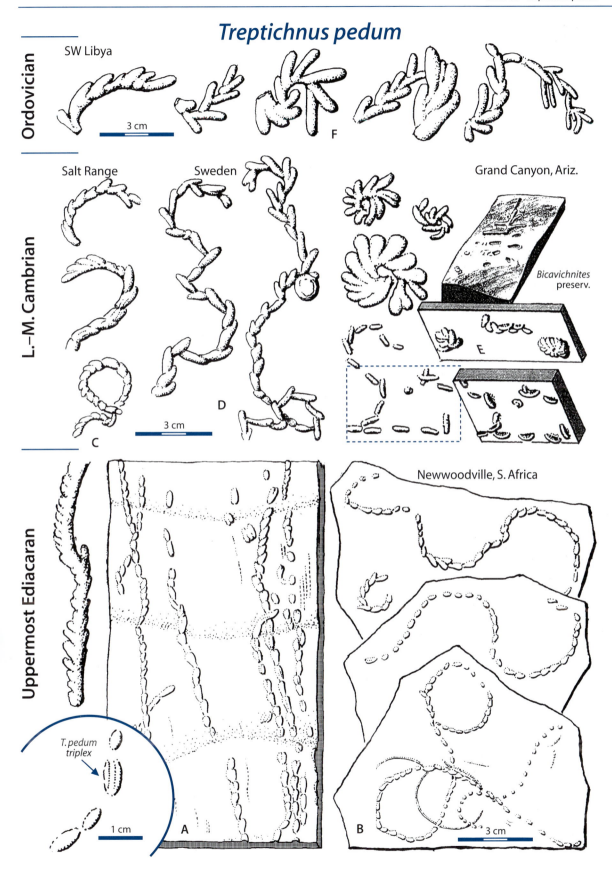

SW Libya

3 cm

Ordovician

Salt Range   Sweden   Grand Canyon, Ariz.

Bicavichnites preserv.

E

L.–M. Cambrian

C   D   3 cm

Newwoodville, S. Africa

Uppermost Ediacaran

T. pedum triplex

1 cm   A

B   3 cm

65

Plate 65
The Cambrian Revolution

The Nature of the Revolution. The term "Cambrian Explosion" refers to the almost sudden appearance and diversification of animal phyla at the beginning of the Cambrian Period. The advent of hard and stiff skeletons favored fossilization potential as well as the differentiation of new designs; but it also changed the trophic structure of communities. One way to escape the new menace of *predation* was to become infaunal. Combined with deposit and sediment feeding, **bioturbation** (in which animals too small to leave recognizable traces were also involved) put an end to **matgrounds** and to biomat-related lifestyles, but not to the microbes. Instead of organizing themselves into mats, they now encrusted individual sand grains, which were moved up and down in the sediment by bioturbation. In such **mixgrounds**, the biologically active zone is much thicker than in matgrounds. Accordingly, microbial productivity could increase to a level that compensated for the loss caused by cropping. Only in environments hostile for metazoans could the microbes continue their Precambrian lifestyles to the present day.

In this perspective, the explosive radiation of metazoan phyla in the Early Cambrian is only one aspect of a more profound ecological turnover. Its effect on the whole biosphere can be compared to the agronomic revolution caused by our own species. In both cases, productivity became increased by sediment mixing. At the same time, intensified cropping reduced the sink of organic matter into the lithosphere. As a consequence, source rocks for hydrocarbons became restricted to sediments that were spared from this development by anoxia. The increased recycling must also have influenced the chemistry of the hydro- and atmosphere. Thus, when we study trace fossils in the Precambrian/Cambrian transition, we register the very processes that contributed to the chief biohistoric event in the last billion years.

Puncoviscana Formation: The Revolution Spreads into Deepsea Environments. Neither the radiation nor the ecological revolution were globally instantaneous. Both started in shallow marine environments, from which they spread into other aquatic realms.

Because there is no reliable biostratigraphic measuring stick, the conquest of rivers and lakes is difficult to pin down. Yet, the occurrence of small crustacean resting tracks (cf. *Isopodichnus*; Pl. 23) in redbeds with salt pseudomorphs (Lower Cambrian, Pakistan) and of crowded worm burrows in a fluvial sequence (Hasawna Formation, Libya) suggests that the conquest happened already by Cambrian times.

The same applies to the conquest of **deepsea** bottoms, as shown by the trace fossil record of the *Puncoviscana*

Formation of **Argentina**. This turbiditic series is probably several thousand meters thick; but tectonization makes it diffcult to measure and correlate continous sections. While the base of the series is ill defined, its Lower Cambrian age is constrained by the trace fossil *Syringomorpha nilssoni* (Pl. 41) that occurs in the unconformably overlying shallow marine sandstones (Mesón Group).

On this background it is remarkable that the Puncoviscana Formation contains two radically different ichnocoenoses. The presumably older one is dominated by the undermat mines of *Oldhamia*, whose burrowing programs are much more complex than in Ediacaran turbidites and differ from one outcrop to the other (Pl. 49). A Cambrian rather than Ediacaran age is also indicated by associated rare arthropod tracks. The second assemblage, shown on the present plate, is not only more diverse and more disparate, but also more closely related to contemporaneous shallow water communities.

The "lasso trail" *Psammichnites gigas* behaves like its neritic counterparts (Pl. 27), although it is smaller and only the bottom relief is preserved. Similarly, the alternating series of burrow openings may be interpreted as a preservational variant (*Bicavichnites* or *Saerichnites*) of *Treptichnus pedum* (Pl. 64). More unusual is a closed circle of tunnel openings with an inner continous ring. Unfortunately the original is cemented into a sidewalk in the beautiful town of Salta, so it is impossible to decide by serial sectioning whether we deal with a variant of *Treptichnus* (*T. coronatus*) or a teichichnoid burrow like **Heliochone** (Pl. 41).

More akin to later flysch ichnocoenoses are the "cautious" meanders of *Psammichnites* ("*Nereites*") *saltensis*. Its bilobed meanders are preserved as negative hyporeliefs. Nevertheless, the associated trilobite footprints (*Diplichnites*) – certainly representing a shallower tier – have not been wiped out by the later meanders. This paradox is explained by a *Psammichnites* mode of burrowing, in which laminae above the burrow become elevated without destroying older signatures (Pl. 27).

Other strange elements of this assemblage are discontinuous trackways (*Tasmanadia*). Like other tracks of "jumping" arthropods (Pl. 7), they consist of a distinctive pattern of footprints that repeats in a disjunct series. While these tracks cannot yet be referred to a particular maker, they indicate that the radiation of strange arthropods in the Cambrian Explosion (as indicated by Chengjiang body fossils) was not restricted to shallow marine habitats.

In summary, the trace fossils of the Puncoviscana Formation carry several important messages: (1) Metazoans invaded deepsea bottoms already in the late Proterozoic. The lifestyle of the characteristic burrowers (*Oldhamia recta*, Pl. 49) suggests that at that time these bottoms were sealed by microbial mats, even though photosynthesis ▶

Plate 65 · The Cambrian Revolution 185

Agronomic Revolution

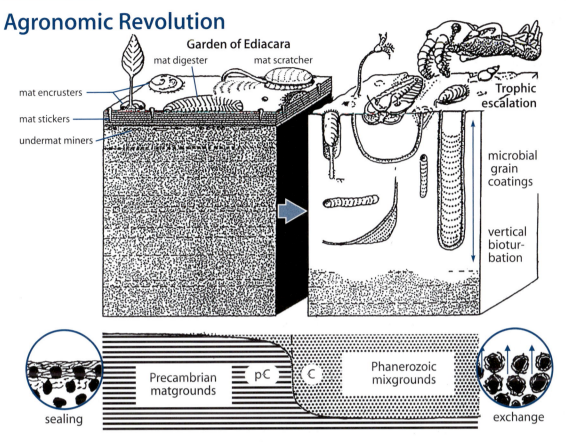

Garden of Ediacara

mat encrusters

mat stickers

undermat miners

mat digester

mat scratcher

Trophic escalation

microbial grain coatings

vertical biotur- bation

sealing

Precambrian matgrounds

p€ €

Phanerozoic mixgrounds

exchange

Invasion into Deepsea Bottoms: L. Cambrian, Argentina

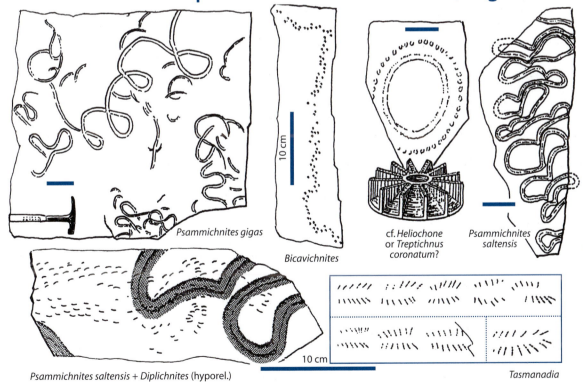

Psammichnites gigas

10 cm

Bicavichnites

cf. *Heliochone* or *Treptichnus coronatum?*

Psammichnites saltensis

10 cm

Psammichnites saltensis + *Diplichnites* (hyporel.)

Tasmanadia

could not take place below the photic zone. (2) The *Oldhamia* deepsea community radiated in the earliest Cambrian with the same characteristics (small size, intensive coverage), but without the disparity of mature post-turbidite associations of later flysch ichnocoenoses. (3) The delayed arrival of the *Agronomic Revolution* in deepsea environments is marked by a richer ichnocoenosis in which burrows are still large and closely related to contemporaneous shallow-water faunas. Yet sediment feeders show a tendency towards more complex search behaviors. (4) Graphoglyptid farming burrows (Pls. 52–55) of the pre-turbidite association (which would be preserved on turbidite soles) are still lacking. They occur in shallow marine ichnocoenoses of Cambrian times, but make their appearance in deepsea deposits only in the Ordovician.

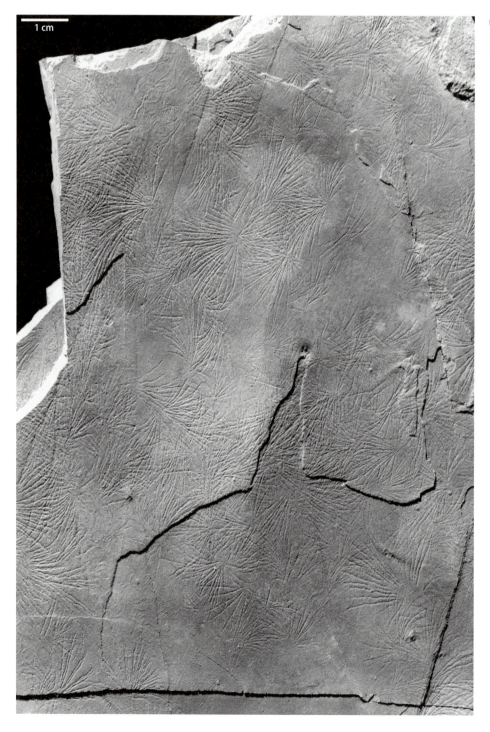

1 cm

■
Oldhamia antiqua
(L. Cambr., Maine)

Cruziana Stratigraphy

The origin of paleontology as a science, as well as its traditional affiliation with earth sciences, derives from the use of fossils as stratigraphic markers. Index species define biozones and stages, while series are marked by the appearance or disappearance of higher taxa and era boundaries by mass extinctions. Although biostratigraphy is not implicitly based on evolutionary theory, the taxa used should be short-ranged and phyletically coherent. As we have seen, most trace fossils fail to meet these requirements, because different organisms may produce similar traces. Nor does current ichnotaxonomy capitalize makership: similar trace fossils, it is claimed, should bear the same name, no matter what kinds of animals made them. As such an attitude is unacceptable for a paleobiologist, I have tried throughout this book to focus on the biological identity of the trace maker as far as possible. Where this could not be done (e.g., in arthrophycids, Chap. IX), biologically coherent "ichnofamilies" have nevertheless been singled out as more or less coherent higher units. On the level below ichnogenera (which have practically become the fundamental units) it would be desirable to take the ichnospecies level more seriously. As this would have required a much more profound review of all the literature (papers that are otherwise irrelevant in this context may still contain new names!) and since many ichnogenera simply do not have enough diagnostic features to justify subdivision, weeding at the ichnospecies level has not been attempted.

Yet, biological identity is not only of academic interest. For biostratigraphic purposes it is necessary to select groups of trace fossils that are not only genetically coherent, but also provide sufficient morphological complexity to make meaningful distinctions at the ichnospecies or even an ichnosubspecies level. One such group are vertebrate tracks, which can be reasonably correlated with taxa known from skeletal remains (Pl. 1). Another example are the arthrophycids (Chap. IX). While floating taxonomically as a whole, they are generically distinct from other worm burrows and evolved sufficiently diverse burrowing behaviors to allow ichnospecific subdivision. The same could be said of graphoglyptids (Pls. 52–55); but their behaviorally distinctive ichnospecies are too long-ranging compared to associated body fossils to serve as index fossils.

In the present chapter, we come back to *trilobite burrows* (Chap. III), now with an emphasis on the distinction of stratigraphically useful ichnospecies. Only few of them can be referred to particular kinds of trilobites (Pl. 14), while others were probably made by non-trilobite arthropods (Pl. 11). Nevertheless, *Cruziana* stratigraphy has become an important tool, because

1. due to taphonomic screening the burrows are very rarely associated with skeletal remains of trilobites (Pl. 14) or of other animals. This is true not only at the bed scale, but may apply to whole series of "unfossiliferous" sandstones. Therefore correlation by trace fossils becomes most useful in cratonic sandstone sequences, whose high porosity makes them prime reservoirs for oil and gas (Pl. 66).

2. like all trace fossils, trilobite burrows are undisplaceable, i.e. they can be neither transported into other facies, nor reworked as "ghost fossils" into younger sediments.

3. determinations can be made directly in the field and recorded by photographs, particularly under the ideal lighting conditions of desert climates.

It is no coincidence that research in this direction started with *Cruziana rugosa* and its allies, because these large burrows call the eye of any observer. They are also most common. Wherever they occur, from the Andes to south China (Pl. 66), they cover the bedding planes, so that the collector must either use cranes and trucks instead of a rucksack or be satisfied with photographs or large-scale rubber casts.

When Alcide d'Orbigny described these spectacular fossils from Bolivia and J. F. N. Delgado did the same in Portugal, they were still interpreted as seaweed impressions. It took the fervor of people like A.G. Nathorst and J. W. Dawson and a fierce (often polemic) discussion, to convince colleagues that these were trace fossils probably made by trilobites. This long history of research and the quantity of material are reflected in the large number of names given. For ichnostratigraphic purposes, most of these "Armorican" trilobite burrows should better be treated as behavioral and preservational variants of a single ichnospecies, *Cruziana rugosa*, because all share a similar (Arenigian) age and characteristic combed fingerprints (Pls. 12 and 66). This does not exclude the distinction of behavioral modifications (*furcifera*; *goldfussi* etc,) at the level of ichnosubspecies. To my knowledge, *Cruziana rugosa* is also the only invertebrate trace fossil that led to a recognized geopark. The place (Penha Garcia) is in Portugal and well worth visiting.

Literature

Chapter XIV

Belka Z (ed) (2000) Excursion guidebook, The Holy Cross Mountains. Joint Meeting of Europrobe (TESZ) and PACE Projects, Zakopane/Holy Cross Mountains, Poland. Warszawa, pp 5–38

Crimes TP (1968) *Cruziana*: A stratigraphically useful trace fossil. Geol Mag 105(4):360–364

Crimes TP (1969) Trace fossils from the Cambro-Ordovician rocks of North Wales and their stratigraphic significance. Geol J 6(2): 333–338

Crimes TP (1975) The stratigraphical significance of trace fossils. In: Frey RW (ed) The study of trace fossils. Springer-Verlag, New York, pp 109–130 (Discussion of *Cruziana* stratigraphy)

Crimes TP (1987) Trace fossils and correlation of late Precambrian and early Cambrian strata. Geol Mag 124(2):97–119

Magwood JPA, Pemberton SG (1990) Stratigraphic significance of *Cruziana*: New data concerning the Cambrian-Ordovician ichnostratigraphic paradigm. Geology 18:729–732 (Critical analysis of *Cruziana* stratigraphy documenting supposed *C. rugosa* from the Lower Cambrian of Canada)

Orbigny A d' (1835–1847) Voyage dans l'Amérique méridionale le Brésil, la République orientale de l'Uruguay, la République Argentine, la Patagonie, la République du Chili, la République de Bolivia, la République du Pérou exécuté pendant les années 1826, 1827, 1828, 1829, 1830, 1831, 1832 et 1833. Pitois-Leverault, Paris & Leverault, Strasbourg, 3(4) (Paléontologie), 188 p (Diagnosis of the ichnogenus *Cruziana*, although interpreted as an algal impression)

Seilacher A (1970) *Cruziana* stratigraphy of non-fossiliferous Palaeozoic sandstones. In: Crimes TP, Harper JW (eds) Trace fossils. Geol J, Special Issue 3, pp 447–476 (Proposal of *Cruziana* ichnostratigraphy)

Seilacher A (1983) Paleozoic sandstones in southern Jordan: Trace fossils, depositional environments and biogeography. In: Abed AM, Khaled HM (eds) Geology of Jordan. Proceedings of the First Jordanian Geological Conference, Jordanian Geologists' Association, pp 209–222 (*Cruziana* ichnostratigraphy in Jordan)

Seilacher A (1990) Paleozoic trace fossils. In: Said R (ed) The geology of Egypt. AA Balkema, Rotterdam, pp 649–670 (*Cruziana* stratigraphy and description of new ichnospecies from Sinai)

Seilacher A (1992) An updated *Cruziana* stratigraphy of Gondwanan Paleozoic sandstones. In: Salem MJ, Hammuda OS, Eliagoubi BA (eds) The geology of Libya, 4, Elsevier, Amsterdam, pp 1565–1581 (Revision of *Cruziana* ichnostratigraphy)

Seilacher A (1996) Evolution of burrowing behavior in Silurian trilobites: Ichnosubspecies of *Cruziana acacensis*. In: Salem MJ, Busrewil MT, Misallati AA, Sola M (eds) The geology of Sirt Basin, 3, Elsevier, Amsterdam, pp 523–530 (Diagnosis of *Cruziana acacensis* ichnosubspecies and their biostratigraphic significance)

Plate 66: Stratigraphic and Paleogeographic Distribution

Bergström J (1976) Lower Palaeozoic trace fossil from eastern Newfoundland. Can J Earth Sci 13(11):1613–1633 (*C. leiferikssoni*)

Bergström J, Peel JS (1988) Lower Cambrian trace fossils from northern Greenland. Grønlands Geologiske Undersøgelse, Rapport 137, pp 43–53 (Documentation of a rich Laurentian ichnofauna, including trilobite trace fossils)

Seilacher A (1994) How valid is *Cruziana* stratigraphy? Geol Rundsch 1994:752–758 (Reinterpretation of Lower Cambrian *C. rugosa* as a new ichnospecies, *C. pectinata*)

Seilacher A, Crimes TP (1969) "European" species of trilobite burrows in eastern Newfoundland. In: Kay M (ed) North Atlantic geology and continental drift. Am Assoc Petr Geol Mem 12:145–148 (Discovery of "European" trilobite trace fossils in Newfoundland, supporting the idea of a Gondwanan or Avalonian terrane)

Plate 67: Cambrian Trilobite Burrows

Seilacher A (1990) Paleozoic trace fossils. In: Said R (ed) The geology of Egypt. AA Balkema, Rotterdam, pp 649–670 (*Cruziana* stratigraphy and diagnosis of the Cambrian ichnospecies *C. nabataeica*, *C. salomonis* and *C. aegyptica*)

Plate 68: Ordovician Trilobite Burrows

Borrello AV (1968) Vocabulario Ichnologico. pp 1–10, Provincia de Buenos Aires

Ctyroky P (1973) *Cruziana* traces from the Khabour quarzite-shale formation in Northern Iraq. Vestník Ustredního Ustavu Geologického 48:281–284, 2 pls. (*C. rugosa*)

Mángano MG, Buatois LA (2003) Trace fossils. In: Benedetto JL (ed) Ordovician fossils of Argentina. Universidad Nacional de Córdoba, Secretaría de Ciencia y Tecnología, pp 507–553 (Description of several ichnospecies of Ordovician *Cruziana* and proposal of ichnosubspecies of *C. rugosa*)

Mángano MG, Buatois LA, Moya MC (2001) Trazas fósiles de trilobites de la Formación Mojotoro (Ordovícico inferior medio de Salta, Argentina): implicancias paleoecológicas, paleobiológicas y bioestratigráficas. Rev Esp Paleontol 16:9–28 (Description of ichnospecies from the *rugosa* group)

Seilacher A (2003) Arte fóssil. Divulgações do Museu de Ciências e Tecnologia – UBEA/PUCRS, Porto Alegre, Publicação Especial 1, pp 1–86 (Illustration of variants of *Cruziana rugosa*)

Plate 69: Silurian to Carboniferous Trilobite Burrows

Seilacher A (1970) *Cruziana* stratigraphy of non-fossiliferous Palaeozoic sandstones. In: Crimes TP, Harper JW (eds) Trace fossils. Geol J, Special Issue 3, pp 447–476 (Discussion of Silurian-Carboniferous trilobite trace fossils)

Seilacher A (1996) Evolution of burrowing behavior in Silurian trilobites: Ichnosubspecies of *Cruziana acacensis*. In: Salem MJ, Busrewil MT, Misallati AA, Sola M (eds) The geology of Sirt Basin, 3, Elsevier, Amsterdam, pp 523–530 (Definition of *Cruziana acacensis* ichnosubspecies and their biostratigraphic significance)

Plate 70: Trans-Gondwanan Seaway

Borrello A (1966) Trazas, restos tubiformes y cuerpos fósiles problemáticos de la Formación La Tinta. Sierras Septentrionales Provincia de Buenos Aires. Paleontografía Bonaerense, Fascículo V, Provincia de Buenos Aires, Gobernación, Comisión de Investigaciones Científicas, La Plata, 42 p (First description of trilobite trace fossils from Argentina, including *C. bonariensis*)

Seilacher A (2005) Silurian trace fossils from Africa and Argentina mapping a trans-Gondwanan seaway? Neues Jahrb Geol P M 3, pp 129–141

Seilacher A, Cingolani C, Varela C (2003) Ichnostratigraphic correlation of early Paleozoic quartzites in central Argentina. In: Salem MJ, Oun KM, Seddig HM (eds) The geology of Northwest Libya. Earth Science Society of Libya 1, Tripoli, pp 275–292 (Proposal of a Silurian Trans-Gondwanan Seaway based on a correlation between strata of Argentina and North Africa)

■ *Cruziana rugosa* (L. Ordov., Iraq)

66

Plate 66
Stratigraphic and Paleogeographic Distribution

The range chart includes only ichnospecies that are morphologically well defined by size, burrowing behavior (stationary = *rusophyciform*, or plowing = *cruzianaeform*) and scratch patterns (claw formula; angle of scratches). With the exception of *Cruziana aegyptica* (Pl. 12) and *C. nabataeica* (Pl. 67), traces smaller than about 1 cm have been excluded because they do not preserve distinctive scratch patterns. This also avoids confusion with the similar, but consistently small, burrows made by nonmarine phyllopod crustaceans (*Isopodichnus*, Pl. 23).

You may also wonder why this scheme contains only forms found in fragments of the ancient Gondwana continent? There are two answers. (1) Compared to other marine organisms, trilobites appear to have had a limited *range of dispersal*; i.e. the planktonic larvae of benthic species (for only these produced burrows) were unable to reach the shelves of other continents if they were more than a few hundred kilometers away. Therefore hardly any of the listed ichnospecies have ever been found in other paleocontinents of the time. (2) The other reason is *monographic*. Gondwana was an area in which cratonic sandstones ("Nubian Facies") were deposited and redeposited over vast areas during periods of transgression. By their high porosity (no silification or cementation!) these sandstones also have a high reservoir potential for oil and gas. Thus there was an economic interest in developing alternative correlation schemes (palynomorphs; trace fossils) that could be used in the absence of standard index fossils. Unfortunately the two schemes cannot easily be combined. Palynomorph stratigraphy is based on well cores that are too small in diameter to show trilobite burrows. In contrast, *Cruziana* is found in surface exposures that are too deeply weathered to preserve palynomorphs. In order to link the two stratigraphies, it would be necessary to core outcrops to depths where the organic-walled microfossils have not yet been eliminated by weathering. Without economic instigation (and without the desert sun emphasizing bedding plane relief), comparable studies have not yet been conducted in other areas. As trilobites as a whole had a global distribution and since burrowing was hardly an exclusively Gondwanan habit, it should be possible to develop similar schemes also for other paleocontinents.

It should also be emphasized that *stratigraphic resolution* is understated in this scheme. In the absence of established index fossils, and of paleomagnetic or radiometric dating, the succession of different trace fossils within a given sandstone sequence is commonly the only age criterion. Only through drilling tests (as discussed above) and the linkage of ichnospecies with body fossils in lithologically more diverse sections could their age be better calibrated. But unfortunately stratigraphers working in such areas tend to neglect trace fossils in favor of the more reliable body fossils.

The sparse data available from *Cruziana* occurrences outside Gondwana suggest that the evolution of burrowing appendages and of burrowing behaviors followed similar but independent lines. This is illustrated by burrows from the Canadian Rocky Mountains, whose multi-clawed scratch pattern is strikingly similar to that of the Lower Ordovician *Cruziana rugosa*. Yet the Canadian forms are clearly Early Cambrian in age and occur in an area that undoubtedly belonged to **Laurentia** rather than Gondwana. The original conclusion, that *Cruziana* stratigraphy is after all not as reliable as we thought, was dispelled by preparation with a metal brush: in the area of the median furrow (which is usually hidden under adhering shale) specimens revealed a series of stronger scratches that are never seen in the Ordovician form. We thus deal with a case of *homeomorphy* caused by the convergent evolution of comblike appendages in two unrelated clades of burrowing trilobites. However, it was the *exopodites* in the Cambrian lineage (*Cruziana pectinata* and *C. jenningsi*, Pl. 13), and the *endopodites* in the Ordovician one, that became transformed into such shovels. Interestingly, the same trend is observed in another Laurentian form (*Cruziana marginata*), from the Lower Cambrian of Greenland), in which scratches of simple endopodites and multiclawed exopodites are combined with rhythmic impressions of the head shield, suggesting a head-down style of burrowing (Pl. 13).

For *palinspastic* purposes, trilobite burrows can so far only be used to identify terranes of Gondwanan origin that happened to dock at other paleocontinents. However, with increased knowledge of developments outside Gondwana it may also be possible to document the switch of terranes in other provinces. For instance, it may be no coincidence that trilobite burrows of the Holy Cross Mountains (Poland) have affinities with Baltica in the Lower Cambrian, while the ichnospecies appearing in the Middle and Upper Cambrian of the same area are Gondwanan.

Plate 66 · Stratigraphic and Paleogeographic Distribution   191

Gondwanan *Cruziana* Stratigraphy and Paleogeography

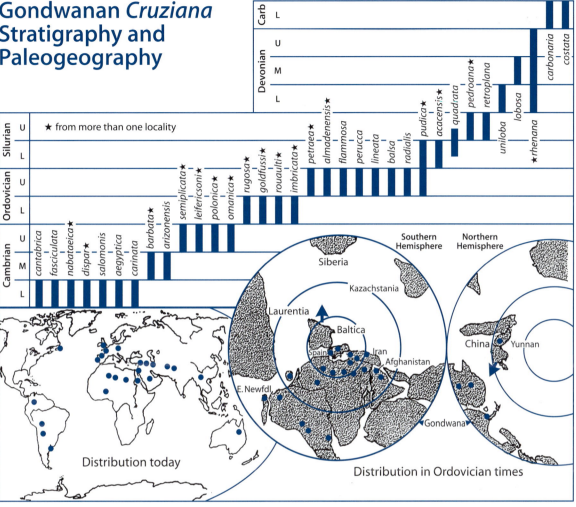

★ from more than one locality

| | | cantabrica | fasciculata | nabataeica ★ | dispar ★ | salomonis | aegyptica | carinata | barbata ★ | arizonensis | semiplicata ★ | leifericsoni ★ | polonica ★ | omanica ★ | rugosa ★ | goldfussi ★ | rouaulti ★ | imbricata ★ | petraea ★ | almadenensis ★ | flammosa | perucca | lineata | balsa | radialis | pudica ★ | acacensis ★ | quadrata | pedroana ★ | retroplana | uniloba | lobosa | rhenana ★ | carbonaria | costata |

(Stratigraphic chart columns: Cambrian L/M/U, Ordovician L/U, Silurian L/U, Devonian L/M/U, Carb L)

Southern Hemisphere · Northern Hemisphere

Siberia · Kazakhstania · Laurentia · Baltica · Spain · Iran · Afghanistan · China · Yunnan · E.Newfdl. · Gondwana

Distribution today

Distribution in Ordovician times

homeomorph scratch patterns

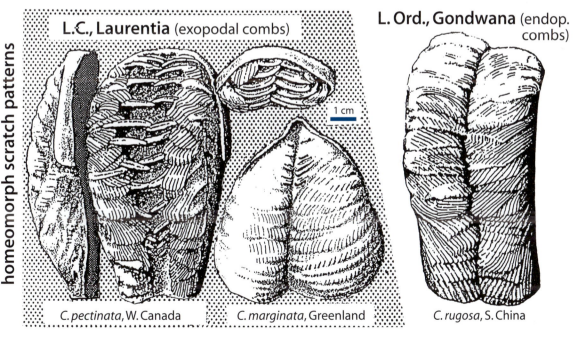

L. C., Laurentia (exopodal combs)

L. Ord., Gondwana (endop. combs)

1 cm

C. pectinata, W. Canada   *C. marginata*, Greenland   *C. rugosa*, S. China

67

Plate 67
Cambrian Trilobite Burrows

In Pls. 67–69 representative ichnospecies are shown with plowing (cruzianaeform) traces on the left and stationary (rusophyciform) variants on the right. If both forms co-occur, a sign indicates which is more common. Obliquity of the dashed lines between ichnospecies indicates which one occurs lower in a local section. Compared to the extension of Gondwanan shelves, samples are as yet too spotty to reflect differences in latitudinal provinces, which should have existed around such a large continent.

In the **Lower Cambrian**, forms from Sweden and Poland (Pl. 13) have been excluded because they belong to Baltica. Trilobite burrows from the Salt Range of Pakistan (where the grazing trackway of Pl. 9 was found) are not distinctive enough. The most important sampling areas are the Middle East (Iran, southern Jordan, Sinai, northwestern Argentina, and the Cantabrian Mountains, Spain).

In the *Middle East*, the earliest form is *Cruziana nabataeica*, named after the type locality in the ancient Nabataean city of Petra. Its rusophyciform burrows are relatively small. Nevertheless one can recognize multiclawed digging traces that comprise about a dozen equal scratches per set. These sets are imbricated, reflecting the sequence in which the appendages became metachronically activated from rear to front. In their active stroke the legs were inclined backwards, so that transition into the recovery stroke is sometimes visible near the midline. The angle between scratches on opposite sides is very wide in front and narrows to about 30° in the rear part of the body. This ichnospecies has also been found in the Lalun Sandstone of Iran.

Cruziana salomonis, found further up in the Jordanian section near the Dead Sea (where king Solomon had his copper mines) and in Sinai, is considerably larger. In the plowings, scratches run almost transversally, but become angular to 120° in the rearmost sets of the rusophyciform version. Endopodal scratches contain three to four sharp claw marks. Between them, introvert exopodal brushings may appear in deep undertraces towards the rear part of the stationary burrow (Pl. 12).

Cruziana aegyptica occurs in other areas of the Eastern Desert of *Egypt*. Therefore its age relationship to the previous ichnospecies is uncertain. The small stationary burrows show angular endopodal scratches and lateral zones of exopodal brushings, whose fine scratches are extrovert (Pl. 12).

The Lower Cambrian of *Europe (N. Spain)* has yielded a different set of distinctive trilobite burrows whose ages are gauged by trilobite body fossils preserved in shale intercalations.

Cruziana cantabrica is most similar to *C. nabataeica*. Except for its much larger size it shows the same imbrication of combed diggings. Proverse scratches are due to a head-down attitude of the body, whereby only the diggings of the anterior legs are preserved in the undertrace (Pl. 13).

In *Cruziana fasciculata*, higher in the same section along the Porma River, endopodal scratches are also combed, but their imbrication is less pronounced. Rusophycifom versions taper pronouncedly towards the rear end.

Cruziana carinata from the same section resembles the footprint of a giant deer, because it is always rusophyciform. Endopodal scratches in the inner field are too poorly preserved to determine the claw formula, but the legs were certainly not comb shaped. The smoother lateral fields could have been produced by the exopodites; but in the casts the boundary to the endopodal field forms a prominent keel rather than a recess.

The name of *Cruziana barbata* from the Middle Cambrian of the same section derives from its resemblance to a moustache and a goat's beard in the rusophyciform version. Obviously, the frontal endopodites dug proversely and thereby removed the sediment up the anterior slope (as in *Cruziana dispar*, Pl. 13). But, as we know from a shallower and more complete undertrace (Pl. 13), the maker of *C. barbata* was much larger than the beard-shaped central depression. Moreover, scratches in the moustache and goat's-beard sections differ not only in direction and penetration depth, but also in claw size. The ichnospecies has also been recognized in the Middle Cambrian of Poland, Turkey, and the Dead Sea area. Cruzianaeform versions are less characteristic.

Cruziana arizonensis, in contrast, is as wide as its maker (note the pleural lobes!). The prominence of exopodal lobes (with introvert brushings, Pl. 12) make it similar to the Upper Cambrian *C. semiplicata* (see below). However, cruzianaeform versions of *C. arizonensis* are not only less common, but also shorter and never form long, continuous plowings. The name refers to a (possibly convergent) *Laurentian* form of the same age.

Forms here referred to the **Upper Cambrian** are less well constrained stratigraphically and may in part already belong to the Ordovician (Tremadocian).

The type material of *Cruziana omanica* was sampled during early oil explorations. It is characterized by endopodal scratches whose three blunt grooves reflect a trifid leg with a stronger claw in the middle (Pl. 12). *Cruziana lata* from the Pacoota Sandstone of central **Australia** is similar, but rusophyciform. Its broad outline suggests that the maker burrowed with a strong dorsal body flexure. If the two occurrences relate to the same or closely related trilobite species, the plowings from Oman must be oriented with the V of the scratches opening tailward, because only the proverse frontal scratches would be depicted. Recently, the Pacoota Sandstone has been dated as lowermost Ordovician from trilobite body fossils. In the future, this spectacular occurrence and a similar one in northwest Argentina may not only clarify the possible synonymy with *Cruziana omanica* and its age, but also identify the maker. That such identifications are possible by cooperation of trilobite and trace fossil experts has already been shown in *Cruziana semiplicata* (Pl. 14). Its maker was a notorious plower. The associated rusophyciform *Cruziana polonica* is probably a molting burrow. In eastern Newfoundland (which is a displaced fragment of Gondwana), *C. semiplicata* is associated with another rusophyciform burrow, *Cruziana leiferikssoni* (Pl. 11), which occurs also in Argentina.

Plate 67 · Cambrian Trilobite Burrows   193

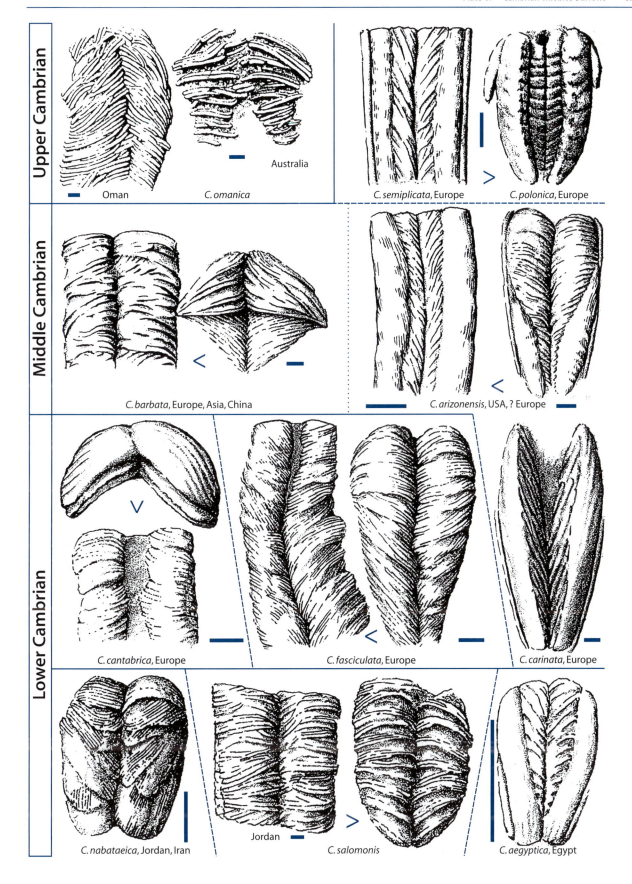

Upper Cambrian

Oman   *C. omanica*   Australia

C. semiplicata, Europe   *C. polonica*, Europe

Middle Cambrian

C. barbata, Europe, Asia, China   *C. arizonensis*, USA, ? Europe

Lower Cambrian

C. cantabrica, Europe   *C. fasciculata*, Europe   *C. carinata*, Europe

C. nabataeica, Jordan, Iran   Jordan   *C. salomonis*   *C. aegyptica*, Egypt

68

Plate 68
Ordovician Trilobite Burrows

When discussing its homeomorphy with *Cruziana pecti-nata* (Pl. 66), we have already come across the distinctive fingerprints of *Cruziana rugosa*: combed dig marks with up to twelve equal and sharp-crested claw scratches. They are most individualized in *Cruziana rugosa rugosa*. This "bathtub" variant is too long for a truly stationary burrow, but short and deep enough to become arched in lateral view (Pl. 66). Still there are no impressions of the pleural edges. Also, each multiclawed dig mark ends with a transversal shelf (hence the name *rugosa*), at which point the leg returned to the recovery stroke. In contrast, continuous plowing is recorded by the long ribbons of *Cruziana rugosa furcifera* and *C. rugosa goldfussi* with the only difference being that marginal furrows left by the pleural edges or spines are present only in the latter ichnosubspecies. Because in these variants all scratches run in the same direction and in the same plane, boundaries between subsequent sets are harder to recognize.

Two undescribed behavioral variants may be added, if they turn out to be more than caprices of extravagant individuals. The one figured on Pl. 13 is from France. The other deviation, observed near Zaragoza (Spain) and in Penha Garcia (Portugal), is a large *C. rugosa furcifera* that circles like *Cruziana semiplicata* (Pl. 14), but with an even smaller turning radius relative to the size of the maker. In view of this variability, it is all the more surprising that the maker of *C. rugosa* never produced truly stationary burrows, which would tell us more about its length and changes in scratch orientation towards the rear end.

Two smaller forms occurring in the Armorican Sandstone deserve the status of separate ichnospecies. *Cruziana imbricata* has scale-like "segments" instead of the usual scratches. They cannot be referred to flaplike appendages resembling the abdominal legs of chelicerates because their shingling is towards the front end, rather than towards the narrower rear part, in the more common rusophyciform version. This shingling can thus be used to orient the rarer cruzianaeform version. The other associate is *Cruziana rouaulti*. Being ribbon-shaped and bounded by pleural grooves, it looks like a minute version of *C. rugosa goldfussi*. The two lobes, however, are usually perfectly smooth. Only specimens from southeast Libya show combed scratches, but with lower claw numbers than in *C. rugosa*.

In contrast to the Lower Ordovician, the **Middle** and **Upper Ordovician** is represented by only a few trace fossil localities. The most diverse ichnofauna comes from an area east of Wadi Rum, Jordan. In this assemblage, only *Cruziana petraea* corresponds to the familiar model of trilobite burrows (Pl. 11). In its more common rusophyciform version, the blunt endopodal scratches show an angular discordance between the front and the rear part of the burrow, although all scratches are retroverse. *Cruziana petraea* has also been found in Chad, Benin and Spain, where the stratigraphic context points to a Middle Ordovician age (personal communication of J. A. Gámez Vintaned 1996).

The remaining forms of the Sabellarifex Sandstone (Jordan) are predominantly stationary burrows, in which the maker was not dorsally flexed, but inclined towards the head end (procline). They also show signs of heteropody: coarser and deeper scratches in frontal and lateral areas, while the inner lobes are either smooth or transversally segmented and show a very regular longitudinal striation. Therefore the trace makers were possibly chelicerates rather than trilobites (Pl. 11). *Cruziana almadenensis* is not only most representative for this problematic group; it also has the widest distribution (Almadén in central Spain; Algeria; Amanos Mountains of Turkey; Saudi Arabia). It is because of this ichnospecies that the assemblages in the Middle East have been tentatively assigned to the Caradocian, the time when Gondwana started to become extensively glaciated.

In the most complete version, *C. almadenensis* presents four prominent features: (1) a vertical front wall left by the edge of a vaulted *head shield*; (2) three pairs of strong scratch bundles that *converge* towards the position of the mouth in a palmate pattern and may protrude beyond the contour of the head shield impression; (3) segmented *lateral lobes* bearing parallel longitudinal striae and merging at the rear end into the sole face of the sandstone; (4) two narrow *central lobes*, sunken in the cast, that look like impressions of coxae. This differentiation goes beyond the endopod/exopod divergence and points to a degree of heteropody unknown in trilobites. Associated plowings show only equivalents of the lateral lobes. As these are not the deepest elements in the stationary version, the animal appears to have plowed in a head-up position.

A similar relationship is observed in *Cruziana flammosa*, whose deep frontal scratches do not converge. The broad lateral lobes are less vaulted than in *C. almadenensis*. They meet along the midline, but are similarly segmented and longitudinally striated. The finely lineated lateral lobes of *Cruziana lineata* cannot be clearly correlated with those of the last two ichnospecies, because there are as yet no corresponding plowings. On the other hand there is a great similarity to the Upper Silurian *C. pedroana* (Pl. 69). The much smaller *Cruziana perucca* occurs in current-aligned clusters. In stationary burrows, one can distinguish (1) the outline of the head shield, (2) two lateral lobes, whose segments appear as series of small tubercles if they just touch the interface, and (3) two sunken and smooth median lobes. Median as well as lateral lobes are expressed in the rare plowings.

Two kinds of stationary burrows (*Cruziana balsa*; *C. radialis*, Pl. 13) differ (1) by a long *oval* outline that does not taper towards the rear end; and (2) by violating the rule that the angle between scratches narrows towards the tail end of the burrow. On the other hand, neither of the two allows us to decipher the claw formula of the endopodites. *C. radialis* is probably a molting burrow.

Plate 68 · Ordovician Trilobite Burrows   195

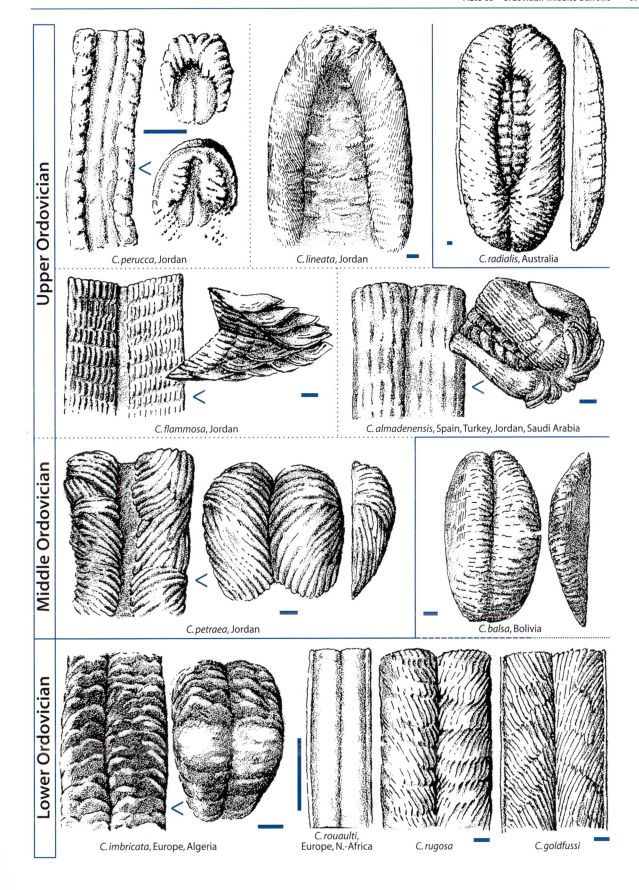

Upper Ordovician

C. perucca, Jordan

C. lineata, Jordan

C. radialis, Australia

C. flammosa, Jordan

C. almadenensis, Spain, Turkey, Jordan, Saudi Arabia

Middle Ordovician

C. petraea, Jordan

C. balsa, Bolivia

Lower Ordovician

C. imbricata, Europe, Algeria

C. rouaulti, Europe, N.-Africa

C. rugosa

C. goldfussi

69

Plate 69
Silurian to Carboniferous Trilobite Burrows

Silurian Forms. The end of the Ordovician is marked by the great Gondwanan glaciation. The following **Early Silurian** transgression resulted, first, in the deposition of bituminous "hot shales" in the depressions of the glacial relief and, then, of prograding marine sandstones. This situation – and its economic interest – have provided us with a rich ichnocoenosis in which piperocks (*Skolithos* = "*Tigillites*"), arthrophycids (Pls. 42, 43), and trilobite burrows are the chief elements.

In the Libyan section, the first representative, *Cruziana kufraensis*, has been found in a sandstone that directly underlies the "hot shales". It is a straight cruzianaeform "bathtub" burrow four to five times as long as wide. Its sharp endopodal scratches are transversal and separated throughout by a median furrow; but the lobes remain relatively flat, so that they form a corner with the lateral walls without pleural markings. In contrast to *C. balsa* (Pl. 68), the outline is not boat-shaped.

In the Acacus Sandstone, prograding over the bituminous shales, dominates *Cruziana acacensis* and its behavioral variants (Pl. 15). Its endopodal scratch sets are combed by five blunt claws. More delicate sets nearer the burrow margins were originally referred to exopodites. But as they run parallel to the endopodal ones, they were more probably made by setae on a more proximal podomer of the same endopodite (reconstruction in Pl. 15).

In the upper parts of the Acacus Sandstone, *C. acacensis* overlaps with another group, in which the two kinds of scratch sets must have been made by different kinds of appendages. In *Cruziana quadrata*, only the coarse endopodal scratches were originally known. Assuming that they are retroverse, this led to a wrong orientation. Additional specimens (some of them rusophyciform) later showed that the seemingly smooth shelves were made by broad and finely setate **exopodites** beating in retroverse direction. This brings *C. quadrata* closer to *C. almadenensis* (Pl. 68) and the large *Cruziana pedroana*, whose rear lobes bear similarly fanned exopodal patterns. The latter ichnospecies was originally described from the San Pedro Sandstone (Upper Silurian) of northern Spain, but occurs also in North Africa.

Not known in Libya is another giant trace of the San Pedro Sandstone, *Cruziana retroplana*. Its rusophyciform version has sharp, uncombed endopodal scratches that run transversely and suppress the median furrow in the rear part.

Devonian Forms. Compared to the Silurian forms, Devonian occurrences of *Cruziana* are even more restricted. They all have transverse scratches without a recognizable claw formula (certainly not combed!).

Cruziana rhenana from the Lower Devonian of Germany shows pleural impressions bounding the short rusophyciform lobes. In the small *Cruziana uniloba* from the Algerian Tassili, the heavy endopodal scratches of the two sides interfinger along the midline, so that the lobes are not separated by a median furrow. In the rear part, the scratches meet at a smaller angle. *Cruziana lobosa* from the Devonian of Libya resembles *C. quadrata*; but the endopodal scratches are not as sharp.

Two undescribed specimens from the Lower Devonian near Marrakech (*Cruziana isp.*; courtesy of Jobst Wendt) show that large trilobites still produced deep "bathtub" burrows at that time.

Carboniferous Forms. The continuing decline of the trilobites in the Carboniferous is also reflected in their meager trace record in Gondwana. In Egypt, *C. carbonaria* represents bath-tub burrows in the style of Devonian predecessors from Morocco, but at a still smaller size. A newly discovered bathtub *Cruziana* from the *Tournaisian* of Morocco has sharp endopodal scratches, whose chevrons are overprinted in opposite directions (D. Korn, personal communication, 2005). Possibly it is a larger variant of *C. carbonaria*.

So far, no trilobite burrows are known from Permian rocks. As a whole, the trace fossil record reflects the evolutionary history of these strange arthropods. Their decline in the Late Paleozoic is bearable from a stratigraphic point of view, because there was a parallel decline of the "Nubian Sandstone" facies, in which substitute guide fossils are most needed. When this facies returned in the Cretaceous (the real Nubian Sandstone), trilobites were gone; but impressions of angiosperm leaves provide another substitute in sandstones deprived of classical index fossils.

■ *Cruziana* cf. *reticulata* (L. Carb., Morocco)

Plate 69 · Silurian to Carboniferous Trilobite Burrows   197

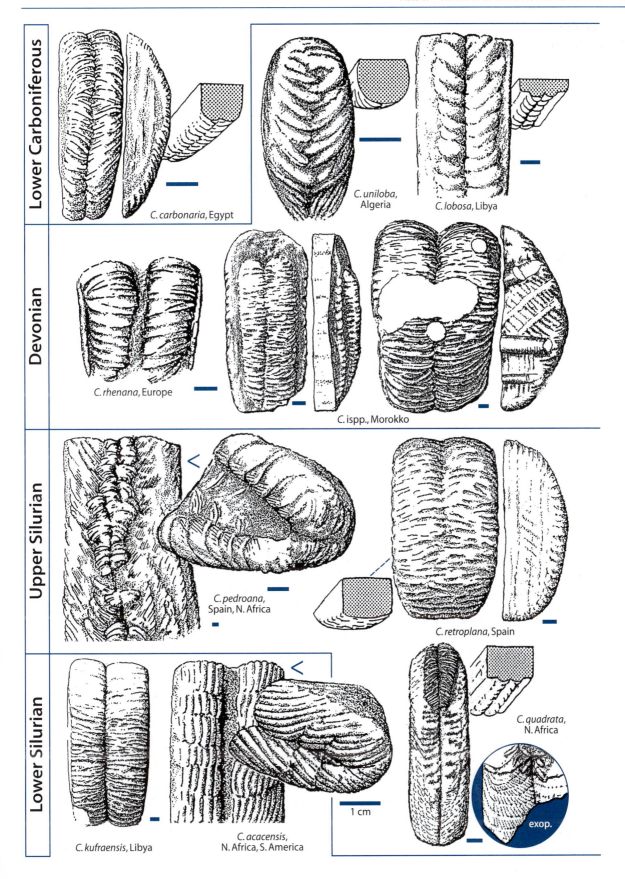

Lower Carboniferous

C. carbonaria, Egypt

C. uniloba, Algeria

C. lobosa, Libya

Devonian

C. rhenana, Europe

C. ispp., Morokko

Upper Silurian

C. pedroana, Spain, N. Africa

C. retroplana, Spain

Lower Silurian

C. kufraensis, Libya

C. acacensis, N. Africa, S. America

C. quadrata, N. Africa

exop.

1 cm

70

Plate 70
Trans-Gondwanan Seaway

Trilobite body fossils provide an unusually high resolution not only in a biostratigraphic, but also in a paleogeographic sense. This is due to the fact that they were largely shallowmarine benthic animals whose planktonic larval stages could not be dispersed across oceans wider than a few hundred kilometers. This rule of thumb can also be applied to trilobite burrows.

A familiar example is Newfoundland, where Cambrian and Early Ordovician trilobites and trilobite burrows in the eastern part of the island resemble European forms, while those in the western part have American and Pacific affinities. In former times one had to hypothesize an isthmus like present Panama to separate the two provinces. Now it is clear that eastern Newfoundland was originally situated on the eastern coast of the protoatlantic Iapetus Ocean (Pl. 66). It then docked, with all its alien fossils, at Laurentia and remained there as an exotic terrane when the modern Atlantic opened again.

The Holy Cross Mountain area of Poland is another exotic terrane. The time of its docking is marked by a switch in the ages of detrital zircons, and also by trilobite burrows. Lower Cambrian sandstones still contain Baltic forms related to *Cruziana dispar*, while Gondwanan ichnospecies (*Cruziana barbata* and *Cruziana semiplicata*) characterize the Middle and Upper Cambrian parts of the sandstone series.

Our plate illustrates another paleogeographic puzzle that could be solved by trace fossils in otherwise unfossiliferous sandstones. It began with the recognition of a characteristic Silurian ichnocoenosis in the Jebel Akakus of southwest Libya and a similar one in the Tandil Range south of Buenos Aires. In principle this was not surprising, as both areas were at that time parts of Gondwana. But migration around the western side of the paleocontinent was difficult, because there was an active margin with no shelf habitats to speak of. Instead, the idea of a *transcontinental seaway* emerged, similar to the one that divided the eastern and western parts of North America in Cretaceous times.

During the years, more and more occurrences of the Silurian ichnofauna were discovered not only in North Africa, but also in the Ennedi region of Chad, and south of the Sahara in the Kandi Basin of Benin. In the same direction, some elements of the fauna disappeared and others replaced them (see symbols on the map). As none of the *Cruziana* ichnospecies was shared by the end stations (Akakus and Tandil), the result was a kind of *relay correlation*.

This situation changed dramatically with the recognition that *Cruziana bonariensis*, originally based on large, but ill defined **cruzianaeform** specimens from the Tandil Range, is in fact very distinctive. Apart from its unusual size (more than 10 cm wide), it resembles *Cruziana acacensis* by scratch sets consisting of five heavy and blunt endopodal claw marks. In addition, a faint central groove can be observed in many scratch casts.

The same fingerprint was subsequently found in an Akakus specimen that otherwise bears little resemblance to the Tandil material. In the front part of this **rusophyciform** burrow, scratches of the four pairs of front legs are well separated. They show that the claws folded together during the medio-posterior digging stroke and that the leg became tilted as it turned to the recovery stroke in a smooth curve (the scratch in the center, drawn in broken lines, belongs to another system, the other side of which appears in the lower right corner). In the sets following, the notch in each claw is also clearly expressed.

The confusing scratch pattern from Jujuy (Museum of La Plata, labeled as Ordovician) is even less *Cruziana*-like. Yet, its analysis suggests a sidling trilobite (Pl. 9) that had at least ten pairs of five-clawed legs and was about the same size as the maker of *C. bonariensis* (**dimorphichniform** version).

On the base of the Jujuy specimen it was possible to recognize another one from the Akakus sandstone as a fourth (**diplichnitiform**) version of *C. bonariensis*. In contrast to true *Diplichnites* (or *Petalichnus*), however, it does not express simple walking. Rather it represents a kind of *probing* in which the endopodites penetrated deeper than usual and spread their five claws in the active backward swing. Notably the sets were not made synchronously on both sides.

The geographic gap between Jujuy and North Africa was eventually bridged by the discovery of another subichnospecies of *Cruziana acacensis* in the Furnas Formation of Paraná, Brazil. The makers were smaller than in Africa and did not dig in a head-down position, but made either short rusophyciform burrows or straight scoopings, whose length is about three times the width (*C. acacensis elongata*; Pl. 15). In Brazil, these variants are poorly preserved, but in Tandil they show the typical five-clawed scratch sets. Similar, but more delicate scratch sets closer to the margins run parallel to the endopodal ones. Therefore they were probably made by setal claws on a proximal section of the endopodite rather than by exopodites.

It is a nice thought that in Silurian times one could have traveled from Tripolis across the Sahara to Buenos Aires by boat. More important, however, is the economic implication of this hypothesis. In North Africa, the most prolific sources of hydrocarbons are Lower Silurian hot shales. They formed in depressions that had been carved out by the Late Ordovician glaciation and became flooded with the eustatic sea level rise in the Early Silurian. As large parts of South America were also glaciated, petroleum exploration could follow the same play – provided the rocks have reached maturity and are correctly dated, which is possible only through *Cruziana* stratigraphy.

Plate 70 · Trans-Gondwanan Seaway   199

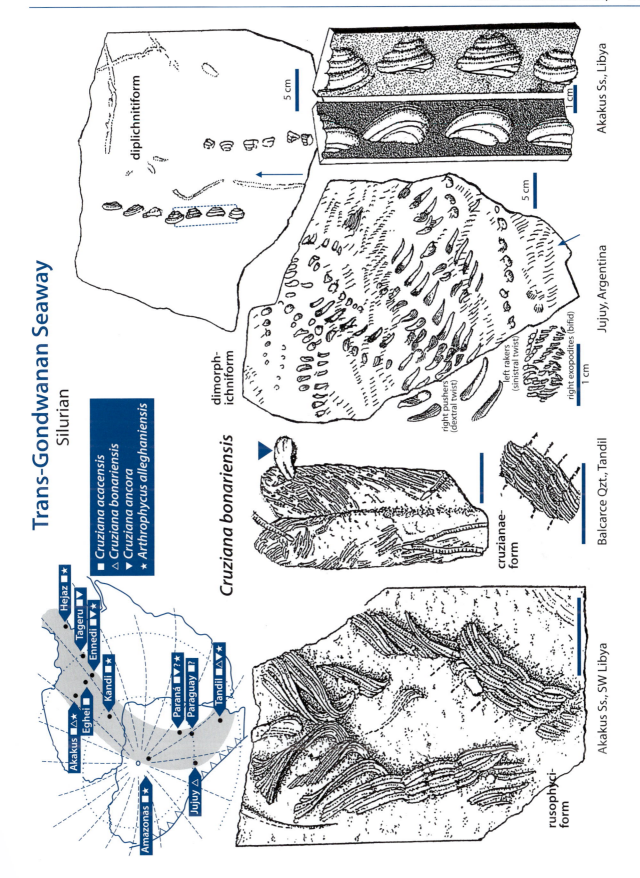

Trans-Gondwanan Seaway

Silurian

■ *Cruziana acacensis*
△ *Cruziana bonariensis*
▼ *Cruziana ancora*
★ *Arthrophycus alleghaniensis*

Cruziana bonariensis

diplichnitiform

Akakus Ss., Libya

dimorph-
ichniform

right pushers
(dextral twist)

left rakers
(sinistral twist)

right exopodites (bifid)

Jujuy, Argentina

cruzianae-
form

Balcarce Qzt., Tandil

rusophyci-
form

Akakus Ss., SW Libya

Hejaz ■ ★
Tageru ■ ▼
Ennedi ■ ▼ ★
Kandi ■ ★
Eghei ■ △ ★
Akakus ■ △ ★
Paraná ■ ▼ ? ★
Paraguay ■ ?
Tandil ■ △ ★
Amazonas ■ ★
Jujuy ■ △ ★

2 cm

■ *Cruziana bonariensis* (Sil., Akakus Mts., SW Libya)

Ichnofacies

Being unreworkable (except for rare and easily recognizable fragments of washed-out tube walls) and reflecting the response to the environment on the spot, trace fossils are ideal facies indicators. This relationship can be studied at different levels, from the global scale, independent of age, to the regional analysis of a particular formation and to the vertical or lateral variations within single beds.

The following plates explicitly exclude a discussion of *ichnofabric*. This omission may appear strange in view of the importance that the ichnofabric approach has recently gained in the context of permeability estimates, sequence stratigraphy and in the study of critical intervals during the history of the biosphere. There are three excuses for this: (1) The ichnofabric concept is adequately covered in other texts (Bromley 1996; Goldring 1999), as well as reports. (2) In an attempt to increase and quantify the data set, ichnofabric analysis emphasizes less the identification of individual trace fossils than degrees of bioturbation and tiering patterns at the bed and sequence level. (3) Ichnofabrics are best studied on vertical rock faces in the field as well as in core samples. In contrast, the approach presented in this book focuses on bedding planes and the interpretation of individual morphologies. It is thus difficult to apply to cores – whether they come from oil wells or from ocean floor drilling.

The two approaches, ichnofabric and ichnomorphology, should be seen as complementary. Which is to be preferred depends on the job to be done and the questions to be answered.

Instead, the following plates emphasize connections between different kinds of trace fossils. This starts with their distribution in different realms or substrates (Pl. 71). What can they tell us about water depth, salinity or substrate conditions and how did these relationships change through Earth history? Plate 72 focuses on deepsea ichnocoenoses, which show a low degree of provincialism, but react strongly to the import of sediments by rare turbidity currents. Temporal change at the local scale is also involved in the shallow marine sequences of Pl. 73. On the one hand hostile background conditions allow rare colonization events to be preserved with frozen tiers. On the other hand, burrows of subsequent generations may become telescoped into a single stratigraphic unit in a storm regime, with low levels of long-term sedimentation, diagenesis, and erosional reactivation being responsible for major changes in substrate conditions. This interplay is discussed in the following paragraph.

Telescoping in Condensed Intervals

A common expression of the interplay between sedimentation, bioturbation, diagenesis, and event erosion are *hardgrounds*, which serve as marker horizons in carbonate mudstone series (Middle Triassic Muschelkalk; Cretaceous Chalk). Their history can be divided into four steps.

(1) An erosion event strips off the *softground* on top and exposes a deeper level, in which the mud was already compacted. (2) The resulting *firmground* is colonized by an association of burrows (mainly those of crustaceans) that were able to penetrate into such a stiff substrate. (3) The transition of this firmground into a *hardground* by cementation, which can be induced either by emergence (beachrock) or, more commonly, by intermittent burial under a carbonate-rich softground. Typically, the cementation front extends downward from the sharp surface of the previous firmground and along the walls of open burrows, so that the lower boundary of the cemented zone becomes very irregular. (4) Now lithified, the re-exposed surface becomes colonized by still another guild of rock *borers* (*Trypanites*; bivalves) and rock *encrusters* (oysters, bryozoans, serpulids, crinoids, etc). Repeated burial and re-exposure of the same hardground is often reflected by successive generations of encrusters that are separated by phosphatized microbial coatings.

Upon final burial, the whole complex history passes into the fossil record. Therefore, hardground faunas are a prime target for synecological studies (Pl. 73); but it should always be remembered that they do not represent census populations.

Hardgrounds have also become important for a *taphonomic* side effect. Its open tunnels (in the firmground as well as the hardground stage) acted as *trap lagerstaetten*, in which delicate fossils such as bryozoans were protected from mechanical abrasion as well as diagenetic dissolution. Such burrow fills often represent time intervals that are not represented in the bed sequence.

Literature

Chapter XV

Bromley RG, Asgaard U (1991) Ichnofacies: A mixture of taphofacies and biofacies. Lethaia 24:153–163 (Discussion of taphonomic controls on ichnofacies)

Buatois LA, Mángano MG (2004) Ichnology of fluvio-lacustrine environments: Animal-substrate interactions in freshwater ecosystems. In: McIlroy D (ed) The application of ichnology to palaeoenvironmental and stratigraphic analysis. Geological Society of London, Special Publication 228, pp 311–333 (Discussion of the ichnofacies concept with emphasis on its application to the study of freshwater ichnofaunas)

D'Alessandro A (1981) Processi tafonomici e distribuzione delle tracce fossili nel Flysch di Gorgolione (Appennino meridionale). Riv Ital Paleontol S 7(3):511–560

Droser ML, Bottjer DJ (1986) A semiquantitative field classification of ichnofabric. J Sediment Petrol 56(4):558–559 (Practical scheme)

Droser ML, Bottjer DJ (1988) Trends in depth and extent of bioturbation in Cambrian carbonate marine environments, western United States. Geology 16:233–236 (Ichnofabrics)

Goldring R (1993) Ichnofacies and facies interpretation. Palaios 8:403–405 (Critical view on the ichnofacies concept and proposal of its replacement by the ichnofabric approach)

Pemberton SG, MacEachern JA, Frey RW (1992) Trace fossil facies models: Environmental and allostratigraphic significance. In: Walker RG, James N (eds) Facies models and sea level change, 3rd ed. Geological Association of Canada, Reprint Series, pp 47–72 (An introduction to the ichnofacies model and its application)

Plate 71: Global Ichnofacies

Archer AW, Maples CG (1984) Trace-fossil distribution across a marine-to-nonmarine gradient in the Pennsylvanian of southwestern Indiana. J Paleontol 58(2):448–466 (Highest diversities in fully marine and fully freshwater facies)

Bradshaw MA (1981) Paleoenvironmental interpretations and systematics of Devonian trace fossils from the Taylor Group (Lower Beacon Supergroup), Antarctica. N Z J Geol Geophys 24:615–652

Bromley RG (1996) Trace fossils: Biology, taphonomy and applications, 2nd edn. Chapman & Hall, London, 361 p (Discussion of the ichnofacies model)

Bromley RG, Asgaard U (1993) Two bioerosion ichnofacies produced by early and late burial associated with sea-level change. Geol Rundsch 82:276–280 (Subdivision of the *Trypanites* ichnofacies)

Bromley RG, Pemberton SG, Rahmani RA (1984) A Cretaceous woodground: The *Teredolites* ichnofacies. J Paleontol 58:488–498 (Introduction of the *Teredolites* ichnofacies)

Buatois LA, Mángano MG (1995) The paleoenvironmental and paleoecological significance of the lacustrine *Mermia* ichnofacies: An archetypical subaqueous nonmarine trace fossil assemblage. Ichnos 4:151–161

Chamberlain CK (1971) Bathymetry and paleoecology of Ouachita Geosyncline of Southeastern Oklahoma as determined from trace fossils. Am Assoc Petr Geol B 55(1):34–50

Chamberlain CK, Clark DL (1973) Trace fossils and conodonts as evidence for deep-water deposits in the Oquirrh Basin of Central Utah. J Paleontol 47(4):663–682

Chaplin JR (1980) Stratigraphy, trace fossil associations, and depositional environments in the Borden Formation (Mississippian), Northeastern Kentucky. Ann. Field Conference of the Geol. Soc. Kentucky, pp 1–114

Ehrenberg K (1941) Über einige Lebensspuren aus dem Oberkreideflysch von Wien und Umgebung. Palaeobiologica 7(4):282–313 (Does not use scientific names)

Ekdale AA, Bromley RG (1984) Comparative ichnology of shelf-sea and deep-sea chalk. J Paleontol 58(2):322–332

Frey RW, Pemberton SG (1984) Trace fossil facies models. In: Walker RG (ed) Facies models, 2nd edn. Geoscience Canada Reprint Series 1, pp 189–207 (*Zoophycos* not only in outer shelf areas but also in shallow marine, protected environments such as lagoons)

Frey RW, Pemberton SG (1987) The *Psilonichnus* ichnocoenose, and its relationship to adjacent marine and nonmarine ichnocoenoses along the Georgia coast. B Can Petrol Geol 35:333–357 (Introduction of the *Psilonichnus* ichnofacies)

Genise JF, Mángano MG, Buatois LA, Laza J, Verde M (2000) Insect trace fossil associations in paleosols: The *Coprinisphaera* ichnofacies. Palaios 15:33–48 (Discussion of paleosol ichnofacies)

Gibert JM de, Martinell J, Domènech R (1998) *Entobia* ichnofacies in fossil rocky shores, Lower Pliocene, northwestern Mediterranean. Palaios 13:476–487 (Detailed analysis of rockground ichnofacies)

Goldring R (1999) Field palaeontology, 2nd edn. Longman, Harlow, 191 p (Proposal to replace the ichnofacies approach by the ichnofabric approach)

Kennedy WJ, Sellwood BW (1970) *Ophiomorpha nodosa* Lundgren, A marine indicator from the Sparnacian of South-East England. P Geologist Assoc 81(1):99–110

Lockley MG, Hunt AP, Meyer CA (1994) Vertebrate tracks and the ichnofacies concept: Implications for palaeogeography and palichnostratigraphy. In: Donovan SK (ed) The palaeobiology of trace fossils. Johns Hopkins University Press, Baltimore, pp 241–268 (Attempt to introduce ichnofacies based on vertebrate trace fossils. Ichnofacies in this context represent ichnocoenoses rather than global ichnofacies)

Pemberton SG (ed) (1992) Applications of ichnology to petroleum exploration. Society of Economic Paleontologists and Mineralogists (SEPM) Core Workshop 17, 429 p, Tulsa (A collection of papers using the ichnofacies model mostly relating to the Mesozoic of Alberta)

Seilacher (1967) Bathymetry of trace fossils. Mar Geol 5:413–428 (Introduces *Cruziana*, *Zoophycos* and *Nereites* ichnofacies)

Plate 72: Post-turbidite Ichnocoenoses

Bromley RG, Ekdale AA (1986) Composite ichnofabrics and tiering of burrows. Geol Mag 123:59–65

Ekdale AA (1985) Paleoecology of the marine endobenthos. Palaeogeog Palaeoclim Palaeoecol 50:63–81 (Excellent review paper on paleoecological aspects of trace fossils, including a discussion of the role of oxygen)

Leszczyński S (1991) Oxygen-related controls on predepositional ichnofacies in turbidites, Guipúzcoan Flysch (Albian-lower Eocene), northern Spain. Palaios 6:271–280 (Proposal of a model that relates trace fossils and oxygenation in turbidite systems)

Orr P (1994) Trace fossil tiering within event beds and preservation of frozen profiles: An example from the Lower Carboniferous of Menorca. Palaios 9:202–210 (Detailed analysis of tiering in Carboniferous turbidites)

Savrda CE (1992) Trace fossils and benthic oxygenation. In: Maples CG, West RR (eds) Trace fossils. Paleontological Society, Knoxville, Short Courses in Paleontology 5:172–196 (Review of the relationships between trace fossils and oxygen in shallow marine environments)

Uchman A (1991) Diverse tiering patterns in Paleogene flysch trace fossils, Magura Nappe, Carpathians, Poland. Ichnos 1:287–292 (Tiering analysis of Paleogene deep marine ichnofaunas from the Polish Carpathians)

Wetzel A, Uchman A (1999) Deep-sea benthic food context recorded by ichnofabrics: A conceptual model based on observations from Paleogene flysch, Carpathians, Poland. Palaios 13:533–546 (Discussion of the food supply as a main controlling factor on deep sea trace fossils)

Wetzel A, Uchman A (2001) Sequential colonization of muddy turbidites in the Eocene Belove·a Formation, Carpathians, Poland. Palaeogeog Palaeoclim Palaeoecol 168:171–186 (Detailed analysis of tiering patterns in an Eocene fine-grained turbidite)

Plate 73: Frozen Tiering and Telescoping in Shallow-Marine Settings

Bottjer DJ, Ausich WI (1986) Phanerozoic development of tiering in soft substrata suspension-feeding communities. Paleobiology 12(4):400–420 (Tiering above and below sediment surface)

Brenner K, Seilacher A (1978) New aspects about the origin of the Toarcian Posidonia Shales. Neues Jahrb Geol P-A 157:11–18 (Frozen tiers in bioturbation events)

Leszczyński S (1991) Oxygen-related controls on predepositional ichnofacies in turbidites, Guipúzcoan Flysch (Albian-lower Eocene), northern Spain. Palaios 6:271–280 (Proposal of a model to link turbidite trace fossil assemblages with oxygen content)

Savrda CE (1995) Ichnologic applications in paleooceanographic, paleoclimate, and sea-level studies. Palaios 10:565–577 (Overall review of trace fossil applications in various paleoenvironmental fields, including references to chondritids)

Savrda CE, Bottjer DJ (1989) Trace-fossil model for reconstructing oxygenation histories of ancient marine bottom waters: Application to Upper Cretaceous Niobrara Formation, Colorado. Palaeogeog Palaeoclim Palaeoecol 7:49–74 (Proposal of a model to link ichnocoenoses with oxygen content in pelagic environments)

Wetzel A (1984) Bioturbation in deep-sea fine-grained sediments: Influence of sediment texture, turbidite frequency and rates of environmental change. In: Stow DAV, Piper DJW (eds) Fine-grained sediments: Deep water processes and facies. Geological Society of London, Special Publication 15, pp 595–608 (Study of bioturbation in modern deep sea sediments that analyzes the response of the infauna to environmental perturbations)

Wetzel A (1991) Ecologic interpretation of deep-sea trace fossils communities. Palaeogeog Palaeoclim Palaeoecol 85:47–69 (Discussion of the controlling factors in deepsea endobenthonic tiering)

Plate 74: Interactions between Trace Fossils

Gaillard C, Olivero D (1993) Interprétation paléoécologique nouvelle de *Zoophycos* Massalongo, 1855. C R Acad Sci (II)316:823–830, Pl. 74 (Cf. *Chondrites* intruding *Zoophycos*)

Jensen S (1990) Predation by Early Cambrian trilobites on infaunal worms: Evidence from the Swedish Mickwitzia Sandstone. Lethaia 20(1):29–42 (Trilobites hunting worms in the Lower Cambrian of Sweden as evidence of predation on worms)

Krejci-Graf K (1938) Ein Grabgang mit Chondriten-Füllung. Senckenberg 20:463–464 (Nucleocave "Chondrites")

Rydell J, Hammarlund J, Seilacher A (2001) Trace fossil associations in the Swedish Mickwitzia sandstone (Lower Cambrian): Did trilobites really hunt worms? Geol Foren Stock For (GFF) 123:247–250 (Critical reanalysis of the Lower Cambrian trilobite traces previously interpreted as representing predation)

Seilacher A (1964) Biogenic sedimentary structures. In: Imbrie J, Newell N (eds) Approaches to paleoecology, John Wiley & Sons, New York, pp 296–316

Plate 75: Solnhofen Mortichnia

Ehrenberg K (1954) Zum Begriff "Lebensspuren" und zur Frage ihrer Benennung. Neues Jahrb Geol P M 1954, pp 141–144 (Trace fossil nomenclature, including the term "Todesspuren")

Schweigert G (1998) Die Ichnofauna des Nusplinger lithographischen limestone (Upper Jurassic, Swabian Alb) Stuttgarter Beitr Naturkd, Ser B 262:1–47 (Mortichnia of shrimp in equivalent of Solnhofen limestones)

71

Plate 71
Global Ichnofacies

Half a century ago it occurred to me that there is a general divergence between ichnocoenoses throughout the Phanerozoic. This was at the time when sedimentologists discovered the process of *turbidity currents*, by which sand and other shelf sediments are episodically transported into deepsea environments. Not much later, the revolutionary concept of *plate tectonics* also explained why the occurrence of such turbidite sequences (the *flysch facies* of earlier geologists) is linked with orogenic events: having formed on the continental slopes, the light turbiditic deposits are preferentially thrust onto continental crust during plate or continent collisions. Thus, the distinction between a shallow-marine **Cruziana ichnofacies** and a deepsea **Nereites ichnofacies** fitted nicely into the new scenario.

Other types of global ichnofacies have since been added.

1. The **Zoophycos ichnofacies** characterizes quiet bottoms, which were disturbed neither by storms nor by turbidity events. Under this condition, bioturbation had time to completely destroy the original depositional fabric. Only the deepest tier of burrows (*Zoophycos*; Pl. 38) is well preserved.
2. *Skolithos* **ichnofacies**: nearshore sands dominated by protective vertical shafts (*Skolithos* and *Diplocraterion*; Pl. 70), expressed as *piperocks* in vertical section.
3. *Glossifungites* **ichnofacies**: muddy *firmgrounds* in fairly turbulent waters that were too stiff for ordinary bioturbation, while protective domiciles could be excavated by *scratching* (e.g., *Spongeliomorpha* modification of *Ophiomorpha*, Pls. 16, 17; *Glossifungites* modification of *Rhizocorallium*, Pl. 19) or by *drilling* (pholadid bivalves, not treated in this volume). Scratches (bioglyphs) are commonly preserved on the tunnel walls.
4. *Trypanites* **ichnofacies**: calcareous hardgrounds, in which domiciles can be excavated only by *drilling* (e.g., rock-chiseling pholadids) or a combination of mechanical abrasion and acid *etching* (*Polydora*, Pl. 18; lithophagid bivalves). The name is derived from *Trypanites*, needle-thin perforations of unknown authorship that are common in fossil hardgrounds (see also Pl. 73).
5. *Teredolites* **ichnofacies**: similar to the *Glossifungites* ichnofacies, except that the substrate is *driftwood*, which can also be used as a food source with the help of endosymbiotic microbes, such as in the bivalve *Teredo* (shipworm).
6. *Scoyenia* **ichnofacies**: softground ichnocoenoses, whose diversity is reduced by nonmarine salinities (freshwater and salt lakes). Aside from *Scoyenia* (Pl. 32, probably made by earthworms or insect larvae), *Cruziana*-like phyllopod burrows (*Isopodichnus*, Pl. 23), swimming tracks of fishes (*Undichna*, Pl. 5) and arthropod undertracks (Pl. 23) are characteristic members.
7. *Oldhamia* **ichnofacies**: microbial matgrounds, in which the decaying lower part of the mat is exploited by undermat miners (Pl. 49).

Even further subdivisions have been made in nonmarine or marine sediments of a particular age; but such distinctions approach the level of time-restricted paleocommunities and dilute the original concept of ichnofacies as a time-independent complex of depositional and biogenic sedimentary structures.

Our original scheme was in fact not independent of time either.

1. It assumed that all aquatic environments, including deepsea bottoms, have been inhabited by trace makers from early on (but see Pl. 65).
2. It neglected the fact that members of one or the other realm have secondarily extended (or shifted) their ranges in an onshore-to-offshore direction (*Zoophycos*, Pls. 37, 38; *Ophiomorpha*, Pl. 17) or in the opposite sense (*Phycosiphon*, Pl. 40).
3. The scheme also neglects the role of *matgrounds* before the Cambrian Revolution (Pl. 65) and traces reflecting mat-related lifestyles (undermat mining, Pl. 50; mat scratching, Pl. 63), which locally persisted into the Phanerozoic.

Nevertheless, the basic message of the tabulation made 40 years ago (modified from a report for the Humble Petroleum Company) remains valid: deepsea ichnocoenoses retain their distinctiveness through time in spite of changing actors. As we have seen in Chap. XI, the remarkable diversity and disparity of deepsea benthos is mainly caused by the limited, but predictable nutrient supply and by the general stability through time of this low-productivity environment, rather than by bathymetry as such.

■ Schistosity sparing traces in concretion (Puncoviscana Fm., Argentina)

Plate 71 · Global Ichnofacies   205

Phanerozoic Ichnofacies

Softgrounds

turbiditic:	*Nereites* ichnofacies
non-event:	*Zoophycos* ichnofacies
tempestitic:	*Cruziana* ichnofacies
littoral:	*Skolithos* ichnofacies
non-marine:	*Scoyenia* ichnofacies
low-oxygen:	*Chondrites* ichnofacies

Other substrates

firmgrounds:	*Glossifungites* ichnofacies
hard- and rockgrounds:	*Trypanites* ichnofacies
woodgrounds:	*Teredolites* ichnofacies
matgrounds:	*Oldhamia* ichnofacies

Abyssal - Bathyal (turbidites) — *Nereites* Facies

Intermediate (no event layers) — *Zoophycos* Fac.

Shelf (tempestites) — *Cruziana* Facies

facies-breaking Ichnogenera

★ BURROWS SHOWN IN UPSIDE - DOWN POSITION

(SEILACHER 1962)

72

Plate 72
Post-turbidite Ichnocoenoses

In contrast to the time-stability of deepsea habitats advocated in the previous plate, the *lithology* of flysch sequences reflects anything but time-constancy: pelagic sedimentation is regularly interrupted by turbiditic event beds. This contradiction dissolves, however, when the rock record is translated into real time: the boredom of years is interrupted on a single day when a turbidity current comes down the continental slope. Since only a limited area was affected by each event, it is also clear that its impact on the global deepsea community was as negligible as the impact of the historical Vesuvius eruption was for the population of Italy as a whole.

On the other hand, the effect of a turbidity current on the local deepsea fauna lasts much longer than one would expect. When mapping modern deepsea floors, biologists discovered that communities on turbidite fans remain distinct from those of unaffected areas for hundreds of years after the event. In contrast to the dramatic picture of a turbidity current killing local populations and importing doomed pioneers from shallower environments, the post-turbidite community is an ecological complex of its own.

In the geologic record, this community is reflected by the **post-turbidite association** of feeding burrows. In contrast to the pre-turbidite association, of which only erosional replicas have survived on the turbidite soles, post-turbidite burrows are recognized by being filled with darker sediment in muddier facies. They also preserve their original three-dimensionality, which can be studied by mechanical preparation or serial sectioning.

Deep Tiers preserved on Turbidite Soles. At a time, when it was still necessary to demonstrate that most turbidites are single-event deposits, sole traces in relation to bed thickness were registered in the sandy turbidites weathering out in the steep coastal cliffs of northern Spain. The result was as expected: while the casts of pre-turbidite burrows are independent of bed thickness (except in the thickest beds, where erosion was too strong to preserve them), the post-turbidite burrows reach the base only in beds up to a certain thickness.

In the present context it is interesting, (1) that the small crustacean that made *Granularia* (a miniaturized ophiomorphid; Pl. 17) outcompeted all the others, (2) that *Phycosiphon* reached much deeper levels than in the Carboniferous, and (3) that search patterns were not as complex as in the pre-turbidite muds.

Tiering in a Carboniferous Post-Tubidite Community. If we read a flysch sequence in terms of depositional history, the lithological alternation between sandy turbidites and intercalated muds must be modified. Every turbidity current erodes, and picks up, muddy sediment on its way down the slope. In the depositional phase, the finest fraction is deposited last and may take days or weeks to settle. Nevertheless, sedimentation rates for the "muddy tail" are much higher than for the following pelagic mud. Although sediments remain virtually the same, the boundary between turbiditic and pelagic mud is marked by the top of the post-turbidite bioturbation. Even though the post-event community may have persisted for hundreds of years, its original tiering has been "frozen" by the negligible rate of pelagic sedimentation.

In our example from a **Lower Carboniferous** flysch (Austrian Alps), tiering was studied by serial sectioning parallel to the bedding plane. The resulting glass model (here divided into three packages) shows three tiers. Upper-most is a level riddled by systems of small *Phycosiphon* (Pl. 40). Next comes a level with fairly large meanders of *Nereites* (Pl. 34), while only the dense spiral system of *Dictyodora* (Pl. 29) reach the deepest level at about 3 cm. If one takes compaction into account, this would have been 10 to 15 cm below the corresponding sediment/water interface.

Tiering in Late Cretaceous/Early Tertiary Post-Turbidite Communities. If one compares examples from Alpine (**Upper Cretaceous to Eocene**) flysches with the Variscan ones, some general changes are noticed. (1) Due to the emergence of calcareous plankton, the mud was micritic. (2) The number of ichnospecies in the community has risen. (3) Penetration has increased by competition.

Interestingly, **Helminthoida** has not only become more efficient and much smaller than its Paleozoic predecessor (*Nereites*, Pl. 33), but also moved to a shallower tier. Also the general correlation between size and penetration depth (Pl. 73) is violated. **Chondrites intricatus** is more delicate than other associated chondritids (Pl. 49); still it manages to reach the level of large *Zoophycos* before radiating into root-like probings.

In summary, *tiering* in the post-turbidite ichnocoenosis resulted probably from a long competitive coevolution. Whether or not the pre-turbidite ichnocoenosis evolved a similar niche splitting is difficult to say, because there is only a two-dimensional record.

Through the Phanerozoic, *diversity* increased in both communities, but by different modes. Graphoglyptids (*Paleodictyon*) have their oldest record in Cambrian shallowmarine deposits. Later they are restricted to deepsea environments. Still they radiate tremendously in the Late Cretaceous (Chap. XI), possibly in response to increased food supply due to the rise of angiosperms in terrestrial as well as shallowmarine environments (eel grass). Of the later immigrants, only a few were able to enter the pre-turbidite club (echinoid burrows in *Taphrhelminthopsis* preservation, Pl. 26; *Spirophycus*, Pl. 33) and only after they had been behaviorally conditioned in the post-turbidite association. The main barrier was perhaps the low level, and poor quality, of food in areas that were excluded from turbidity fertilization.

In contrast, the Upper Cretaceous/ Lower Tertiary *post-turbidite* association appears to have been less exclusive. Of the 30 forms figured, 10 are Mesozoic immigrants from shallow habitats (black arrows), 6 were already present in the Paleozoic (black dots), 5 evolved in place from immigrants (thin arrows) and 9 have no obvious outside relationship. Remarkably, no member of the pre-turbidite community has ever been found in the post-turbidite association.

Plate 72 · Post-turbidite Ichnocoenoses   207

Post-turbidite Association (U.Cret./Eoc.)

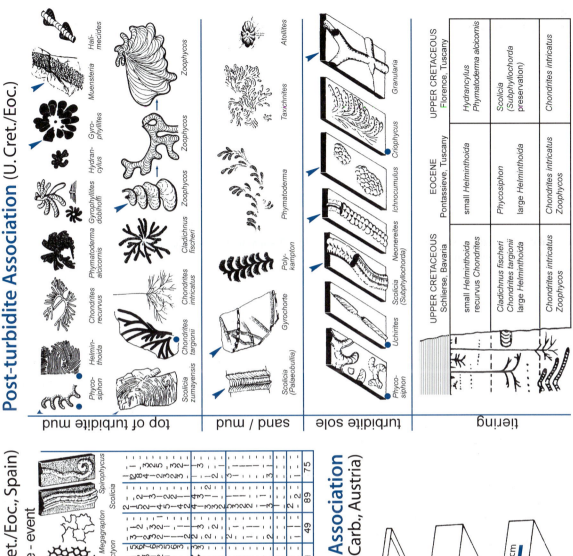

| | | | mud / sand | turbidite sole | | | |

mud of turbidite mud — top of turbidite mud — sand / mud — turbidite sole — tiering

	UPPER CRETACEOUS Schlierse, Bavaria	EOCENE Pontassieve, Tuscany	UPPER CRETACEOUS Florence, Tuscany
	small Helminthoida recurvus Chondrites	small Helminthoida	Hydrancylus Phymatoderma alcicornis
	Cladichnus fischeri Chondrites targionii large Helminthoida	Phycosiphon large Helminthoida	Scolicia (Subphyllochorda) preservation
	Chondrites intricatus Zoophycos	Chondrites intricatus Zoophycos	Chondrites intricatus

Burrows on Turbidite Soles (U.Cret./Eoc., Spain)

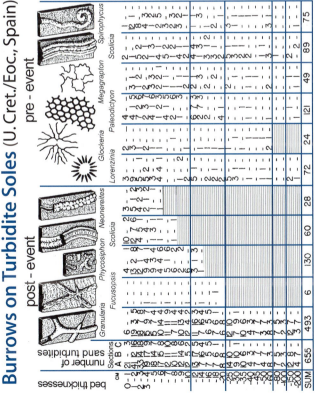

pre - event

post - event

Tiering in Post-turbidite Association (L. Carb., Austria)

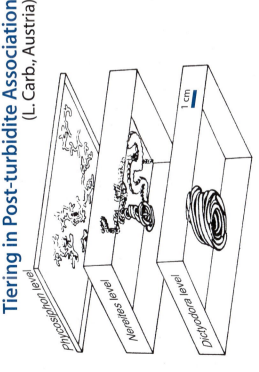

Phycosiphon level

Nereites level

Dictyodora level

1 cm

73

Plate 73
Frozen Tiering and Telescoping in Shallow-Marine Settings

After having zoomed on within-bed patterns in deepsea sediments, we now return to shallow settings. Here, the dominant events are storms on open shelves and river floods in protected estuaries. The corresponding event beds (tempestites and inundites) resemble turbidites sedimentologically (e.g., by grading). But not only do these events occur more frequently; also their impacts on the local bottom fauna are very different. (1) Individual tempestites and inundites may be too thick to become completely penetrated by the post-event generation of burrows, so that only their upper parts are bioturbated. (2) As the intervals between such events are too short for a particular post-event community to become established, the old inhabitants either rose to the new surface (escape burrows; Pl. 24) or were replaced by a new generation, unless the event had changed substrate conditions.

In the absence of such events, different tiers became mixed as they shifted in response to gradual sedimentation. In this case, the lower-tier burrows always intersect the ones of the upper tiers (*upward telescoping*). The responses to negative or oscillating sedimentation (*downward telescoping*) are more complex – particularly if they interfere with a third group of processes that can be referred to as *sediment maturation*. This will be shown in the second example.

Colonization Windows in Anoxic Environments.
In anoxic parts of a basin, high-energy events are replaced by **oxic events**. They leave no sedimentological signature; but they allow opportunistic species to invade otherwise unhabitable bottoms and to tap the organics that accumulated in the sediment during the long anoxic intervals. The Lower Jurassic Posidonia Shales of southern Germany have their name from such an opportunist, the small bivalve *Posidonia*. It covers particular laminae, but is virtually absent otherwise. Since shells within an assemblage have the same size and are largely still articulated ("butterfly position"), we probably deal with a single pioneer generation that was able to complete its life cycle before anoxic conditions returned and killed the whole population.

The rare *bioturbation horizons* in these bituminous shales tell a similar story. Dominated by chondritids (Pl. 49), they show a **frozen tiering**. But in contrast to the frozen tiers of post-turbidite burrows (Pl. 72) we deal only with a few generations of burrows, whose makers explored the shallow layers as juveniles and penetrated deeper as they grew up. A still deeper tier is occupied by the tunnels of crustaceans (*Thalassinoides*, Pl. 17), whose large size suggests that their makers immigrated as adults to take part in the feast. While tiering patterns are essentially the same in all studied horizons (levels are indicated in the stratigraphic column), each one has its peculiarities. Thus the crustacean in horizon **A** systematically searched for buried ammonite

shells and excavated their body chambers. This suggests that remains of the ammonite's soft body were still present. However, as the maker of the figured burrow left across the wall of the body chamber, the aragonitic ammonite shell must already have been demineralized.

In horizon **B**, the chondritid level is occupied by *Phymatoderma granulosa* (Pl. 50) and the crustacean burrows are well-sized *Thalassinoides* (Pl. 18) with typical Y-shaped branchings.

Horizon **D**, at the very top of the bituminous shales, is the world of *Chondrites bollensis* (Pl. 50), while the crustacean burrows show globular swellings – a feature that is also found in black shales of the next stage and resembles *Halimedides* in the post-turbidite association of a Cretaceous flysch (Pl. 72).

Strangely, burrow horizons of the Posidonia Shales do not coincide with benthic shell layers. So the two groups of pioneers may have used different kinds of windows to invade this otherwise hostile environment. The model may also be applied to burrow horizons in the Upper Jurassic limestones of the Solnhofen area. There, *Chondrites* is missing, while it fills particular horizons in non-lithographic marls of similar age. Did it not find the H_2S for its chemosymbionts?

Storm Condensation.
Hardground histories reach maximum complexity in stratigraphically condensed intervals, i.e. in times when long-term net sedimentation approached zero. In such horizons, ammonites of different biozones are found together in a single bed. If there is any succession, it is grading by shell size rather than by geologic age. In a first approach, one would tend to explain this phenomenon by very low sedimentation rate: shells lying on the sea floor became mixed for lack of a separating matrix. However, the aragonitic shells are excellently preserved, not corroded and not encrusted as one would expect them to be after a million-year exposure.

Franz Fürsich dissected such a case in coastal exposures near Caen (northern France). There, a condensed horizon (**Calcaire à Oolithes Ferrugineuses**) of the *Bajocian* (Middle Jurassic) has for long been searched by collectors for its beautifully preserved ammonites. The key to understanding the problem, however, came not from the condensed oolite itself, but from the underlying mudstone (**Couche Verte**), which is less fossiliferous but preserves a lot of bioturbational structures. Their study in polished sections revealed the complex history shown in the diagram.

This story teaches us how historical information becomes stored in the stratigraphic record by the combination of physical, chemical, and biological processes. Whereas the initial mud layer probably contained a minimum of information, the final sediment package of similar thickness was stuffed with a rich and varied historical record. Also, one would have considered the *Calcaire à Oolithes Ferrugineuses* as just another mappable rock unit, had it not preserved the ammonite shells with their high stratigraphic resolution and the trace fossils recording changes in substrate consistency.

Telescoping By Storm Condensation

a primary softground (deposit feeders)

original fabric

b erosional firmground (*Thalassinoides*, 1st generation)

c same surface scoured as diagenetic hardground (encrusters, borers)

storm

d again exhumed with firmground fillings (2nd *Thalass.* generation)

e burial under winnowed layer of small oncoids

f stromatolitic encrustation into biogenic hardground

g accumulation & event winnowing of locally produced loose particles (ooids; large „snuff box" oncoids; mollusc shells mixed from 3 ammonite zones)

final fabric

"Calcaire à Oolithes Ferrugineuses"

Conglomerat de Bayeux

Couche verte

Top of Malière

modified from Fürsich 1971

Frozen Tiers in Jurassic Oil Shales

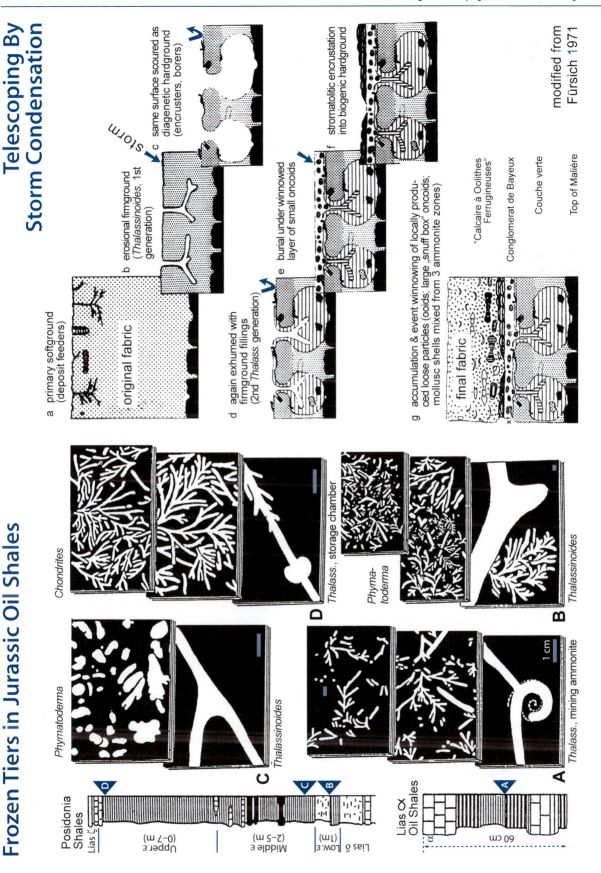

Chondrites

Phymatoderma

Thalassinoides

C

Thalass., storage chamber

Phymatoderma

D

Thalassinoides

B

Thalass., mining ammonite

A

1 cm

Posidonia Shales

Lias ζ

Upper ε (0–7 m)

Middle ε (2–5 m)

Low. ε (1m)

Lias δ

D

C

B

Lias α Oil Shales

α

A

60 cm

74

Plate 74
Interactions between Trace Fossils

Trace fossils provide a record of activities that took place millions of years ago at the very spot where we find them today. They can be analyzed in the same way in which a detective reconstructs past happenings at the site of a crime. So it is understandable that ichnologists dream of the smoking gun: two trails approaching the point of encounter and only one of them leading away. The reaction of the sidling trilobite in Pl. 9 to the track of a larger competitor comes close to this scenario. But can we be certain that the two trace makers actually met? Much more common, and more preservable, are interactions *within* the sediment, because older burrows are part of the "infaunal landscape", just like other sedimentary structures.

We have already come across such interactions with insects preferring the sites of preexisting tetrapod footprints for their pupa chambers (Pl. 3). Similarly, the trackways of large arthropods became deviated by a *Climactichnites* trail in an intertidal environment (Pl. 10). In all these cases we deal with reactions to other traces, not to the animals that produced them.

Graphoglyptids. The avoidance reaction of *Spirorhaphe* towards its own and conspecific tunnels (but not to associated systems of earlier generations) is repeated here from Pl. 52. The congruent course of *Belorhaphe* and the echinoid burrow (*Taphrhelminthopsis*, Pl. 26) is certainly not coincidental; but it remains uncertain who came first. In any case, the echinoid newcomer to the exclusive pre-turbidite community burrowed at a shallower level than its post-event relatives.

Nucleocave *Chondrites*. In *Diplocraterion* from a dark limestone in the Lower Lias of Germany, *Chondrites* burrows penetrated twice as deep as in the surrounding sediment. They preferred the passive infill of the U-tube as well as the actively constructed spreite. A larger chondritid from the Middle Jurassic shows the same preference with regard to *Zoophycos*.

Our next example is an **ammonite steinkern** from Cretaceous chalks of Poland (Jagellonian University, Kraków). In the glauconitic matrix, *Chondrites* systems are not restricted to the body chamber, but extend into what would have been the phragmocone; nevertheless they keep protruding in an apical direction. The animals probably hugged the sediment-filled periostracal bag only after the aragonitic shell and septa had been diagenetically dissolved. Thus the case resembles the small *Thalassinoides* probing systematically into ammonite body chambers in Lower Jurassic oil shales (Pl. 73D).

A similar "nucleocave" behavior is observed in steinkerns of larger Jurassic *Thalassinoides*. The mud around these shrimp tunnels (Pl. 18) was originally stiffer than the sandy fill sediment; but the latter became subsequently lithified into a pressure-shadow concretion, so that it could freely weather out. Note that the branches of *Chondrites* in the tunnel fill are as widely spaced as in bituminous shales (Pl. 73). Smaller shrimp burrows in deepsea deposits (*Granularia*, Pl. 17) were similarly penetrated by widely spaced *Chondrites* from **Romania**. This contrasts with their close packing in the second example from **Germany**, where uniformly directed *Chondrites* is restricted to the wall lining. Probably it was the binding mucus that made this layer attractive for sediment feeding.

Trilobite "Hunts". In the previous cases one clearly deals with reactions to preexisting burrows. Specimens of rusophyciform *Cruziana* associated with large worm burrows, however, might suggest that the trilobite preyed on the worm. In fact, the worm burrow runs more or less along the midline of the trilobite burrow in all published "hunting scenes". Yet there are doubts. First, burrowing trilobites were probably sediment processors rather than predators, as shown by the lack of differentiated mouthparts (Pl. 11). Second, the published specimens may not be representative, but result perhaps from accidental superposition and biased sampling.

In the classic case of *Cruziana dispar*, teichichnoid burrows are common enough on the adjacent bedding plane for accidental superposition. Also the worm burrow (blank) appears to intersect trilobite scratches. Intersection of *Cruziana acacensis* by *Arthrophycus* is even clearer; i.e. the worm burrows were produced only after their tiers had shifted upwards due to additional sedimentation. Yet their alignment with trilobite burrows is not random. Probably the worm simply followed the depression that the trilobite had made in the mud/sand interface. This leaves *Cruziana carleyi* as the only candidate for a real hunting scene. But as this is probably a molting burrow, it is unlikely that the trilobite was in the mood for hunting

These are only a few examples. Others can be found throughout the book, such as the usurpation of Miocene *Rhizocorallium* (*Glossifungites*) tunnels by thalassinid shrimp in an Eocene firmground (Pl. 19), the mutual avoidance reactions in chondritids (Pl. 50), other "fucoids" (Pl. 48), and the phobotactic behavior of *Nereites* (Pl. 34).

In general, it should always be remembered that associations of trace fossil *are* strictly autochthonous, but that they do not necessarily represent isochronous snapshots. Buried bedding planes served as writing pads for many generations of trace makers!

Plate 74 · Interactions between Trace Fossils   211

Interactions

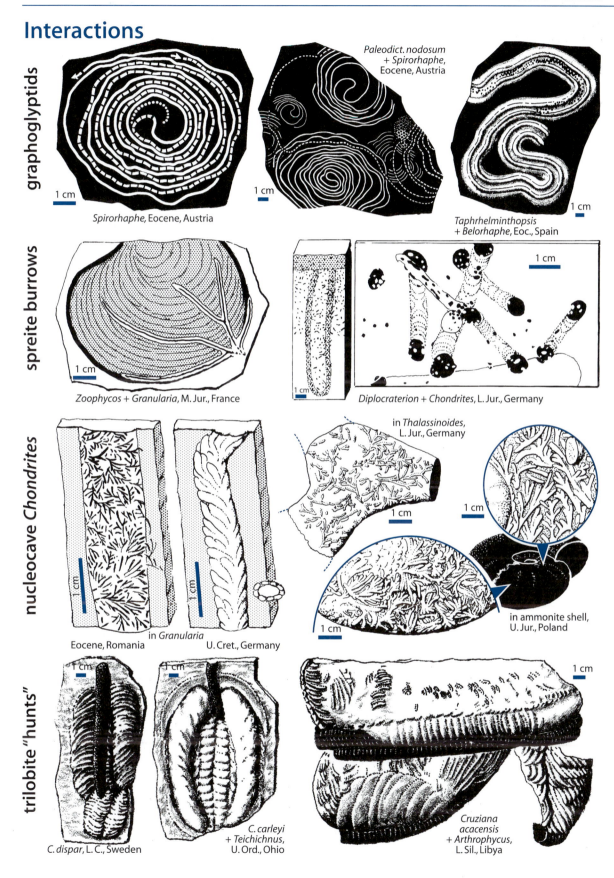

graphoglyptids

Spirorhaphe, Eocene, Austria

Paleodict. nodosum + Spirorhaphe, Eocene, Austria

Taphrhelminthopsis + Belorhaphe, Eoc., Spain

spreite burrows

Zoophycos + Granularia, M. Jur., France

Diplocraterion + Chondrites, L. Jur., Germany

nucleocave *Chondrites*

in *Granularia*
Eocene, Romania

U. Cret., Germany

in *Thalassinoides,* L. Jur., Germany

in ammonite shell, U. Jur., Poland

trilobite "hunts"

C. dispar, L. C., Sweden

C. carleyi + Teichichnus, U. Ord., Ohio

Cruziana acacensis + Arthrophycus, L. Sil., Libya

75

Plate 75
Solnhofen Mortichnia

In aquatic *Konservat-Lagerstaetten*, the absence of trace fossils stands for hostile conditions in the bottom water. Thus, whole skeletons could be preserved without the destructive interference of scavengers. Conversely, burrowed horizons in the Lower Jurassic Posidonia Shales (Pl. 73) signal oxic events in an otherwise anoxic setting. On the other hand, tracks or undertracks would remain undiscovered even if they had been present, because the bituminous rock neither splits nor weathers along bedding planes.

In the Upper Jurassic lithographic limestones of Solnhofen and Nusplingen (southern Germany) the situation is different. Due to selective calcification of turbiditic event layers (*Flinz*) versus background sediments (*Fäulen*), quarried slabs expose the most delicate bedding-plane features, including low-relief *elephant skin* structures caused by microbial mats (Pl. 58). There are also rare surfaces covered with *roll marks* of ammonite shells that had been picked up by muddy turbidity currents (Pl. 57). Also preserved are drag marks of jellyfish, driftwood, or other dead objects that ripped away the microbial scum along their paths.

All the more remarkable are biogenic structures that hardly deserve to be called *Lebensspuren* (= traces of life), because they record the last movements (or even postmortem convulsions) of the trace makers preserved together with them. So they may not record regular activities; but at least it is certain who was responsible.

Most common are **limulids** at the end of their trackways (*Kouphichnium*; Pl. 6). An experienced quarryman recognizes the pattern and knows exactly in which direction to dig for the prized body fossil. Nor does he have to dig far, because even a sturdy horseshoe crab made it no more than a few meters after touching bottom in the asphyxiating environment. There is only a minor drawback: limulids thus preserved never reach the body sizes recorded by trackways in marginal facies. Probably only juveniles swam up and became swept into the anoxic zone when a turbidity current passed by. A very small individual described by the late Rudolf Mundlos tried to swim up again and died upside-down.

The figured specimen of *Mecochirus* had a similar fate; but it tried to escape backwards and did not manage to crawl as far as the limulids.

In contrast, *Antrimpos* lived nektobenthically, like modern shrimps. As Günther Schweigert found out, most specimens in the Nusplingen Basin are exuviae that sank down through the water column. The figured individual, however, was still alive. As shown by its trail (*Telsonichnus*), it did not crawl after having landed, but moved forward by strokes of the abdomen before it died.

Death marches with the animal at the end are also found in displaced benthic mollusks. The burrowing bivalve *Solemya* (Pl. 6) loves low-oxygen muds as an energy source for its endosymbionts. Still, it survived only briefly in this harsher environment. The same was true for *gastropods*, whose winding trails and crushed shells are too small for identification.

Fish were rarely alive when reaching the bottom. In the figured case, however, the tail still made some swimming motions. This is shown by the body impressions and the blown-off biomat around the head, which was already sticking in the mud. As the curved impressions fail to align with the fish tail, it is also possible that they stem from smaller scavengers that managed to reach the otherwise toxic bottom.

For completeness, one should also mention *postmortem traces*. Dead fish came down belly-up, as shown by the **landing marks** of the head, the dorsal fin and the tail fin. But after having fallen to the side, the carcass suffered a dorsal bend that is found only in a few other *Konservat-Lagerstaetten*. Such bends are well known in all skeletons of the famous *Archaeopteryx*, as well as in carcasses of modern birds after desiccation in the sun. They are due to shrinkage of the ligaments that connect the dorsal spines of the vertebral column. The same phenomenon may occur underwater, if the ligaments are dehydrated by osmosis. Thus one may assume that anoxia was combined with hypersalinity in the bottom waters of the Solnhofen basins. In the bending process, the tail fins of fish carcasses commonly broke off because they were held in place by the sticky matground.

Shrimp carcasses also reached the bottom in a belly-up attitude, producing landing marks of the rostrum, the stalked eyes and the buckled abdomen. When the carcass tilted, the eye mark on the tilting side became more deeply impressed. The body also suffered taphonomic bending, but in this case to the ventral side. In the figured specimen, the bending tail fan scraped off a heap of mud, so the deformation must have occurred well after the animal had died.

Ammonite shells (*Perisphinctes*) reached the bottom in swimming attitude, which can be reconstructed from the landing mark. As they remained in place until the shell tilted, the animals were already dead upon landing, but with the soft parts still in place. Yet no one has ever found impressions of the arms, which were probably reduced to a delicate filter fan.

Although such markings are exceptional, I would like to call them mortichnia – perhaps an adequate act upon finishing this book and a happy affair that has lasted for sixty-five years!

Plate 75 · Solnhofen Mortichnia   213

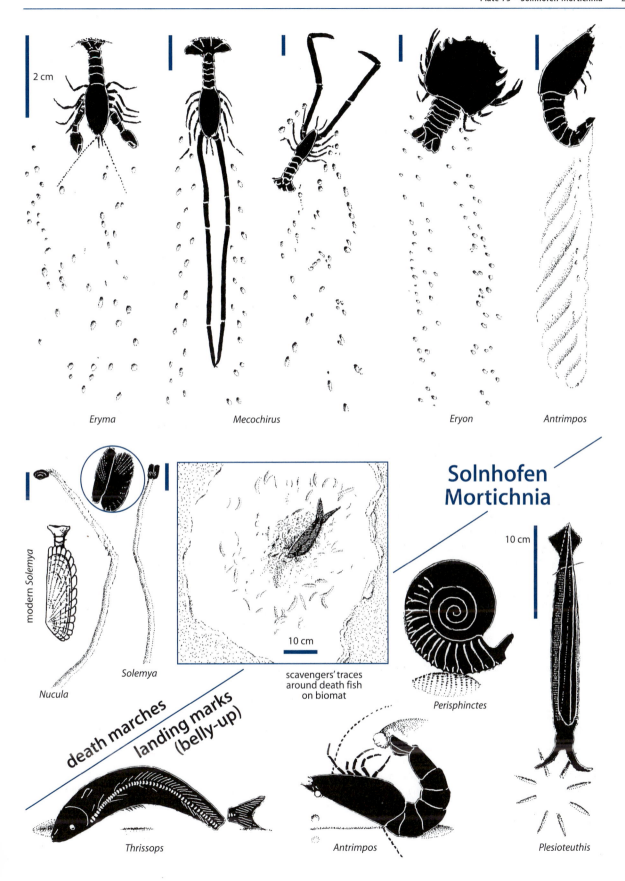

Eryma

Mecochirus

Eryon

Antrimpos

modern *Solemya*

Nucula

Solemya

scavengers' traces
around death fish
on biomat

10 cm

Solnhofen
Mortichnia

10 cm

Perisphinctes

death marches

landing marks
(belly-up)

Thrissops

Antrimpos

Plesioteuthis

2 cm

■ Death march of *Limulus* (U. Jur., Solnhofen)

Glossary

agglutination	Reinforcement of an external skeleton or a tube by sand or other sedimentary particles.
backfill phantom	Part of a backfill burrow in which the general shape and length of the trace maker is reflected.
ballast shell	Shell thickened beyond the requirements of protection to make it heavier, for instance in burrowing.
Bauplan	German term to describe the basic construction and architecture of an animal and its skeleton.
bioglyphs	Scratches on burrows left by digging appendages and other non-intentional "fingerprints".
cephalon	Head shield of trilobites produced by fusion of segments (tagmatization).
cololites	Fossilized contents of the intestine. Due to very early microbial cementation, they depict the morphology of these soft organs and commonly have a higher preservation potential than any other part of a vertebrate carcass.
coxae	Basal elements of arthropod legs; in trilobites used for transporting food particles to the mouth.
death masks	Impressions of soft organisms stabilized by mineralization of microbial films coating carcasses (typical for Ediacaran body fossils; Pl. 62).
doublure	Ventral rims of calcified dorsal carapaces of trilobites, providing strength and keeping sediment from sliding into filter chamber by outwardly ratcheted terrace lines.
draft fill	Passive filling of a cavity through openings of a smaller diameter with a terminal fill channel of the same diameter preserved along the crest of the burrow.
epichnial (epirelief)	See Pl. 31.
farming	Term for ectosymbiosis with micro-organisms that can use nutrients not exploitable by the host.
flysch facies	Old term in Alpine geology for sedimentary sequences now interpreted as deep-sea turbidite facies.
Furchensteine	Cobbles from lakes corroded by modern insect larvae.
Golden Age	Term for periods in which equable climates allowed evolution to reach unusually high levels of diversification and specialization. Golden Age biota were thus unusually vulnerable towards global catastrophies.
hypichnial	Preservation of trace fossils on the soles of sandstone beds (Pl. 31).
hypostome	Mineralized ventral element in front of trilobite mouth (Pl. 11), providing purchase for muscles of the suctional stomach.
ichnotope	Site, in which a characteristic suite of ichnofossils is preserved in a particular kind of rocks.
inglutination	Internal skeletons consisting of sand or finer clastic particles.

mantle burrows	Tunnels, the walls of which are actively reinforced either by mucus impregnation or a lining of marginal backfill (see Plate 37).
meniscoid	Watchglass-shaped. Used for terminal backfill structures in burrows.
parataxonomy	A classification in which parts of an animal, or its work (traces), with uncertain relationships, are classified and named separately.
pereiopods	Arthropod appendages specialized for walking.
proboscis	Evertible mouth, particularly in worm-like animals.
pygidium	Fused tail shield of trilobites fitting against cephalon in enrolled state for protection of the non-mineralized ventral side (Plate 11).
Red Queen filtrators	Planktonic animals that reach almost neutral buoyancy and compensate the rest by constant beat of swimming organs acting as strainers.
rhachis	Axial part of trilobite carapace, containing intestines.
seleniform	Halfmoon-shaped. Used for the cross-sectional aspect of terminal and transversal backfill structures (crescentic).
shadow trace	A less distinct trace pressed through to an adjacent bedding plane not touched by the trace maker (see Plate 2 for a corresponding vertebrate expression).
spicate	Chevroned trail pattern resembling an ear of wheat.
spreite	Term derived from early interpretations when trace fossils were regarded as plants and the webs between the shafts of rhizocoralliid burrows were taken for the photosynthesizing areas of a leaf. Now used for transverse backfill structures in general; see Plates 18, 37.
stylolitization	Process in late diagenesis, by which a pressurized water film dissolves the rock on either side of a former bedding plane and thereby obscures trace fossil reliefs or projects tubes into one plane.
taphonomy	Science of the post-mortem fate of biological remains, including decay, transport, burial, and diagenetic alteration or elimination.
telson	Arrow-like tail spine of horse shoe crabs, used for rear support in locomotion and for righting-up.
terrace lines	Ratcheted sculpture on arthropod outer skeletons, providing purchase in burrowers and crevice dwellers.
thigmotaxis	Behavioral tendency to keep in touch with, or maintain a certain distance from, previous parts of the same or another burrow without overcrossing.
toponomy	Spatial relationship of traces to layers of different grain size. Coarser beds serve as reference.
undertrace	Expression of surface track on bedding plane within the sediment.
wiggling phantom	Directional changes too abrupt and to closely spaced for a long worm-like trace maker.

Index